# The Mechanics of Inhaled Pharmaceutical Aerosols
## An Introduction

# The Mechanics of Inhaled Pharmaceutical Aerosols

## An Introduction

### Warren H. Finlay

*University of Alberta*
*Edmonton, Canada T6G 2G8*

## ACADEMIC PRESS

A Harcourt Science and Technology Company

San Diego   San Francisco   New York   Boston
London   Sydney   Tokyo

Academic Press
*A Harcourt Science and Technology Company*
Harcourt Place, 32 Jamestown Road, London NW1 7BY, UK
http://www.academicpress.com

Academic Press
*A Harcourt Science and Technology Company*
525 B Street, Suite 1900, San Diego, California 92101-4495, USA
http://www.academicpress.com

ISBN 0-12-256971-7

Library of Congress Catalog Number: 2001090350

A catalogue record for this book is available from the British Library
Transferred to Digital Printing 2005

Typeset by Paston PrePress Ltd, Beccles, Suffolk, UK

01 02 03 04 05 06 BC 9 8 7 6 5 4 3 2 1

# Contents

# Contents

# Contents

# Contents

# Contents

# Preface

The field of inhaled pharmaceutical aerosols is growing rapidly. Various indicators suggest this field will only expand more quickly in the future as inhaled medications for treatment of systemic illnesses gain popularity. Indeed, worldwide sales of inhalers for treating respiratory diseases alone are expected to nearly double to $22 billion by 2005 from the estimated 1997 value of $11.6 billion. However, this is only the start of what is likely to be a much larger period of growth that will occur because of the increasing realization that inhaled aerosols are ideally suited to delivery of drugs to the blood through the lung. Indeed, in the future, inhaled aerosols are expected to be used for vaccinations, pain management and systemic treatment of illnesses that are currently treated by other methods.

With the explosive growth of inhaled pharmaceutical aerosols comes the need for engineers and scientists to perform the research, development and manufacturing of these products. However, this field is interdisciplinary, requiring knowledge in a diverse range of subjects including aerosol mechanics, fluid mechanics, transport phenomena, interfacial science, pharmaceutics, physical chemistry, respiratory physiology and anatomy, as well as pulmonology. As a result, it is difficult for newcomers (and even experienced practitioners) to acquire and maintain the knowledge necessary to this field. The present text is an attempt to partially address this fact, presenting an in-depth treatment of the diverse aspects of inhaled pharmaceutical aerosols, focusing on the relevant mechanics and physics involved in the hope that this will allow others to more readily improve the treatment of diseases with inhaled aerosols.

Chapter 1 supplies a brief introduction for those unfamiliar with the clinical aspects of this field. Chapter 2 is a short introduction to particle size concepts, which is important and useful to those new to the field, but which is standard in aerosol mechanics. Chapter 3 lays down the basic equations and concepts associated with the motion of aerosol particles through air, including the effects of electrical charge. The complications added by considering particles that may evaporate or condense, as commonly occurs with liquid droplets in nebulizers and metered dose inhalers, are dealt with in detail in Chapter 4.

Chapter 5 introduces some basic aspects of breathing and respiratory tract anatomy, while Chapter 6 introduces the concepts of fluid motion in the respiratory tract. Both chapters are necessary for understanding subsequent chapters, particularly Chapter 7 which delves deep into the details of aerosol particle deposition in the respiratory tract, one of the most important aspects of inhaled pharmaceutical aerosols.

The last three chapters of the book each introduce basic aspects of the mechanics of the three major device types currently on the market: nebulizers, dry powder inhalers, and metered dose inhalers. From a traditional engineering point of view the mechanics of these devices has not been well studied, so that a reasonable part of this material is speculative, drawing on work done in related engineering applications and extrapolating in an attempt to gain some understanding of the mechanics of existing aerosol delivery devices.

Any book will have its shortcomings, and the present one is no exception. In particular, there are several topics that I would have liked to include, but have chosen not to because of time and energy limitations. Some of these neglected topics include nasal administration of aerosols, the mechanics of several new and promising delivery devices (including various novel powder and liquid systems about to be launched on the market), various aspects of formulation, as well as particle sizing methods. My apologies to those who had hoped for coverage of these topics. However, this book has taken far longer to complete than I had planned, and the time has come to send it to the presses.

Edmonton                                                                    W. H. F.
September 2000

# Acknowledgments

This book would not have been possible without the help of many individuals. Thanks are due to those who read and suggested changes to drafts of various chapters, including, in alphabetical order, Tejas Desai, Werner Hofmann, Martin Knoch, Carlos Lange, Edgar Matida, Antony Roth, David Wilson, and Austin Voss. Thanks also to my many colleagues, graduate students, postdoctoral fellows, research associates, technicians and collaborators too numerous to list here, that I have worked with in this field and from whom I have learned so much. Finally, I thank my parents for instilling a boundless curiosity in me, and my wife and children for the kind patience and support they showed while I wrote this book.

# Dedication

To Susan, Chris, Paul and Jenise.

# 1
# Introduction

Aerosols (gasborne suspensions of liquid or solid particles) are commonly used as a means of delivering therapeutic drugs to the lung for the treatment of lung diseases. Indeed, most people probably know someone who is taking asthma medication from an inhaler or nebulizer. However, inhaled pharmaceutical aerosols (IPAs) can also be used to deliver drugs to the bloodstream by depositing the drug in the alveolar regions (where the deposited drug crosses the alveolar epithelium and enters the blood in the capillaries). This latter approach allows treatment of the entire body by inhaled aerosols, and has opened up the field of inhaled pharmaceutical aerosols tremendously, since drugs traditionally administered by injection (such as vaccines) can potentially be administered by inhalation.

There are many successful inhaled pharmaceutical aerosol delivery systems available. Traditional systems include propellant (pressurized) metered dose inhalers based on aerosol container technology, dry powder inhalers (in which a small amount of powder is dispersed into a breath), and nebulizers (which produce a 'mist' that is inhaled through a mouthpiece or mask). Various new devices and technologies continue to be developed and introduced. Understanding and designing these various delivery methods requires an understanding of inhaled pharmaceutical aerosol mechanics (for which we introduce the abbreviation IPAM).

However, before understanding IPAs, it is useful to briefly mention the more traditional drug delivery routes. Probably the most familiar are the oral route (i.e. swallowing a pill, tablet or elixir) or needle administration (which includes subcutaneous injection, intramuscular injection and intravenous administration). Other less familiar delivery routes are also used (e.g. transdermal, buccal, nasal, etc.). There are advantages and disadvantages to each of these routes of administration, some of which are summarized in Table 1.1 for the aerosol route and the two most common nonaerosol routes.

If the aerosol route is chosen as the delivery method for a new drug under development, it is not easy to develop an appropriate aerosol delivery system. There are a number of reasons for this, including the difficulty in efficiently producing small aerosol particles, and the difficulty in consistently delivering a reliable dose to the appropriate parts of the respiratory tract (partly because of different inhalation patterns and lung geometries among different individuals). In addition, ergonomic considerations eliminate many possible designs, since cost, portability, delivery times and ease-of-use are important issues that can dramatically affect marketability and patient compliance (i.e. whether patients take the prescribed dose at the prescribed frequency). However, the advantages of the aerosol route given in Table 1.1 are often enough to warrant its use for a particular medication. It is then imperative that an understanding of inhaled pharmaceutical aerosol mechanics be invoked in designing and using the delivery

1

**Table 1.1**   Advantages and disadvantages of inhaled pharmaceutical aerosols compared to oral delivery and injection

| Route | Advantages | Disadvantages |
|---|---|---|
| Oral | • Safe<br>• Convenient<br>• Inexpensive | • Unpredictable and slow absorption (e.g. foods ingested with drug can affect drug)<br>• For lung disease: drug not localized to the lung (systemic side effects may occur)<br>• Large drug molecules may be inactivated |
| Needle | • Predictable and rapid absorption (particularly with i.v.) | • Requires special equipment and trained personnel (e.g. sterile solutions)<br>• Improper i.v. can cause fatal embolism<br>• For lung disease: drug not localized to the lung (systemic side effects may occur)<br>• Infection |
| Inhaled aerosol | • Safe<br>• Convenient<br>• Rapid and predictable onset of action<br>• Decreased adverse reactions<br>• Smaller amounts of drug needed (particularly for topical treatment of lung diseases) | • May have decreased therapeutic effect, e.g. in severe asthma other routes may be more beneficial<br>• Unpredictable and variable dose<br>• For systemic delivery: some drugs poorly absorbed or inactivated |

system, since otherwise a suboptimal delivery system usually results, reducing the effectiveness and market potential of the drug.

Several mechanical parameters are important in determining the effectiveness of an IPA. One of the most important is the size of the inhaled aerosol particles, since this determines where aerosols will deposit in the lung. If the inhaled particles are too large, they will tend to deposit in the mouth and throat (which is undesirable if the lung is the targeted organ), while if the particles are too small, they will be inhaled and then exhaled right back out with little deposition in the lung (again undesirable since this wastes medication and results in unintended airborne release of the drug). The number of particles $m^{-3}$ and surface properties of the inhaled aerosol can also affect the fate of the particles as they travel through the respiratory tract via evaporation or condensation effects; very high inhaled concentrations may also induce coughing and prevent proper inhalation. Inhalation flow rate, as well as particle surface properties, can affect aerosol generation, and the former can also strongly affect aerosol deposition in the lung. Finally, respiratory tract geometry affects the location of deposition, which can alter the effect of an inhaled aerosol.

Understanding how these various factors can affect an IPA requires combining aspects of a wide range of traditional science and engineering areas, including aerosol mechanics, single-phase and multiphase fluid mechanics, interfacial science, pharmaceutics, respiratory physiology and anatomy, and pulmonology. It is the purpose of this book to introduce the relevant aspects of these various disciplines that are needed to yield an introductory understanding of inhaled pharmaceutical aerosol mechanics.

# 2
# Particle Size Distributions

Most inhaled pharmaceutical aerosols are 'polydisperse', meaning that they contain particles of different sizes. (Under controlled laboratory conditions, 'monodisperse' aerosols that consist of particles of a single size can be produced, but such aerosol generation methods are usually not practical for inhaled pharmaceutical purposes.) Because particle size is one of the most important attributes of inhaled pharmaceutical aerosols, it is important to characterize their particle size distributions. The measurement of inhaled pharmaceutical aerosols has received considerable treatment in the literature (see Mitchell and Nagel 1997 for a recent perspective) and large parts of readily available aerosol texts are devoted to methods of measuring particle sizes (Willeke and Baron 1993). The reader is referred to these references for more information on this topic. Instead, the purpose of the present chapter is to give the reader a basic theoretical understanding of particle size distributions.

## 2.1 Frequency and count distributions

One way to describe the size distribution of an aerosol is to define its frequency distribution $f(x)$, defined such that the fraction of particles that have diameter between $x$ and $x + dx$ is $f(x)dx$, as shown schematically in Fig. 2.1.

Note that $f$ by itself has no physical meaning. It is only when $f(x)$ is integrated between two values of $x$ that its meaning occurs, i.e. $\int_a^b f(x)dx$ is the fraction of particles with diameter between $a$ and $b$.

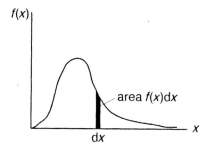

**Fig. 2.1** The fraction of aerosol particles between diameter $x$ and $x + dx$ in an aerosol is given by $f(x)dx$, where $f(x)$ is the frequency distribution.

A size distribution definition that is related to the frequency distribution is the count distribution $n(x)$, defined so that the number of particles having diameters between $x$ and $dx$ is $n(x)dx$. The count distribution $n(x)$ is related to $f(x)$ by

$$n(x) = N_{\text{particles}} f(x) \tag{2.1}$$

where $N_{\text{particles}}$ is the number of particles in the aerosol. Note that $n(x)$ is often defined instead on a unit volume basis, in which case $N_{\text{particles}}$ is the number of particles per unit volume (e.g. particles m$^{-3}$).

If we integrate $f(x)dx$ over all particle diameters, we obtain the total fraction of aerosol in all sizes, which is of course equal to one, so we have

$$\int_0^\infty f(x)dx = 1 \tag{2.2}$$

Equations (2.1) and (2.2) imply that if we integrate $n(x)dx$ over all particle diameters we obtain the number of particles, i.e.

$$\int_0^\infty n(x)dx = N_{\text{particles}} \tag{2.3}$$

The mean diameter (specifically, the 'count mean diameter') of an aerosol can be obtained by considering the case where we have a discrete count distribution such that there are $n_i$ particles of different sizes $x_i$. Then, the average diameter is defined in the usual way as

$$\bar{x} = \frac{\sum x_i n_i}{\sum n_i} \tag{2.4}$$

This can be generalized to the case where we have a continuous count distribution $n(x)$, so that the count mean diameter, $\bar{x}$, is given by

$$\bar{x} = \frac{\int_0^\infty x n(x)dx}{\int_0^\infty n(x)dx} \tag{2.5}$$

Substituting Eqs (2.1) and (2.3) into Eq. (2.5) we have

$$\bar{x} = \frac{\int_0^\infty x N_{\text{particles}} f(x)dx}{N_{\text{particles}}} \tag{2.6}$$

or simply

$$\bar{x} = \int_0^\infty x f(x)dx \tag{2.7}$$

### 2.1.1 The log-normal distribution

Experimentally, the functions $f$ or $n$ are only known at a finite number of points, but we may be able to curve fit an analytical formula through these points to give us a continuous distribution. A commonly used form for $f$ that works well for many inhaled pharmaceutical aerosols is the so-called log-normal distribution, defined as

$$f(x) = \frac{1}{x\sqrt{2\pi}\ln\sigma_g}\exp\left[\frac{-(\ln x - \ln x_g)^2}{2(\ln\sigma_g)^2}\right] \tag{2.8}$$

This is simply a normal (Gaussian) distribution in which the $x$-axis is replaced with $\ln x$ (thus the name log-normal). The log-normal distribution is uniquely specified if we know the two parameters $x_g$ and $\sigma_g$, where $x_g$ is the count median diameter (also called the geometric mean diameter) and $\sigma_g$ is the geometric standard deviation (sometimes abbreviated to GSD). These two parameters are defined as follows:

The value of $x_g$ is simply the diameter at the median value of $f$, i.e. $\int_0^{x_g} f(x)dx = 1/2$ where $f(x)$ is given by Eq. (2.8). It can be shown that for the log-normal distribution $x_g$ satisfies

$$\ln x_g = \int_0^{\infty} \ln x\, f(x)dx \tag{2.9}$$

Note that for mass distributions (defined in the next section) that are log-normal, $x_g$ would be the mass median diameter (MMD) instead of the count median diameter.

The parameter $\sigma_g$ is called the geometric standard deviation (GSD), defined as the standard deviation of the logarithm of particle diameter, i.e.

$$\ln \sigma_g = \left[ \int_0^{\infty} \left(\ln x - \ln x_g\right)^2 f(x)dx \right]^{1/2} \tag{2.10}$$

By drawing parallels with the normal (Gaussian) distribution, it can be shown that for a log-normal distribution, 68% of the particles have a diameter between $x_g/\sigma_g$ and $x_g \times \sigma_g$ (recall that for a normal distribution, 68% of the distribution lies in the region defined by the mean $\pm$ standard deviation). For a monodisperse aerosol, $\sigma_g = 1$.

### 2.1.2 Cumulative distributions

Instead of defining an aerosol size distribution in terms of the number or fraction of particles that have a certain diameter, it is often convenient to instead define a size distribution that gives the number or fraction of particles that are smaller than a given size. Such a distribution is referred to as a cumulative size distribution. For example, the cumulative frequency distribution $F(d)$ gives the fraction of particles, $F$, having diameter less than $d$, and is related to the frequency distribution $f(x)$ by

$$F(d) = \int_0^{d} f(x)dx \tag{2.11}$$

Similarly, the cumulative number distribution $N(d)$ gives the number of particles $N$ having diameter less than diameter $d$, and is defined in terms of the count distribution $n(x)$ by

$$N(d) = \int_0^{d} n(x)dx \tag{2.12}$$

## 2.2 Mass and volume distributions

Although frequency and count distributions may be a useful way of thinking of aerosol size distributions, a much more commonly used way of presenting data for pharmaceutical aerosols is the mass distribution, since it is the mass of drug delivered by a particle that determines its therapeutic effect.

The mass distribution $m(x)$ is defined such that $m(x)dx$ is the mass of particles having diameters between $x$ and $x + dx$. The normalized mass distribution $m_{normalized}(x)$ is defined as the fraction of aerosol mass contained in particles having diameters between $x$ and $x + dx$, and is given by

$$m_{normalized}(x) = m(x)/(\text{total mass of aerosol}) \tag{2.13}$$

For a log-normal distribution,

$$m_{normalized}(x) = \frac{1}{x\sqrt{2\pi}\ln\sigma_g}\exp\left[\frac{-(\ln x - \ln MMD)^2}{2(\ln\sigma_g)^2}\right] \tag{2.14}$$

where $MMD$ is the mass median diameter, defined so that half of the mass of the aerosol is contained in particles with diameters $\leq MMD$, as we shall see in the next section.

The normalized volume distribution $v_{normalized}(x)$ and volume distribution $v(x)$ are defined just like $m_{normalized}(x)$ and $m(x)$ but in terms of volume rather than mass.

Note that the above definitions of mass and volume distributions are often defined on a unit volume basis, e.g. $m(x)dx$ can be defined as the mass of particles *per unit volume* with diameters between $x$ and $x + dx$.

If we have an aerosol consisting of particles that all have the same density, then $m_{normalized}(x)$ and $v_{normalized}(x)$ are identical since in this case mass and volume are directly proportional with the constant of proportionality, the density, canceling out in the normalization.

For spherical particles, we can obtain $v_{normalized}(x)$ directly from the frequency $f(x)$ distribution as follows. First, the volume of a single spherical particle of diameter $x$ is $\pi x^3/6$, so that the volume of $n$ particles of this size is then

$$v(x) = n(x)\pi x^3/6 \tag{2.15}$$

The total volume of particles is then

$$\int_0^\infty v(x)dx = \int_0^\infty n(x)\frac{\pi x^3}{6}dx \tag{2.16}$$

The fraction of volume occupied by particles of diameter $x$ is simply $v_{normalized}(x)$, which is the volume of $n$ particles of size $x$ divided by the total volume, i.e.

$$v_{normalized}(x) = \frac{n(x)\dfrac{\pi x^3}{6}}{\int_0^\infty n(x)\dfrac{\pi x^3}{6}dx} \tag{2.17}$$

Using Eq. (2.1) and simplifying gives

$$v_{normalized}(x) = \frac{f(x)x^3}{\int_0^\infty f(x)x^3dx} \quad \text{(for spherical particles only)} \tag{2.18}$$

Equation (2.18) allows us to specify the normalized volume distribution if we know the frequency distribution. Note that for spherical particles, a log-normal frequency or count distribution will give rise to a log-normal volume distribution with the same geometric standard deviation. This result follows from the fact that if $f(x)$ is log-normal, then it can be shown that $x^k f(x)$ is also log-normal (with the same $\sigma_g$ but different mean).

As noted by Crow and Shimizu (1988), the above volume and mass distributions are different from those defined in mathematical statistics where, for example, the volume distribution is defined in terms of volume (rather than diameter as used here). This difference in definitions results in some differences in the meaning of various mean diameters, but we have chosen to follow the conventions of the aerosol literature.

## 2.3 Cumulative mass and volume distributions

Just as we defined cumulative distributions for the frequency and count distributions, we can define cumulative distributions for mass and volume distributions. Thus, the cumulative mass distribution $M(d)$ gives the mass $M$ contained in all particles having diameter less than $d$, and is related to the mass distribution $m(x)$ by

$$M(d) = \int_0^d m(x)dx \qquad (2.19)$$

The normalized cumulative mass distribution is defined as

$$M_{\text{normalized}}(d) = \int_0^d m_{\text{normalized}}(x)dx \qquad (2.20)$$

Equations (2.19) and (2.20) imply

$$M_{\text{normalized}}(d) = M(d)/(\text{total mass of aerosol particles})$$

Similarly, the cumulative volume distribution $V(d)$ is defined as the volume $V$ of the total aerosol volume contained in particles having diameter less than $d$, and is related to the volume distribution $v(x)$ by

$$V(d) = \int_0^d v(x)dx \qquad (2.21)$$

The normalized cumulative volume distribution is defined as

$$V_{\text{normalized}}(d) = \int_0^d v_{\text{normalized}}(x)dx \qquad (2.22)$$

from which it follows that

$$V_{\text{normalized}}(d) = V(d)/(\text{total volume of aerosol particles}) \qquad (2.23)$$

These definitions are often made on a unit volume basis, e.g. $M(d)$ can be defined as the mass of particles with diameter $\leq d$, per m$^3$ of aerosol.

Note that for aerosols having particles all of the same density $M_{\text{normalized}}(d) = V_{\text{normalized}}(d)$.

An important definition that is made using the normalized cumulative mass distribution is the mass median diameter ($MMD$), which we defined earlier as the diameter such that half the mass of the aerosol particles is contained in particles with larger diameter and half is contained in particles with smaller diameter, i.e. $M_{\text{normalized}}(MMD) = 1/2$.

### 2.3.1 Obtaining $\sigma_g$ for log-normal distributions

For log-normal distributions, the geometric standard deviation can be determined approximately from the normalized mass distribution using the fact that 68% of the particle mass is contained between $MMD/\sigma_g$ and $MMD \times \sigma_g$. This implies that 34% of the aerosol mass must lie between $MMD/\sigma_g$ and $MMD$ and 34% must lie between $MMD$ and $MMD \times \sigma_g$. However, since 50% of the aerosol mass is contained in particles with diameters less than $MMD$, $50\% + 34\% = 84\%$ of the aerosol mass must lie between diameters $x = 0$ and $x = MMD \times \sigma_g$ and similarly, $50\% - 34\% = 16\%$ must lie between diameters $x = 0$ and $x = MMD \sigma_g$. In other words

$$\sigma_g = d_{84}/MMD \qquad (2.24)$$

where $d_{84}$ is the diameter at which 84% of the aerosol mass is contained in diameters less than this diameter. Also,

$$\sigma_g = MMD/d_{16} \qquad (2.25)$$

where $d_{16}$ is the diameter at which 16% of the aerosol mass is contained in diameters less than this diameter.

Multiplying Eqs (2.24) and (2.25), we obtain

$$\sigma_g^2 = d_{84} \, d_{16} \qquad (2.26)$$

or

$$\sigma_g = (d_{84} \, d_{16})^{1/2} \qquad (2.27)$$

which is a simple, commonly used expression to estimate the geometric standard deviation, $\sigma_g$, of a log-normal distribution. Note the mass median diameter and $\sigma_g$ for experimentally measured size distributions are usually estimated using a nonlinear regression fit of Eq. (2.14) to the data (for example, using methods described in Chapter 7 of Albert and Gardner 1967). However, Eqs (2.24)–(2.27) provide a commonly used first approximation for $\sigma_g$.

### 2.3.2 Obtaining the total mass of an aerosol from its *MMD*, $\sigma_g$ and number of particles/unit volume

A useful equation can be derived that gives the mass of particulate matter of an aerosol for a log-normal aerosol, as follows.

Consider an aerosol consisting of $N$ spherical particles per m³ with uniform particle density $\rho$ and having a log-normal size distribution with known mass median diameter $MMD$ and geometric standard deviation $\sigma_g$. The total volume, $V$, of particles per unit volume is given from Eq. (2.21) as

$$V = \int_0^\infty v(x)dx \qquad (2.28)$$

For spherical particles we can substitute Eq. (2.15) for $v(x)$, and we have

$$V = \int_0^\infty n(x)\frac{\pi x^3}{6}dx \qquad (2.29)$$

where we interpret $n(x)$ as the total number of particles per unit volume. From Eq. (2.1), we have $n(x) = N f(x)$ where $N$ is the total number of particles per unit volume, so that Eq. (2.29) becomes

$$V = \int_0^\infty N f(x)\frac{\pi x^3}{6}dx \qquad (2.30)$$

The total mass, $M$, of particles per unit volume is then simply $\rho V$, i.e.

$$M = \frac{\rho N\pi}{6} \int_0^\infty f(x)x^3 \, dx \tag{2.31}$$

Using the definition of the log-normal function $f(x)$ from Eq. (2.8), we have

$$M = \frac{\rho N\pi}{6\sqrt{2\pi} \ln \sigma_g} \int_0^\infty x^2 \exp\left[\frac{-(\ln x - \ln x_g)^2}{2(\ln \sigma_g)^2}\right] dx \tag{2.32}$$

Defining $u = \ln x$, which implies $x = e^u$ and $dx = e^u du$, a substitution of variables in Eq. (2.32) gives

$$M = \frac{\rho N\pi}{6\sqrt{2\pi} \ln \sigma_g} \int_{-\infty}^\infty e^{3u} \exp\left[\frac{-(u - \ln x_g)^2}{2(\ln \sigma_g)^2}\right] du \tag{2.33}$$

which can be written as

$$M = \frac{\rho N\pi}{6\sqrt{2\pi} \ln \sigma_g} \int_{-\infty}^\infty \exp\left[\frac{6u(\ln \sigma_g)^2 - u^2 + 2u \ln x_g - (\ln x_g)^2}{2(\ln \sigma_g)^2}\right] du \tag{2.34}$$

Completing the square inside the exponential gives

$$M = \exp\left[3 \ln x_g + \frac{9}{2}(\ln \sigma_g)^2\right] \frac{\rho N\pi}{6} \left\{\frac{1}{b\sqrt{2\pi}} \int_{-\infty}^\infty \exp\left[\frac{(u - c)^2}{2b^2}\right] du\right\} \tag{2.35}$$

where $c = \ln x_g + 3 \ln \sigma_g$. The term in curly brackets is simply the integral of the normal distribution from $-\infty$ to $\infty$, which is 1, so Eq. (2.35) can be written

$$M = \frac{\rho N\pi}{6} \exp\left[3 \ln x_g + \frac{9}{2}(\ln \sigma_g)^2\right] \tag{2.36}$$

which simplifies to

$$M = \frac{\rho N\pi}{6} x_g^3 \exp\left[\frac{9}{2}(\ln \sigma_g)^2\right] \tag{2.37}$$

Using the log-normal function, it can be shown that

$$MMD = x_g \exp[3(\ln \sigma_g)^2] \tag{2.38}$$

so that we can rewrite Eq. (2.37) as

$$M = \frac{\rho N\pi}{6} (MMD)^3 \exp\left[-\frac{9}{2}(\ln \sigma_g)^2\right] \tag{2.39}$$

which gives the mass of aerosol per unit volume from its $MMD$, geometric standard deviation, $\sigma_g$, number of particles per unit volume, $N$, and mass density, $\rho$, of the material making up the particles.

## 2.4 Other distribution functions

Aerosol size distributions are not always log-normal, and different continuous functions have been developed to describe such distributions (see Hinds 1982 for brief descriptions of a few of these). One of the most commonly used distributions other than the log-normal one is the Rosin–Rammler distribution (Rosin and Rammler 1933), given by the cumulative mass distribution

$$M_{\text{normalized}}(d) = 1 - \exp[-(d/\delta)^n] \tag{2.40}$$

where $\delta$ and $n$ are specified constants (just as the $MMD$ and $\sigma_g$ specify the log-normal distribution). Crowe et al. (1998) discuss several aspects of this distribution.

One shortcoming of the log-normal and Rosin–Rammler distributions is the lack of an upper and lower limit on particle sizes. In reality, there is zero probability of having particles larger or smaller than a certain size in an aerosol, whereas the log-normal and Rosin–Rammler distributions give nonzero probabilities. To overcome this limitation, various other distributions have been suggested, including the log-hyperbolic distribution (Xu et al. 1993) and the beta distribution (Popplewell et al. 1988). These distributions have not been widely used, partly because many inhaled pharmaceutical aerosol distributions are reasonably well-approximated by the log-normal distribution, and partly because of the long history of assuming log-normal distributions with inhaled pharmaceutical aerosols.

## 2.5 Summary of mean and median aerosol particle sizes

Consider an aerosol having count distribution given by $n(x)$, frequency distribution $f(x)$, mass distribution $m(x)$, normalized mass distribution $m_{\text{normalized}}(x)$, cumulative mass distribution $M(x)$, normalized cumulative mass distribution $M_{\text{normalized}}(x)$, normalized volume distribution $v_{\text{normalized}}(x)$, and cumulative volume distribution $V(x)$. Discrete values of these functions at the discrete diameters $d_i$ are given by $n_i$, $M_i$, $f_i$, etc. The following definitions of mean and median particle sizes can then be made:

$$\text{count mean diameter} = \int_0^\infty x f(x) \, dx = \frac{\int_0^\infty x n(x) \, dx}{\int_0^\infty n(x) \, dx} \approx \frac{\sum n_i d_i}{\sum n_i} \tag{2.41}$$

The count mean diameter is also called the arithmetic mean diameter.

$$\text{mass mean diameter} = \int_0^\infty x m_{\text{normalized}}(x) \, dx = \frac{\int_0^\infty x m(x) \, dx}{\int_0^\infty m(x) \, dx} \approx \frac{\sum m_i d_i}{\sum m_i} \tag{2.42}$$

$$\text{volume mean diameter} = \int_0^\infty x v_{\text{normalized}}(x) \, dx = \frac{\int_0^\infty x v(x) \, dx}{\int_0^\infty v(x) \, dx} \approx \frac{\sum v_i d_i}{\sum v_i} \tag{2.43}$$

For an aerosol consisting of particles all having the same density, the volume mean diameter is equal to the mass mean diameter.

The geometric mean diameter $x_g$ is defined such that

$$\ln x_g = \int_0^\infty \ln x f(x) \, dx = \frac{\int_0^\infty \ln x n(x) \, dx}{\int_0^\infty n(x) \, dx} \approx \frac{\sum n_i \ln d_i}{\sum n_i} \tag{2.44}$$

The count median diameter, $CMD$, is the diameter such that half the particles have larger diameter and half have smaller diameter, i.e. $F(CMD) = 1/2$. For log-normal distributions the count median and geometric mean diameter have the same value.

The mass median diameter ($MMD$) is the diameter such that half the mass of the aerosol particles is contained in particles with larger diameter and half is contained in particles with smaller diameter, i.e. $M_{\text{normalized}}(MMD) = 1/2$.

The volume median diameter ($VMD$) is the diameter such that half the volume of the aerosol particles is contained in particles with larger diameter and half is contained in particles with smaller diameter, i.e. $V_{\text{normalized}}(VMD) = 1/2$. A common abbreviation for the volume median diameter used with software for particle measurement devices is Dv0.5, the '0.5' indicating that 1/2 of the volume of the aerosol is contained in particles up to this size.

Assuming spherical particles (which is a necessary assumption in order to relate mass to diameter via $\pi x^3/6$), the following equations are valid:

$$\text{diameter having the average mass} = \left(\int_0^\infty x^3 f(x)\,dx\right)^{1/3}\text{(for constant particle density)} \tag{2.45}$$

$$\text{diameter having the average area} = \left(\int_0^\infty x^2 f(x)\,dx\right)^{1/2} \tag{2.46}$$

$$\text{volume mean diameter} = \frac{\int_0^\infty f(x)x^4\,dx}{\int_0^\infty f(x)x^3\,dx} \tag{2.47}$$

$$\text{Sauter mean diameter } (SMD) = \frac{\int_0^\infty x^3 f(x)\,dx}{\int_0^\infty x^2 f(x)\,dx} \approx \frac{\sum n_i d_i^3}{\sum n_i d_i^2} \tag{2.48}$$

The Sauter mean diameter is often used in the literature on sprays and atomization since such sprays are often found to obey Simmons' universal root normal distribution, which is an empirical observation that for many atomization sprays the ratio $MMD/SMD = 1.2$ (Simmons 1977). For a spray having a Simmons universal root normal distribution, only one parameter is needed to characterize the particle size distribution (e.g. the Sauter mean diameter).

The Sauter mean diameter is a specific form of the general definition of $d_{nm}$ commonly used in the literature on sprays. Here $n$ and $m$ are integers and the definition of $d_{nm}$ is

$$d_{nm} = \left[\frac{\int_0^\infty x^n f(x)\,dx}{\int_0^\infty x^m f(x)\,dx}\right]^{1/(n-m)} \approx \left[\frac{\sum n_i d_i^n}{\sum n_i d_i^m}\right]^{1/(n-m)} \tag{2.49}$$

The Sauter mean diameter is thus $d_{32}$, the volume mean diameter is $d_{43}$ and the count mean diameter is $d_{10}$.

For log-normal distributions of spherical particles all having the same density, the following equations are valid

$$\text{count median diameter} = \text{geometric mean diameter} = x_g \tag{2.50}$$

$$\text{count mean diameter} = x_g \exp[0.5(\ln \sigma_g)^2] \tag{2.51}$$

$$\text{count mode diameter (i.e. diameter at frequency peak)} = x_g \exp[-(\ln \sigma_g)^2] \tag{2.52}$$

$$\text{diameter of particle having average area} = x_g \exp[(\ln \sigma_g)^2] \tag{2.53}$$

$$\text{diameter of particle having average mass} = x_g \exp[1.5(\ln \sigma_g)^2] \tag{2.54}$$

$$MMD \text{ or volume median diameter } (VMD) = x_g \exp[3(\ln \sigma_g)^2] \tag{2.55}$$

$$\text{mass mean diameter or volume mean diameter} = x_g \exp[3.5(\ln \sigma_g)^2] \tag{2.56}$$

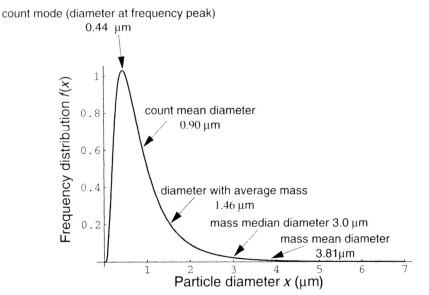

**Fig. 2.2** The frequency distribution $f(x)$ is shown for a log-normal distribution with mass median diameter of 3.0 μm and geometric standard deviation $\sigma_g = 2.0$, along with the values of various other diameter definitions.

Shown in Fig. 2.2 is a log-normal frequency distribution with mass median diameter ($MMD$) = 3.0 μm and $\sigma_g = 2.0$. Also shown are some of the above defined diameters.

Notice that $f(x)$ can be greater than 1, and is indeed greater than 1 near the count mode for the distribution in Fig. 2.2. There is no reason why $f(x)$ cannot be greater than 1, since $f(x)$ only has meaning when it is multiplied by a diameter range $dx$, so that $f(x)dx$ is the fraction of particles between $x$ and $x + dx$. In fact, as $\sigma_g$ approaches one, $f(x)$ at the count mode increases without bound, since we must have $\int_0^\infty f(x)dx = 1$.

See Reist (1984) for further definitions of various diameters and their relations.

## Example 2.1

The aerosol emitted by a salbutamol metered dose inhaler was collected on a cascade impactor, and Table 2.1 shows the amounts of drug, determined by chemical assay, on each impactor plate. Assuming this distribution is log-normal:

(a) Estimate the $MMD$ and geometric standard deviation ($\sigma_g$) of this distribution using simple methods.
(b) Provide an estimate for the particle size that will have the largest number of particles for this distribution.
(c) What factors may cause errors in your estimates?

## *Solution*

(a) The simplest way to calculate the $MMD$ is to linearly interpolate the data to find the value of $d$ at which $M_{normalized} = 50\%$. The points we must interpolate between are

Table 2.1 An experimentally determined cumulative mass distribution for a salbutamol metered dose inhaler

| $d$ (μm) | Cumulative mass under size (i.e. $M_{normalized} \times 100$) (%) |
|---|---|
| 0.4 | 0 |
| 0.7 | 1.85 |
| 1.1 | 6.66 |
| 2.1 | 40.58 |
| 3.3 | 84.06 |
| 4.7 | 96.22 |
| 5.8 | 98.1 |
| 9 | 98.77 |

$M_{normalized} = 40.58\%$ at $d = 2.1$ μm and $M_{normalized} = 84.06\%$ at $d = 3.3$ μm. Linear interpolation gives

$$M_{normalized} = 50\% \text{ at } d \approx 2.1 \text{ μm} + \left( \frac{50\% - 40.58\%}{84.06\% - 40.58\%} \right)(3.3 \text{ μm} - 2.1 \text{ μm})$$

$$= 2.4 \text{ μm}$$

i.e. MMD = 2.4 μm.
To estimate $\sigma_g$, we can use Eq. (2.24)

$$\sigma_g = (\text{diameter with 84\% cumulative mass})/MMD$$

Thus, we have $\sigma_g = 3.3$ μm/2.4 μm, or $\sigma_g = 1.4$.

(b) The largest number of particles will occur at the peak in the number distribution $n(x)$, which will be the same diameter as the peak in the frequency distribution curve $f(x)$. To estimate this diameter, we can use Eq. (2.52), i.e.

$$\text{count mode diameter} = x_g \exp[-(\ln \sigma_g)^2]$$

where $x_g$ is the count median diameter, which we do not know yet. However, we can obtain $x_g$ from Eq. (2.55) relating $x_g$ to the volume median diameter ($VMD$):

$$\text{volume median diameter } (VMD) = x_g \exp[3 (\ln \sigma_g)^2]$$

where $VMD = MMD$ if we assume all the particles in the aerosol have the same density. Thus using

$$MMD = x_g \exp[3 (\ln \sigma_g)^2]$$

we have

$$x_g = MMD/\exp[3 (\ln \sigma_g)^2]$$

and substituting $MMD = 2.4$ μm from above, we obtain

$$x_g = 2.4 \text{ μm}/ \exp[3 (\ln 1.4)^2]$$

$$x_g = 1.7 \text{ μm}$$

We can now use Eq. (2.52) to obtain

$$\text{count mode diameter} = x_g \exp[-(\ln \sigma_g)^2]$$
$$= 1.7\ \mu\text{m} \exp[-\ln (1.4)^2]$$
$$= 1.0\ \mu\text{m}$$

Our estimate for the particle size with the largest number of particles is thus 1.0 μm.

(c)  Several factors may cause errors in the above estimates, including the following.

- The use of simple linear interpolation is somewhat inaccurate – this can be corrected by performing a nonlinear regression fit of the log-normal distribution (Eq. 2.14) to the data, and is commonly done (for example, using the methods described in Chapter 7 of Albert and Gardner, 1967).
- Experimental error in the cascade impactor data will cause all calculations to be approximate (since the size cut-offs for each impactor stage are not perfectly sharp).
- If the distribution is not exactly log-normal, the GSD and all results in (b) will be in error (one can plot the data for $M_{\text{normalized}}$ and compare to an integrated log-normal function with the calculated MMD and GSD to see how different they are).
- The particles are not spherical, so (b) will be in error since it was assumed that the volume of each particle was $\pi d^3/6$ to obtain the count distribution from the volume distribution.
- If the densities of the particles are not all the same then in (b) the VMD and MMD may be different.
- The cascade impactor actually measures aerodynamic diameter, so that our estimate for MMD is actually the mass median aerodynamic diameter (MMAD), where $MMAD = (\text{specific gravity})^{1/2} MMD$ (see Chapter 3), so that if the particles have a specific gravity $\neq 1$, our estimate of MMD will be off by the factor (specific gravity)$^{1/2}$.

# References

Albert, A. E. and Gardner, L. A. (1967) Stochastic approximation and non-linear regression. Research Monograph No. 42, The MIT Press, Cambridge, MA.

Crow, E. L. and Shimizu, K. (1988) *Lognormal Distributions*, Marcel Dekker, New York.

Crowe, C., Sommerfeld, M. and Tsuji, Y. (1998) *Multiphase Flow with Droplets and Particles*, CRC Press, New York.

Hinds, W. C. (1982) *Aerosol Technology: Properties, Behaviour, and Measurement of Airborne Particles*, Wiley, New York.

Mitchell, J. P. and Nagel, M. W. (1997) Medical aerosols: techniques for particle size evaluation, *Particulate Sci. Technol.* **15**:217–241.

Popplewell, L. M., Campanella, O. H., Normand, M. D. and Peleg, M. (1988) Description of normal, log-normal and Rosin–Rammler particle populations by a modified version of the beta distribution function, *Powder Technol.* **54**:119–135.

Reist, P. C. (1984) *Introduction to Aerosol Science*, MacMillan, New York.

Rosin, P. and Rammler, E. (1933) *J. Inst. Fuel* **7**:29.

Simmons, H. C. (1977) The correlation of drop size distributions in fuel nozzle sprays, *J. Eng. Power* **99**:309–319.

Willeke, K. and Baron, P. A. (1993) *Aerosol Measurement: Principles, Techniques and Applications*, Van Nostrand Reinhold, New York.

Xu, T.-H., Durst, F. and Tropea, C. (1993) The three-parameter log-hyperbolic distribution and its application to particle sizing, *Atom. Sprays* **3**:109.

# 3
# Motion of a Single Aerosol Particle in a Fluid

Much of aerosol mechanics can be understood by studying the motion of a single particle in a fluid. For this reason, it is useful to look at the forces and equations that govern the motion of a single, isolated, particle moving through a fluid.

At first it might seem that it should not be too difficult to obtain the trajectory of a particle in a fluid flow, since this is a relatively simple system -- we have a small, rigid particle all by itself moving through air. However, the task is not simple, and large parts of entire books have been written on this subject (Clift *et al.* 1978, Happel and Brenner 1983). Fortunately, for most of our purposes we can examine a simplified version of this problem that arises with the following two major simplifying assumptions: (1) the particle is assumed to be spherical; and (2) the particle density is assumed to be much larger than the surrounding fluid density.

Much of the work in aerosol science is based on these two assumptions. For inhaled pharmaceutical aerosols, assumption (1) is usually reasonable. It is certainly reasonable for liquid inhaled pharmaceutical aerosol droplets, since such small liquid droplets are spherical. For dry powder aerosols and evaporated metered dose inhaler aerosols, an assumption of sphericity is not exact, but most such aerosols consist of reasonably compact particles, so that the drag on these particles is often not far from that on a sphere.

Assumption (2) (i.e. $\rho_{particle} \gg \rho_{fluid}$) is usually quite reasonable for inhaled pharmaceutical aerosols, since the densities of pharmaceutical compounds are typically near that of water, which is 1000 times the density of air, so $\rho_{particle} \approx 10^3 \rho_{fluid}$.

By invoking the first assumption, we can make use of the vast body of work that has been done on the motion of spheres in fluids. It might be thought that this finally makes the problem easy to solve, but this is not necessarily true. In fact, it was not until 1983 (Maxey and Riley 1983) that a relatively complete development of the equation governing the motion of a spherical particle in a flow field was made.

The second assumption, i.e. $\rho_{particle} \gg \rho_{fluid}$, simplifies the analysis because it results in the drag force on the particle being much larger than all the other fluid forces acting on the particle (Crowe *et al.* 1998, Barton 1995). If the particle density is, instead, not much greater than the fluid density, then several fluid forces (buoyancy force, Magnus force, lift force, Basset force, pressure force, Faxen corrections and virtual mass forces) become important and make the analysis more difficult. For such cases, the reader is referred to Kim *et al.* (1998) who developed an equation, valid up to a much higher Reynolds number than previous equations, that includes all but the lift, Magnus and Faxen corrections (and which could be modified to include these forces). However, throughout this text we will largely neglect such forces because we assume that the particle is much denser than the surrounding air. The only exception to this rule is considered in Chapter 9, where powder particles near solid boundaries can experience aerodynamic lift forces.

## 3.1 Drag force

The equation of motion governing the trajectory of a particle is Newton's second law:

$$m\frac{d\mathbf{v}}{dt} = \mathbf{F} \tag{3.1}$$

where $\mathbf{F}(t)$ is the total external force exerted on the particle and $\mathbf{v}$ is its velocity. Assuming that the drag force is the only nonnegligible fluid force on the particle, and assuming that the only body force is gravity, Eq. (3.1) can be written as

$$m\frac{d\mathbf{v}}{dt} = m\mathbf{g} + \mathbf{F}_{drag} \tag{3.2}$$

To solve this equation for $\mathbf{v}(t)$ we must determine the drag force.

From known results on the drag coefficient for flow past spheres, we have the drag coefficient

$$C_d = \frac{|\mathbf{F}_{drag}|}{\frac{1}{2}\rho_{fluid}v_{rel}^2 A} \tag{3.3}$$

where $A$ is the cross-sectional area of the sphere, i.e. $A = \pi d^2/4$ where $d$ is the diameter of the particle. In Eq. (3.3), $v_{rel}$ is the magnitude of the velocity of the particle relative to the fluid, i.e.

$$v_{rel} = |\mathbf{v} - \mathbf{v}_{fluid}| \tag{3.4}$$

where $\mathbf{v}_{fluid}$ is the velocity of the fluid (many diameters away from the particle, i.e. the 'free stream' fluid velocity).

The drag force $\mathbf{F}_{drag}$ acts in the same direction as the velocity of the particle relative to the fluid, i.e. it is parallel to $\mathbf{v} - \mathbf{v}_{fluid}$. Thus, we have

$$\mathbf{F}_{drag} = -\frac{1}{2}\rho_{fluid}v_{rel}^2\frac{\pi d^2}{4}C_d\hat{\mathbf{v}}_{rel} \tag{3.5}$$

where

$$\hat{\mathbf{v}}_{rel} = \frac{\mathbf{v} - \mathbf{v}_{fluid}}{v_{rel}} \tag{3.6}$$

is the unit vector giving the drag force its direction parallel to the relative velocity of the particle, and recall that $v_{rel}$ is the speed of the particle relative to the fluid, given by Eq. (3.4).

The drag coefficient $C_d$ depends on particle Reynolds number $Re$ where

$$Re = v_{rel}\,d\,\nu \tag{3.7}$$

Here, $\nu$ is the kinematic viscosity of the fluid surrounding the particle and is given by

$$\nu = \mu/\rho_{\text{fluid}} \qquad (3.8)$$

where $\mu$ and $\rho_{\text{fluid}}$ are the dynamic viscosity and mass density, respectively, of the fluid surrounding the particle. Various empirical equations for $C_d(Re)$ based on experimental data are normally used (Crowe *et al.* 1998), one such correlation being

$$C_d = 24(1 + 0.15\, Re^{0.687})/Re \qquad (3.9)$$

However, most inhaled pharmaceutical aerosol particles have very small diameters $d$ and low velocities $v_{\text{rel}}$, so that $Re$ is small. If $Re \ll 1$, the drag coefficient of a sphere is given by

$$C_d = 24/Re \qquad (3.10)$$

which for $Re < 0.1$, gives a value of $C_d$ that is accurate to within 1%.

Combining Eqs (3.4)–(3.10), for $Re \ll 1$ we can write

$$\mathbf{F}_{\text{drag}} = -3\pi d\mu(\mathbf{v} - \mathbf{v}_{\text{fluid}}) \qquad (3.11)$$

Equation (3.11) is often referred to as Stokes law[1]. It is derived from the continuum equations of fluid motion (since Eq. (3.10) comes by solving the Navier–Stokes equations), and so is valid only for particle diameters that are much greater than the mean free molecular path (which in air at typical inhalation conditions is near 0.07 μm). Extension of Eq. (3.11) to particles with diameter $d$ near the mean free path is considered later in this chapter, while extension to larger Reynolds number is readily accomplished with correlations such as Eq. (3.9).

## 3.2 Settling velocity

A particle in stationary air will settle under the action of gravity, and reach a terminal velocity quite rapidly. The settling velocity (also referred to as the 'sedimentation velocity') is defined as the terminal velocity of a particle in still fluid.

Because the particle's velocity does not change once it reaches the settling velocity, the acceleration on the particle is zero at this velocity, so that the net force on the particle must also be zero. Assuming the only forces on the particle are the aerodynamic drag and gravity, then for a solid, nonrotating, spherical particle only a vertical drag force will be present, which must balance gravity, i.e.

$$mg = F_{\text{drag}} \qquad (3.12)$$

where $F_{\text{drag}}$ is the magnitude of the drag force. Assuming the Reynolds numbers $Re \ll 1$, we can use Eq. (3.11) for $F_{\text{drag}}$, in which the air velocity is zero ($v_{\text{fluid}} = 0$), so that Eq. (3.11) reduces to

$$F_{\text{drag}} = 3\pi d\mu v_{\text{settling}} \qquad (3.13)$$

Also, the gravity force is

$$mg = \rho_{\text{particle}} V g \qquad (3.14)$$

---

[1]It is named after George Stokes, who first determined the flow field due to a rigid sphere in translational motion through a fluid for very low Reynolds number flow (Stokes 1851).

where $V = \pi d^3/6$ is the volume of the spherical particle and $g$ is the acceleration of gravity. Equation (3.14) can thus be written

$$mg = \rho_{particle}(\pi d^3 6)g \qquad (3.15)$$

Substituting Eqns (3.13) and (3.15) into Eq. (3.12), we have

$$3\pi d\mu v_{settling} = \rho_{particle}(\pi d^3/6)g \qquad (3.16)$$

or

$$v_{settling} = \rho_{particle}\, gd^2/18\mu \qquad (3.17)$$

Equation (3.17) gives the settling velocity for a spherical particle settling under the action of gravity under the condition that $Re \ll 1$ and diameter $\gg$ mean free path. Most inhaled pharmaceutical aerosols readily satisfy the condition diameter $\gg$ mean free path, and many inhaled pharmaceutical aerosols also satisfy the condition that $Re \ll 1$, as seen in the example below. Exceptions to the condition $Re \ll 1$ are uncommon with inhaled pharmaceutical aerosols, but do occur in the entrainment of large carrier particles that occur in dry powder particles (discussed in Chapter 9), and high-speed metered dose propellant droplets (discussed in Chapter 10).

## Example 3.1

What is the Reynolds number of a 10 micron diameter spherical, budesonide powder particle (a drug used in treating asthma, specific gravity = 1.26) settling in room temperature air?

### Solution

We have

$$\rho_{particle} = 1.26 \times \text{density of water} = 1260 \text{ kg m}^{-3}$$
$$\text{viscosity of air } \mu = 1.8 \times 10^{-5} \text{ kg m}^{-1}\text{s}^{-1}$$
$$d = 10 \times 10^{-6} \text{ m}$$

which gives

$$v_{settling} = (1260 \text{ kg m}^{-3})(9.81 \text{ m s}^{-2})(10 \times 10^{-6} \text{ m})^2/(18 \times 1.8 \times 10^{-5} \text{ kg m}^{-1}\text{ s}^{-1})$$
$$= 3.8 \times 10^{-3} \text{ m s}^{-1}$$
$$= 3.8 \text{ mm s}^{-1}$$

This gives us a Reynolds number of

$$Re = U_{rel}d/v$$
$$= (3.8 \times 10^{-3} \text{ m s}^{-1}) \times (10 \times 10^{-6} \text{ m})/(1.5 \times 10^{-5} \text{ m}^2 \text{ s}^{-1})$$

where we have used Eq. (3.8) for the kinematic viscosity of air with the density of air being $\rho = 1.2 \text{ kg m}^{-3}$. Calculating the numbers, we have

$$Re = 0.0025$$

This is very much lower than 1 and so we are quite justified in using Eq. (3.11) for the drag force, and Eq. (3.17) that results from Eq. (3.11).

## 3.2.1 Settling velocities for droplets

The above discussion and Eqs (3.9), (3.10), (3.11) and (3.17) all assume solid spherical particles. If the particle is not solid, but is instead a liquid droplet, then it is possible for the relative motion of the air flowing past the droplet to induce fluid flow (internal circulation) inside the droplet. This lowers the drag force and increases the settling velocity compared to a solid sphere of the same mass and diameter. However, surface impurities on the droplet surface appear to hinder internal circulation for small droplets (see Wallis 1974 for some discussion on this). Even if surface impurities did not prevent internal circulation, the magnitude of the drag force including such circulation can be shown to be given by

$$F_{drag} = 3\pi\mu dv_{rel} \left\{ \frac{1 + 2\mu_{air}/3\mu_{drop}}{1 + \mu_{air}/\mu_{drop}} \right\} \quad (3.18)$$

where $\mu_{air}$ is the viscosity of the air surrounding the drop and $\mu_{drop}$ is the viscosity of the liquid in the drop (this result was derived independently by both Hadamard (1911) and Rybczynski (1911)). This equation differs from Stokes law by the factor in curly brackets. For water droplets in air, as well as HFA 134a propellant droplets in air at their wet bulb temperature (211 K), this factor is 0.994, and is thus negligible for such droplets.

## 3.2.2 Particle–particle interactions in settling of particles

For dense aerosols (i.e. high number concentrations), settling velocities are lower than predicted by the standard analysis (Eq. (3.17)) because the particles travel in each other's wakes, rather than in an undisturbed fluid. This effect is often referred to as 'hindered settling'.

The drag on particles in dense clouds undergoing hindered settling has not been well studied. However, we can obtain an estimate as to when this effect becomes important by using empirical correlations in the literature (e.g. Di Felice 1994, Crowe et al. 1998). These results suggest that for aerosols with particle Reynolds numbers $Re \ll 1$, hindered settling alters the Stokes drag formula by a factor $1/\alpha^{3.7}$, i.e.

$$\mathbf{F}_{drag} = -3\pi d\mu(\mathbf{v} - \mathbf{v}_{fluid})/\alpha^{3.7} \quad (3.19)$$

where $\alpha$ is the volume fraction of the continuous phase (i.e. air), and is always $< 1$. Specifically,

$$\alpha = \text{volume of air}/(\text{volume of air} + \text{volume of particles}) \quad (3.20)$$

in a given total volume of aerosol.

Notice that the drag force in Eq. (3.19) increases as the volume of particles per unit volume is increased (i.e. as the air volume fraction, $\alpha$, is decreased), which is of course why it is called hindered settling.

For the drag force to be 10% more than that for a single particle, $\alpha$ must be 0.975 or less, i.e. the aerosol needs to occupy more than 2.5% of the volume. Thus, in a cubic meter of aerosol, 0.025 m$^3$ would need to be occupied by aerosol. At a particle density of 1000 kg m$^{-3}$, this implies that 25 kg of particles must be present per m$^3$, which is

$25 \, \text{g} \, \text{l}^{-1}$. This is much higher than is normally encountered in inhaled pharmaceutical aerosol applications, and so hindered settling is negligible for such aerosols.

## 3.3 Drag force on very small particles

As mentioned earlier, Stokes law (Eq. (3.11)) is derived from the Navier–Stokes equations, which assume that the fluid surrounding the particle is a continuum. This is valid only if the diameter of the particle is very much greater than the mean free path of the fluid molecules surrounding the particle. For air at room temperature and 1 atmosphere pressure, the mean free path is 0.067 μm. For inhaled pharmaceutical aerosols, particles of interest have diameters down to 0.5 μm or so, which gives radii of 0.25 μm. This is in the range where the particle radius is not very much greater than the mean free path, and so a correction to Eq. (3.11) is required for these small particles. This correction was first suggested by Cunningham in 1910, and is thus referred to as the Cunningham slip correction factor. It is defined so that the drag coefficient for a sphere used to obtain Stokes law is replaced by

$$C_d = \frac{1}{C_c} \times \frac{24}{Re}$$

where $C_c$ is the Cunningham slip correction factor. This is an empirically determined factor. The drag force is then

$$\mathbf{F}_{drag} = -\frac{3\pi d\mu(\mathbf{v} - \mathbf{v}_{fluid})}{C_c} \qquad (Re \ll 1) \qquad (3.21)$$

Here the only restriction is that $Re \ll 1$ in order that we can use the Stokes flow solution for zero $Re$ flow past spheres. Equating the drag force with the weight of the particle as we did before to obtain the terminal settling velocity of a spherical particle, we obtain

$$v_{settling} = C_c \rho_{particle} \, g d^2 / 18\mu \qquad (3.22)$$

A simple, approximate formula for $C_c$ when $d > 0.1$ μm is

$$C_c = 1 + 2.52 \, \lambda/d \qquad (d > 0.1 \text{ μm}) \qquad (3.23)$$

where $\lambda$ is the mean free path of molecules in the fluid. For air, the mean free path at room temperature and 1 atm pressure is 0.067 μm. At other temperatures and pressures it is different, e.g. at body temperature (37°C) $\lambda = 0.072$ μm. More general and complex formulae for $C_c$ and also for $\lambda$ are given in the literature (Willeke and Baron 1993).

Note that since $C_c > 1$, the settling velocity obtained with the slip correction is larger than when this factor is neglected, i.e. noncontinuum effects result in larger settling velocities than predicted with a continuum assumption. For air at typical inhalation conditions, only for particles with diameter smaller than 1.7 μm does the Cunningham slip factor result in a correction to the drag coefficient that is larger than 10%.

### Example 3.2

Calculate the settling velocity in air of a 0.5 μm diameter spherical droplet of nebulized Ventolin[R] respiratory solution (2.5 mg ml$^{-1}$ salbutamol sulfate with 9 mg ml$^{-1}$ NaCl in water) both with and without the Cunningham slip correction factor.

### Solution

Without the Cunningham slip factor, we use Eq. (3.17)

$$v_{settling} = \rho_{particle} \, gd^2/18\mu$$

where the density of the droplet is the same as that of water (the drug and salt have negligible effect on the density). Thus, we have

$$v_{settling} = (1000 \text{ kg m}^3)(9.81 \text{ m s}^{-2})(0.5 \times 10^{-6} \text{ m})^2/18(1.8 \times 10^{-5} \text{ kg m}^{-1} \text{ s}^{-1})$$
$$= 7.6 \times 10^{-6} \text{ m s}^{-1}$$
$$= 0.0076 \text{ mm s}^{-1} \text{ (neglecting Cunningham slip factor)}$$

If we now include the Cunningham slip factor, then when calculating the drag force on the particle we must use Eq. (3.22).

$$v_{settling} = C_c \rho_{particle} \, gd^2/18\mu$$

Since we have $d = 0.5$ µm, we can use Eq. (3.23) for the slip factor, and we have

$$C_c = 1 + 2.52(0.067 \text{ µm})/0.5 \text{ µm}$$
$$= 1 + 0.34$$
$$= 1.34$$

Putting this into (3.22), we then obtain

$$v_{settling} = 1.34 \times (1000 \text{ kg m}^{-3})9.81 \text{ m s}^{-2}(0.5 \times 10^{-6} \text{ m})^2/18(1.8 \times 10^{-5} \text{ kg m}^{-1} \text{ s}^{-1})$$
$$= 1.34(0.0076 \text{ mm s}^{-1})$$
$$= 0.010 \text{ mm s}^{-1}$$

We see that we obtain a 34% increase in the settling velocity when we include the Cunningham slip correction factor.

## 3.4 Brownian diffusion

For very small particles, collisions with the randomly moving air molecules will cause the particle to undergo a nondeterministic random walk called Brownian motion. Consider, for example, the motion of a particle settling in air under the action of gravity, shown in Fig. 3.1.

Very small particles ($d \ll 1$ µm) diffuse readily due to molecular collisions with the gas. These molecular collisions are nondeterministic and so we cannot actually predict the motion of a given particle. However, if we examine the particle motion only over times that are much longer than the time between collisions with molecules, we can use a result developed by Einstein in 1905, which states that the root mean square displacement, $x_d$, of a particle in time $t$ (where $t \gg$ time between molecular collisions) due to Brownian motion is

$$x_d = (2D_d t)^{1/2} \tag{3.24}$$

where $D$ is the particle diffusion coefficient and is given by

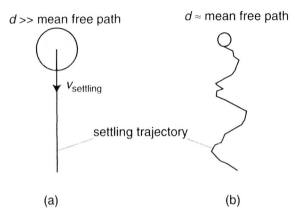

**Fig. 3.1** The trajectory of a spherical particle settling in air for (a) a particle of diameter $d \gg$ mean free path of the air molecules, and (b) a particle with diameter near that of the mean free path.

$$D_d = \frac{kTC_c}{3\pi\mu d} \qquad (3.25)$$

Here, $k = 1.38 \times 10^{-23}$ J K$^{-1}$ is Boltzmann's constant, $T$ is the temperature in Kelvin, $C_c$ is the Cunningham slip factor (Eq. 3.23), $d$ is particle diameter and $\mu$ is the viscosity of the surrounding fluid.

Because the diffusion coefficient $D_d$ increases with decreasing particle size, diffusion becomes important for small particles. To decide at what particle size diffusion starts to become important, we can compare the distance $x_s = v_{settling}t$ that a particle will settle in time $t$ to the distance $x_d$ in Eq. (3.24) that the particle will diffuse in the same time $t$. The ratio $x_d/x_s$ then is a measure of the importance of diffusion compared to sedimentation. Using Eq. (3.22) for $v_{settling}$ we have

$$\frac{x_d}{x_s} = \frac{18\mu\sqrt{2D_dt}}{\rho_{particle}gd^2C_ct} \qquad (3.26)$$

which simplifies, with the definition of $D_d$ in Eq. (3.25), to

$$\frac{x_d}{x_s} = \frac{1}{\rho_{particle}g}\sqrt{\frac{216\mu kT}{\pi t d^5 C_c}} \qquad (3.27)$$

Diffusion can be considered negligible if $x_d/x_s < 0.1$ or so. Thus, substituting the value of $x_d/x_s = 0.1$ into Eq. (3.27) allows us to solve for the time $t$ above which diffusion will be negligible for a given particle diameter $d$. The result is shown in Fig. 3.2.

The residence time $t$ of a particle in a lung airway can be estimated for simplified models of the lung given in Chapter 5, and we find that for an inhalation flow rate of 18 l min$^{-1}$ (typical of a tidal breathing delivery device, such as a nebulizer) the shortest residence time of a particle in any airway is approximately 0.03 s, so that, from Fig. 3.2, we see diffusion can be considered to have negligible effect on a particle's motion in all lung air passages if the particle's diameter is larger than approximately 2.8 μm. For an inhalation flow rate of 60 l min$^{-1}$ (typical of single breath inhalers), residence times decrease to $t > 0.01$ s, and Fig. 3.2 suggests that diffusion is negligible for particles with diameters larger than 3.5 μm. If we include a breath hold of 10 seconds duration (which

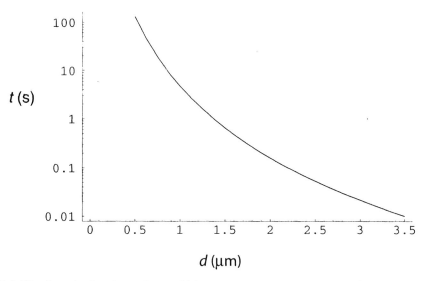

**Fig. 3.2** The time $t$ in Eq. (3.27) above which $x_d < 0.1x_s$, so that particle motion due to Brownian diffusion is estimated to be negligible compared to particle motion due to sedimentation.

is often suggested in the clinical use of single breath inhalers), then Fig. 3.2 suggests that diffusion has negligible effect on the particle's motion compared to sedimentation for particle's with diameters larger than 0.9 μm.

Thus, in deciding whether diffusion is an important mechanism of deposition for inhaled pharmaceutical aerosols, we must decide over what time interval we expect deposition to occur. If deposition occurs mainly during sedimentation with a breath hold, then diffusion is probably negligible for most inhaled pharmaceutical aerosols. However, if deposition occurs mainly during inhalation while the particle is in transit through the lung, then diffusion may need to be included for particles with diameter below a few microns in diameter. For larger particles, diffusion remains unimportant. Further discussion of this issue is given in Chapter 7.

## 3.5 Motion of particles relative to the fluid due to particle inertia

Besides diffusion and gravitational settling, a third mechanism that can cause inhaled pharmaceutical aerosols to move relative to the fluid, and deposit on the walls of the airways in the respiratory tract is due to particle inertia. In particular, if the fluid travels around a bend, a particle that is massive enough may not be able to execute the bend and will deposit on the wall, as shown in Fig. 3.3. Deposition of particles in this manner is called inertial impaction.

In order to determine whether a particle will deposit by impaction, we need to determine its trajectory. This requires solving the equation of motion, Eq. (3.2):

$$m\frac{d\mathbf{v}}{dt} = m\mathbf{g} + \mathbf{F}_{\text{drag}} \tag{3.2}$$

Once we know the particle velocity $\mathbf{v}$, the particle position $\mathbf{x}$ can be obtained by

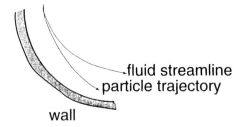

**Fig. 3.3** Particle inertia results in a particle not following the fluid motion.

integration, i.e. by integrating $d\mathbf{x}/dt = \mathbf{v}$, and if this trajectory intersects the wall, the particle will impact.

### 3.5.1 Estimating the importance of inertia: the Stokes number

It is possible to estimate whether inertial impaction is likely to occur without even solving Eq. (3.2). This can be done as follows. First, let us substitute Stokes law for the drag force of the fluid on the particle, corrected for slip (Eq. (3.21)), into Eq. (3.2) and divide by particle mass to obtain

$$\frac{d\mathbf{v}}{dt} = \mathbf{g} - \frac{3\pi d\mu}{mC_c}(\mathbf{v} - \mathbf{v}_{\text{fluid}}) \tag{3.28}$$

But for a spherical particle, we know $m = \rho_{\text{particle}}\pi d^3/6$, so we can write Eq. (3.28) as

$$\frac{d\mathbf{v}}{dt} = \mathbf{g} - \frac{1}{\tau}(\mathbf{v} - \mathbf{v}_{\text{fluid}}) \tag{3.29}$$

where

$$\tau = \rho_{\text{particle}}d^2 C_c/18\mu \tag{3.30}$$

The parameter $\tau$ is called the particle relaxation time (and is an important parameter, as we will see in a later section of this chapter).

Now let us nondimensionalize Eq. (3.29) by introducing $U_0$ as a typical velocity in the fluid flow (e.g. the mean velocity of the fluid in the lung airway the particle is in) and $D$ as a typical dimension of the geometry containing the fluid flow (e.g. the diameter of the lung airway the particle is in). Equation (3.29) can then be rewritten as

$$\frac{U_0}{D/U_0}\frac{d(\mathbf{v}/U_0)}{d(t/(D/U_0))} = \mathbf{g} - \frac{U_0}{\tau}\left(\frac{\mathbf{v}}{U_0} - \frac{\mathbf{v}_{\text{fluid}}}{U_0}\right) \tag{3.31}$$

Introducing the dimensionless variables

$$\mathbf{v}' = \mathbf{v}/U_0 \tag{3.32}$$

$$\mathbf{v}'_{\text{fluid}} = \mathbf{v}_{\text{fluid}}/U_0 \tag{3.33}$$

$$\mathbf{v}'_{\text{rel}} = (\mathbf{v}' - \mathbf{v}'_{\text{fluid}}) \tag{3.34}$$

$$t' = t/(D/U_0) \tag{3.35}$$

$$\hat{\mathbf{g}} = \mathbf{g}/g \tag{3.36}$$

and multiplying (3.31) by $\tau/U_0$, we obtain

$$\frac{\tau U_0}{D}\frac{dv'}{dt'} = \frac{\tau g}{U_0}\hat{g} - v'_{rel} \tag{3.37}$$

The coefficient in front of the time derivative term is called the Stokes number, $Stk$, defined as

$$Stk = \tau U_0/D \tag{3.38}$$

Substituting the definition of $\tau$ from Eq. (3.30) into Eq. (3.38) we have

$$Stk = U_0\rho_{particle}d^2C_c/18\mu D \tag{3.39}$$

Recalling our previous result for the settling velocity given in Eq. (3.22), and using the definition (3.38), we can rewrite Eq. (3.37) as

$$Stk\frac{dv'}{dt'} = \frac{v_{settling}}{U_0}\hat{g} - v'_{rel} \tag{3.40}$$

Equation (3.40) is simply a nondimensionalized version of the equation motion for a spherical particle. Because we have nondimensionalized all the variables, all the terms $dv'/dt'$, $\hat{g}$ and $v'_{rel}$ are expected to be at most of order 1 (i.e. O(1)). For example if $|v'_{rel}| = 1$ then the particle is moving relative to the fluid at a velocity of $U_0$ which is about as large as one would expect the relative velocity to be (e.g. a stationary particle dropped into the moving fluid would have this relative velocity).

With $dv'/dt'$ being of order 1 due to our nondimensionalization, the LHS of Eq. (3.40) will be zero if $Stk \to 0$. In the absence of gravity, then both terms other than $v'_{rel}$ in Eq. (3.40) will be zero if $Stk \to 0$. The only way for Eq. (3.40) to be satisfied in this case is if $v_{rel} = 0$, which implies that $v = v_{fluid}$ and that particle trajectories are fluid streamlines. This leads to the conclusion that particles with $Stk \ll 1$ will follow fluid streamlines (neglecting gravitational settling). By a parallel argument, if $Stk \approx 1$ or larger, a particle that encounters a rapid change in direction of flow (so that $dv'/dt' \approx 1$), will move relative to the fluid. In other words, particles with $Stk \approx 1$ or larger will not follow rapid changes in fluid streamlines. From this discussion we conclude that the value of the Stokes number determines whether a particle will undergo inertial impaction.

## Example 3.3

Assuming an inhalation flow rate of 300 cm$^3$ s$^{-1}$, a tracheal diameter of 1.8 cm and a 16th lung airway generation diameter of 0.06 cm, calculate the Stokes number for a 1 μm diameter particle of specific gravity 1.0:

(a) in the trachea;
(b) in the 16th generation of the lung.
(c) Suggest whether inertial impaction is likely to be an important consideration for this particle.

## Solution

(a) The Stokes number is defined in Eq. (3.39) as

$$Stk = U_0\rho_{particle}d^2C_c/18\mu D \tag{3.39}$$

For the present case we have

$$\rho_{particle} = 1000 \text{ kg m}^{-3}$$
$$d = 1 \times 10^{-6} \text{ m}$$
$$\mu = 1.8 \times 10^{-5} \text{ kg m}^{-1} \text{ s}^{-1}$$
$$C_c = 1 + 2.52 \, \lambda/d = 1 + 2.52(0.07 \text{ μm})/1 \text{ μm} = 1.176$$
$$D = 1.8 \times 10^{-2} \text{ m}$$

We also know that fluid velocity is related to the flow rate and area by

$$U_0 = \text{flow rate/area} \tag{3.41}$$

where the flow rate is $(300 \text{ cm}^3 \text{ s}^{-1})(1 \text{ m}^3/10^6 \text{ cm}^3) = 3 \times 10^{-4} \text{ m}^3 \text{ s}^{-1}$ and the area $= \pi(1.8 \times 10^{-2} \text{ m})^2/4$, so that Eq. (3.41) gives

$$U_0 = 1.18 \text{ m s}^{-1}$$

Putting these numbers into our definition of Stokes number in Eq. (3.39) gives

$$Stk = (1.18 \text{ m s}^{-1})(1000 \text{ kg m}^{-3})$$
$$\times (1 \times 10^{-6} \text{ m})^2 \, 1.176/(18 \times 1.8 \times 10^{-5} \text{ kg m}^{-1} \text{ s}^{-1}) \times (1.8 \times 10^{-2} \text{ m})$$

which gives $Stk = 2.4 \times 10^{-4}$.

(b) In the 16th generation, $D = 0.06$ cm. Since there will be $2^{16}$ airways in the 16th generation, all of which are carrying the air that was in the trachea, the cross-sectional area in Eq. (3.41) is

$$\text{area} = (\pi \, (0.06 \times 10^{-2} \text{ m})^2/4) \times 2^{16} = 0.0185 \text{ m}^2$$

So that Eq. (3.41) gives

$$U_0 = (3 \times 10^{-4} \text{ m}^3 \text{ s}^{-1})/0.0185 \text{ m}^2$$
$$= 0.0162 \text{ m s}^{-1}$$

Putting this into the Stokes number definition we have

$$Stk = 1 \times 10^{-4}$$

(c) Inertial impaction is not an important mechanism of deposition since $Stk \ll 1$. We will see later in Chapter 7 that it is only the larger particles (larger than a few microns) that experience significant inertial impaction in the lung.

### 3.5.2 Particle relaxation time

Because the Stokes number, $Stk$, appears as the coefficient in front of the inertial term in the dimensionless equation of motion for a particle (Eq. 3.40), $Stk$ is a measure of how important inertial effects are in determining particle trajectories. However, the Stokes number can also be interpreted as being a dimensionless version of the particle relaxation time given in Eq. (3.30).

To understand the meaning of this, consider a particle that is placed with zero velocity into a fluid having constant velocity $U_0$. Because of the drag of the fluid on the particle, the particle will start moving, and will be accelerated so that after a while the particle's velocity will be the same as the fluid velocity. We can determine the velocity of the particle as a function of time by solving the equation of motion (Eq. 3.40) neglecting gravity:

$$Stk \frac{dv'}{dt'} = -v'_{rel} \tag{3.42}$$

where $v' = v/U_0$ is the particle velocity nondimensionalized by $U_0$ (where $v$ is the dimensional particle velocity) and $v'_{rel} = (v' - v'_{fluid})$ is the velocity of the particle relative to the fluid, also nondimensionalized by $U_0$, where $v'_{fluid} = v_{fluid}/U_0$.

Since $v'_{rel} = v' - v'_{fluid}$ and $v'_{fluid} = U_0/U_0$ is a constant unit vector, $dv'_{rel}/dt' = dv'/dt'$, and we can write Eq. (3.42) as

$$Stk \frac{dv'_{rel}}{dt'} = -v'_{rel} \tag{3.43}$$

which we can integrate as follows:

$$\int \frac{dv'_{rel}}{v'_{rel}} = -\frac{1}{Stk} \int dt \tag{3.44}$$

to obtain

$$\ln v'_{rel} = \frac{1}{Stk} t' + const. \tag{3.45}$$

Exponentiating both sides, we obtain

$$v'_{rel} = Ae^{-\frac{1}{Stk}t'} \tag{3.46}$$

where $A$ is a constant vector that we can obtain from the initial condition:

$$A = v'_{rel}(t=0) = \frac{[v(t=0) - v_{fluid}(t=0)]}{U_0}$$

$$A = -\frac{U_0}{U_0} \tag{3.47}$$

so that Eq. (3.46) gives

$$|v'_{rel}| = e^{-\frac{t'}{Stk}} \tag{3.48}$$

From Eq. (3.48), we see that a particle moving in a uniform flow and initially having a velocity different from the fluid around it will have its velocity decay exponentially with time until it is eventually moving at the same speed as the fluid surrounding it, as shown in Fig. 3.4.

The particle's velocity relative to the fluid will reach a value of $e^{-1} = 1/e = 37\%$ of its initial relative velocity when

$$t' = Stk \tag{3.49}$$

Thus, the Stokes number can be interpreted as the nondimensional time required for the particle's velocity relative to the fluid to drop to 37% of its initial value when injected into the fluid. From Eq. (3.35), the dimensionless time was related to dimensional time as

$$t' = tU_0/D \tag{3.50}$$

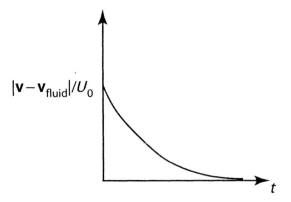

**Fig. 3.4** The relative velocity of a particle initially having different velocity than a surrounding constant velocity fluid decays exponentially with time, as given by Eq. (3.48).

Also, from Eq. (3.38), the particle relaxation time $\tau$ was related to $Stk$ by

$$Stk = \tau U_0 / D \tag{3.51}$$

Examining Eqs (3.49)–(3.51) we have the useful result that the particle relaxation time $\tau$ is the dimensional time (in seconds) required for the particle's velocity relative to the fluid to decay to 37% of its initial value, and the Stokes number is simply a dimensionless particle relaxation time.

### 3.5.3 Particle stopping (or starting) distance

The Stokes number can also be interpreted as a dimensionless version of a distance called the stopping (or starting) distance. This comes from realizing that the particle's position relative to a fluid particle will be given by

$$v'_{rel} = \frac{dx'_{rel}}{dt'} \tag{3.52}$$

where $x'_{rel}$ is the particle's dimensionless position relative to a fluid particle. Combining Eq. (3.52) with the equation we obtained for the particle's relative velocity (Eqs (3.46) and (3.47)), we have

$$\frac{dx'_{rel}}{dt'} = -\frac{U_0}{U_0} e^{-\frac{t'}{Stk}} \tag{3.53}$$

Integrating with respect to time we obtain

$$x'_{rel} = const. + Stk \frac{U_0}{U_0} e^{-\frac{t'}{Stk}} \tag{3.54}$$

Using the initial condition that $x'_{rel} = 0$ at $t = 0$ allows us to evaluate the constant and we obtain

$$|x'_{rel}| = Stk(1 - e^{-\frac{t'}{Stk}}) \tag{3.55}$$

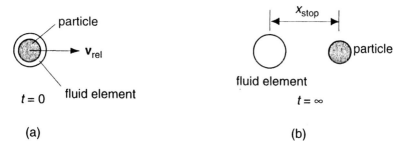

**Fig. 3.5** A particle initially moving with velocity $v_{rel}$ relative the fluid at $t = 0$ will move a distance $x_{stop}$ (the 'stopping distance') relative to the fluid before its motion relative to the fluid stops.

Equation (3.55) can also be shown to be valid for a particle injected at velocity $U_0$ into a stationary fluid, instead of a stationary particle injected into a region with fluid velocity $U_0$.

Now, the 'stopping (or starting) distance' $x_{stop}$ is defined as the separation that will occur at $t = \infty$ between the particle and the element of fluid that the particle started out with, as shown in Fig. 3.5.

Thus, letting $t \to \infty$ in Eq. (3.55) and taking the magnitude of this equation we obtain

$$|x'_{rel}(t = \infty)| = Stk \tag{3.56}$$

Since $x'_{rel} = x_{rel}/D$, Eq. (3.56) implies

$$Stk = x_{stop}/D \tag{3.57}$$

where $D$ is a characteristic dimension in the fluid flow. From Eq. (3.57) we see that the Stokes number is thus the ratio of the stopping distance to a characteristic length $D$ in the fluid.

## Example 3.4

What is the stopping distance given by Eq. (3.57) for a 20 µm diameter droplet of HFA 134a propellant (commonly used in metered dose inhalers) emitted into still air at 30 m s$^{-1}$? Use a density of 1220 kg m$^{-3}$ for liquid HFA 134a. Is this result expected to be accurate for such a droplet emitted from a metered dose inhaler?

## Solution

The stopping distance is given from Eq. (3.57) as:

$$x_{stop} = Stk\ D$$

Using our definition of Stokes number from Eq. (3.39):

$$Stk = U_0 \rho_{particle} d^2 C_c / 18 \mu D$$

we have

$$x_{stop} = U_0 \rho_{particle} d^2 C_c / 18 \mu \tag{3.58}$$

The information we have is

$$U_0 = 30 \text{ m s}^{-1}$$
$$\rho_{\text{particle}} = 1.22 \text{ g cm}^{-3} \times 1000 \text{ kg m}^{-3}/(\text{g cm}^{-3}) = 1220 \text{ kg m}^{-3}$$
$$d = 20 \times 10^{-6} \text{ m}$$
$$C_c \approx 1$$

viscosity of fluid surrounding droplet (air) $\mu = 1.8 \times 10^{-5}$ kg m$^{-1}$ s$^{-1}$.
   Putting these numbers into Eq. (3.58) gives

$$x_{\text{stop}} = 0.045 \text{ m} = 4.5 \text{ cm}$$

Droplets produced by propellant metered dose inhalers have stopping distances different from this value for several reasons, as follows. First, the propellant will rapidly evaporate from the droplet, quickly making it much smaller than 20 microns – this would tend to make the stopping distance smaller (since Eq. (3.58) shows that $x_{\text{stop}}$ varies as $d^2$, smaller $d$ makes $x_{\text{stop}}$ smaller). In addition, the aerosol emitted from a metered dose inhaler consists not of a single aerosol particle, but a jet of propellant vapor plus particles. The aerosol particles will be carried along by the propellant vapor (i.e. the gas surrounding the particles is not stationary as we assumed in our stopping distance analysis). Predicting the distance of travel of a vapor jet that entrains ambient air and that includes aerosol particles is a difficult area in multiphase fluid dynamics. Finally, the formula we used for stopping distance was derived using the equation of motion of a particle that assumes a particle Reynolds number $Re \ll 1$. The Reynolds number of the particle here is

$$Re = \rho_{\text{fluid}} v_{\text{rel}} d / \mu$$
$$= (1.2 \text{ kg m}^3)(30 \text{ m s}^{-1})(20 \times 10^{-6} \text{ m})/(1.8 \times 10^{-5} \text{ kg m}^{-1} \text{ s}^{-1})$$
$$= 40$$

Thus we are in violation of our assumption that $Re \ll 1$. Correction for this fact can be made using, for example, Eq. (3.9), for the drag coefficient. However, the equation of motion is now nonlinear and our simple formula, Eq. (3.58), for stopping distance is no longer valid. Instead, we must resort to numerical methods to solve the equation of motion (see Fuchs 1964, Section 18 for further discussion).

## 3.6 Similarity of particle motion: the concept of aerodynamic diameter

Because the respiratory tract is a difficult geometry to perform measurements in, we often wish to instead perform experiments or numerical simulations in casts or replicas of certain parts of the lung. To make these simulations or experiments easier, we may want to look at scaling the geometry to a different size, or perhaps using larger particles to make their measurement easier, or even using water instead of air as the fluid that flows through our experiment (it is sometimes easier to perform flow visualization in water than in air). However, in order for our experiments to give data that can be scaled to predict what happens in the actual geometry, several factors must be considered as follows.

First, the geometry must be an exact scale replica – this is sometimes neglected and causes interesting effects. For example, the respiratory tract geometry is itself a function of air velocity, so we may not be able to scale data at different flow rates using dimensional analysis. We will see this in Example 3.5.

Second, all aspects of the equations governing the fluid flow must be the same. These equations are the Navier–Stokes equations. For flows in the lung, fluid motion is low speed and is thus incompressible (see Chapter 6), so that if we nondimensionalize the Navier–Stokes equations, the only parameter that appears is the fluid Reynolds number

$$Re_{\text{flow}} = \rho_{\text{fluid}} \, U_0 D / \mu \tag{3.59}$$

from which we conclude that the Reynolds number of the fluid flow must be the same between our scale model and the actual case.

Finally, for particle Reynolds number $\ll 1$, and with gravity as the only external body force, the dimensionless equation of motion of a particle (Eq. (3.40)) is dependent on only two parameters: $Stk$ and $v_{\text{settling}}/U_0$. As a result, $Stk$ and $v_{\text{settling}}/U_0$ must be the same between our scale model and the actual case. Recall, from Eqs (3.39) and (3.22)

$$Stk = U_0 \rho_{\text{particle}} d^2 C_c / (18 \mu D) \tag{3.60}$$

$$\frac{v_{\text{settling}}}{U_0} = g \rho_{\text{particle}} \frac{d^2 C_c}{18 \mu U_0} \tag{3.61}$$

Because many aerosol particles have densities near that of water, it is common to write

$$\rho_{\text{particle}} \, d^2 = \rho_{\text{w}}(sg) d^2 \tag{3.62}$$

where

$$sg = \rho_{\text{particle}} / \rho_{\text{w}} \tag{3.63}$$

is the specific gravity of the particle and $\rho_{\text{w}} = 998 \text{ kg m}^{-3}$ is a constant equal to the density of water. Substituting Eq. (3.62) into Eqs (3.60) and (3.61) we have

$$Stk = U_0 \rho_{\text{w}} (sg \, d^2) C_c / 18 \mu D \tag{3.64}$$

or

$$\frac{v_{\text{settling}}}{U_0} = \frac{g \rho_{\text{w}}(sg \, d^2)}{18 \mu} \tag{3.65}$$

Assuming particle diameters are much greater than the mean free path (so that $C_c \approx 1$), the only particle parameters that appear in the above equations are $\rho_{\text{particle}}$ and $d^2$, which appear together in both $Stk$ and $v_{\text{settling}}$ as $(sg \, d^2)$. We thus come to the conclusion that the only particle property that affects a spherical particle's trajectory is the value of its aerodynamic diameter $d_{\text{ae}}$, defined as

$$\text{aerodynamic diameter, } d_{\text{ae}} = (sg)^{1/2} d \tag{3.66}$$

This conclusion is restricted to the case where particle Reynolds number $Re \ll 1$ and particle diameter $\gg$ mean free path, and where gravity and fluid drag are the only external forces on the particle.

Because of this result, aerosol particle diameters are usually given as aerodynamic diameters. For example, for a log-normal distribution, it is not the mass median diameter ($MMD$) that is of interest, but rather the mass median aerodynamic diameter, i.e. $MMAD$. For particles with densities equal to water, the two are the same of course.

## Example 3.5

Deposition of inhaled pharmaceutical aerosols in the mouth and throat is usually considered a waste of drug if the lung is the target region for the inhaled aerosol. One unorthodox suggestion for reducing deposition in the mouth and throat is to reduce the Reynolds number in the larynx (and hopefully reduce the amount of turbulent deposition) by having patients inhale the aerosol with a gas called Heli-Ox that has a kinematic viscosity which is several times that of air (Heli-Ox is a mixture of helium and oxygen). Assume an 80%–20% He–$O_2$ mixture that has a kinematic viscosity which is 2.6 times that of air, and a density which is 1/2.8 times that of air. For particles with $Re \ll 1$ and $d \gg$ mean free path, and knowing that existing data give deposition as a function of $d_{ae}$ and air inhalation flow rate $Q$:

(a) Is it possible to make use of existing measurements already available on mouth–throat deposition of particles inhaled in air in a cast of the mouth–throat in order to predict the mouth–throat deposition of particles inhaled with Heli-Ox? Assume the only mechanism of deposition in the mouth and throat is inertial impaction.
(b) Same as (a), but now the existing data with air are for actual subjects, i.e. can we predict the mouth–throat deposition of particles inhaled with Heli-Ox in actual subjects based on data on particles inhaled in air in actual subjects?

## *Solution*

(a) If the only mechanism of deposition is inertial impaction, then the only nondimensional parameters involved are the Stokes number and the fluid Reynolds number. Thus, if $Stk$ and $Re_{flow}$ are the same for a particle in Heli-Ox and in air, the particle will have the same trajectory and deposit in the same location as a particle inhaled in air. Thus, if we are to somehow scale the data for deposition with air, we must have

$$Stk_{HeO_2} = Stk_{air} \tag{3.67}$$

$$Re_{HeO_2} = Re_{air} \tag{3.68}$$

Equation (3.67) combined with Eqs (3.64) and (3.66) means we must have

$$(U_0 d_{ae}^2 C_c / 18\mu D)_{HeO_2} = (U_0 d_{ae}^2 C_c / 18\mu D)_{air} \tag{3.69}$$

But the geometry of the cast is the same whether air or Heli-Ox is used, so $D_{air} = D_{HeO_2}$ and we can cancel $D$ on both sides of Eq. (3.69).

Also, since inhalation flow rate $Q$ is directly proportional to $U_0 D^2$, and $D_{air} = D_{HeO_2}$ we have

$$U_0 \propto Q \tag{3.70}$$

Thus, we can replace the $U_0$ in Eq. (3.69) by flow rate $Q$. Finally, if $d \gg$ mean free path, we can approximate the Cunningham slip factor as $C_c \approx 1$ and we can thus rewrite (3.69) as

$$(d_{ae}^2 Q/\mu)_{HeO_2} = (d_{ae}^2 Q/\mu)_{air} \tag{3.71}$$

Fluid Reynolds number equality (Eq. 3.68) means we also must have

$$(\rho_{fluid}\, U_0 D/\mu)_{HeO_2} = (\rho_{fluid}\, U_0 D/\mu)_{air} \tag{3.72}$$

Again, since $D_{air} = D_{HeO_2}$ and $U_0 \propto Q$, we can rewrite this as

$$(\rho_{fluid}\, Q/\mu)_{HeO_2} = (\rho_{fluid}\, Q/\mu)_{air} \tag{3.73}$$

which can be written as

$$Q_{air} = Q_{HeO_2}\, \nu_{air}/\nu_{HeO_2} \tag{3.74}$$

where $\nu = \mu/\rho$ is kinematic viscosity. The ratio of kinematic viscosities is given as $1/2.6$, so we have

$$Q_{air} = Q_{HeO_2}/2.6 = 0.38\, Q_{HeO_2} \tag{3.75}$$

Substituting $Q/\mu$ from Eq. (3.73) into Eq. (3.71) gives

$$(d_{ae})_{air} = (d_{ae})_{HeO_2}\, (\rho_{HeO_2}/\rho_{air})^{-1/2} \tag{3.76}$$

We were given the ratio of densities as $1/2.8$, so we have

$$(d_{ae})_{air} = 1.68\, (d_{ae})_{HeO_2} \tag{3.77}$$

Equations (3.75) and (3.77) imply that a particle of a given aerodynamic diameter inhaled in $HeO_2$ at a given flow rate will behave exactly like a particle inhaled in air with aerodynamic diameter that is 1.68 times as large as that in $HeO_2$, at an air flow rate that is 38% times that of the Heli-Ox flow rate. Data on deposition of particles in air in this cast could then be used to predict deposition of particles with $HeO_2$ using these relations.

(b) For this part of the question, we would like to predict the deposition in the mouth–throat region of actual subjects using data we have on particles inhaled in air in actual people. The analysis done in part (a) would suggest we could go ahead and predict the deposition of a particle inhaled with Heli-Ox by using particle deposition data for air at a flow rate that is 38% that of Heli-Ox with particle diameters that are 68% larger. However, we have assumed that the geometry is the same when a subject inhales at 38% of the original flow rate. This is incorrect, since the geometry of the larynx in the throat changes with flow rate, so that in fact we cannot predict deposition of particles inhaled with Heli-Ox based solely on using dimensional analysis like that in part (a).

## 3.7 Effect of induced electrical charge

So far we have said that the only factors causing deposition of pharmaceutical aerosols in the lung are sedimentation, inertial impaction and diffusion. However, if an aerosol particle has a net electrical charge, then electrostatic forces affect the particle's motion and can affect its deposition.

The force on a particle with charge $q$ (in Coulombs) in an external electric field of strength $\mathbf{E}$ (in V m$^{-1}$) is given by

$$\mathbf{F} = q\mathbf{E} \tag{3.78}$$

In the airspace in the lung, there is generally not an external electric field. However, a charged particle in the lung sets up an electric field around it, which causes the molecules in the tissues in the lung to orient themselves (called the dielectric effect), which in turn

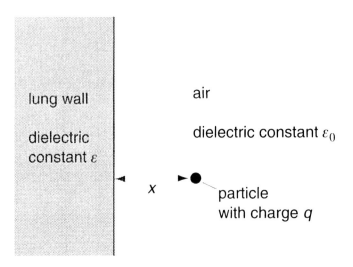

**Fig. 3.6** Geometry of a charged particle in air adjacent to a dielectric plane wall.

alters the electric field around the particle such that there is in fact a net force on the particle. The force of the molecules in the lung tissue on the charged particle appears like an 'induced charge', whose strength can be found by solving the equations of electrostatics[2].

The induced charge force can be found using elementary electrostatics, and requires solving Laplace's equation in the geometry of interest. To demonstrate the concepts involved, consider the induced charge for a particle next to a planar wall, shown in Fig. 3.6.

The solution for the problem associated with Fig. 3.6 is given in undergraduate texts in electrostatics (Reitz *et al.* 1979). It requires introduction of the dielectric constant $\varepsilon$, which is a material property that must be measured and indicates the molecular response to the particle's electric field in the dielectric. For the respiratory tract, the wall tissue typically has a dielectric constant close to that of water, which is very large (approximately 80 times that of air). For such a large value of $\varepsilon$, this problem simplifies so that the net effect of the wall is to supply a force as if the particle was in free space but with a mirror image induced charge of equal strength and opposite in magnitude, as shown in Fig. 3.7.

The magnitude of the force on the particle is then equal to the force of attraction of two point charges in free space of equal but opposite charge separated by a distance $r = 2x$, and is given by

$$|\mathbf{F}| = q^2/(r^2\,4\pi\varepsilon_0) \tag{3.79}$$

where $r = 2x$, so that

$$|\mathbf{F}| = q^2/(x^2\,16\pi\varepsilon_0) \tag{3.80}$$

where $\varepsilon_0 = 8.85 \times 10^{-12}\,\mathrm{C^2\,N^{-1}\,m^{-2}}$ is the permittivity of free space.

---

[2] It might be thought that we should be using the equations of electrodynamics here, but the time scales over which a pharmaceutical aerosol moves around in the air in the lung are so much longer than the time scales involved with electomagnetic wave propagation times and molecular orientation time scales that things are essentially static from the point of view of the electromagnetic phenomena involved.

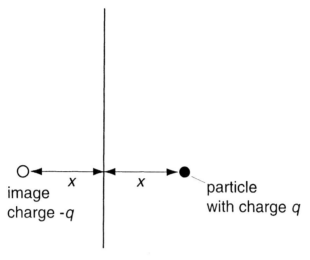

**Fig. 3.7** For a wall with large dielectric constant, the electric field seen by the particle in Fig. 3.6 is the same as that produced by the particle in free space with an induced image charge located as shown.

If we include the electrostatic force in the equation of motion for the particle (Eq. (3.2)), we have

$$m\frac{d\mathbf{v}}{dt} = m\mathbf{g} + \mathbf{F}_{\text{drag}} + \mathbf{F}_{\text{electric}} \tag{3.81}$$

Nondimensionalizing Eq. (3.81) as we did earlier when obtaining Eq. (3.40), and using Eq. (3.80) for the electric force on the particle, we obtain

$$Stk\frac{d\mathbf{v'}}{dt'} = \frac{v_{\text{settling}}}{U_0}\hat{\mathbf{g}} - \mathbf{v}'_{\text{rel}} + \left(\frac{C_c}{3\pi\mu U_0 d}\frac{e^2}{16\pi\varepsilon_0 D^2}\frac{n'^2}{x'^2}\right)\hat{\mathbf{x}} \tag{3.82}$$

where

$n'$ is an integer and is the number of elementary electronic charges on the particle

$x' = x/D$ is the nondimensional distance of the particle from the wall

$\hat{\mathbf{x}}$ is the unit vector pointing from the particle to the nearest point on the wall

$e = 1.6 \times 10^{-19}$ C is the magnitude of the charge on an electron

Let us introduce the symbol *Inc* (for induced charge), defined as

$$Inc = \frac{C_c}{3\pi\mu U_0 d}\frac{e^2}{16\pi\varepsilon_0 D^2}\frac{n'^2}{x'^2} \tag{3.83}$$

so that Eq. (3.82) can be written as

$$Stk\frac{d\mathbf{v'}}{dt'} = \frac{v_{\text{settling}}}{U_0}\hat{\mathbf{g}} - \mathbf{v}'_{\text{rel}} + Inc\,\hat{\mathbf{x}} \tag{3.84}$$

*Inc* is a coefficient that indicates the importance of the electric force relative to the other terms in Eq. (3.84) (although it does not have the same general significance as *Stk* or *Re* because it applies only for a particle attracted to a planar wall and requires specifying *n'* and $x'$). Thus, the ratio $Inc/Stk$ tells us the importance of the induced charge electrostatic term compared to the inertial term, i.e. if $Inc/Stk \ll 1$, the electrostatic effect is expected to be negligible compared to inertial effects. Also, $Inc \times U_0/v_{settling}$ tells us the importance of the electrostatic term compared to gravity so that if $Inc \times U_0/v_{settling} \ll 1$, induced electrostatic effects are expected to be negligible compared to gravitational effects.

As we will see in Chapter 7, inertial effects are most important in the larger airways, so the ratio $Inc/Stk$ is the parameter of most importance in determining the extent of induced charge effects in the larger airways. Deeper in the lung, gravitational settling is the most important deposition mechanism for inhaled pharmaceutical aerosols, so the parameter $Inc \times U_0/v_{settling}$ is the parameter of most importance in determining the extent of induced charge effects in the smaller airways and alveolar region.

Let us look at these two ratios, $Inc/Stk$ and $Inc \times U_0/v_{settling}$, separately. First, consider the ratio $Inc/Stk$. Recall the definition of $Stk$ from Eq. (3.39) is

$$Stk = (\rho_{particle}d^2 C_c/18\mu)\, U_0/D \tag{3.85}$$

Combining this with the definition of $Inc$ in Eq. (3.83), we have

$$Inc/Stk = \frac{3e^2}{8\pi^2 U_0^2 d^3 \rho_{particle}\varepsilon_0 D}\frac{n'^2}{x'^2} \tag{3.86}$$

In the airways, this coefficient will be largest when airway diameter $D$ is smallest, i.e. in the smallest airways where significant impaction might occur. In addition, this coefficient will be largest when $U_0$ is smallest (again in the smaller airways) and when particle diameter $d$ is smallest. Thus, if there is any chance that the induced charge force is of any consequence, it will be for a small particle in the smaller airways.

Thus, using $D = 5$ mm (a typical diameter of an airway in about the 12th generation, beyond which impaction is not usually important), $U_0 = 0.005$ m s$^{-1}$ (a typical velocity in such an airway for $Q = 300$ cm$^3$ s$^{-1}$, as can be calculated from $Q = vA$), and $d = 3$ μm (a typical pharmaceutical aerosol particle for which inertial impaction is still important), we obtain

$Inc/Stk$

$$= \frac{3 \times (1.6 \times 10^{-19}\text{C})^2}{(8 \times \pi^2\ 0.005^2\ \text{m}^2\ \text{s}^{-2}\ (3 \times 10^{-6})^3\ 1000\ \text{kg m}^{-3}\ 8.85 \times 10^{-12}\ \text{C}^2\ \text{N}^{-1}\ \text{m}^{-2}\ 0.005\ \text{m})}$$
$$\times \frac{n'^2}{x'^2} \tag{3.87}$$

which gives

$$Inc/Stk = 3 \times 10^{-8}\, n'^2/x'^2 \tag{3.88}$$

For induced charge effects to be nonnegligible, this ratio should be greater than or equal to 0.1 or so, and we then must have

$$3 \times 10^{-8}n'^2/x'^2 \geq 0.1 \tag{3.89}$$

or

$$n' \geq 1700\ x' \tag{3.90}$$

Equation (3.90) gives us a condition under which we expect electrostatic effects to be nonnegligible relative to inertial impaction, where recall $n'$ is the number of elementary charges on the particle and $x' = x/D$ is the dimensionless distance of the particle from the wall. In a circular tube, 10% of the cross-sectional area is contained within a distance of $0.025\,D$ from the outer wall, so a reasonable value of $x'$ to use is $x' = 0.025$, and Eq. (3.90) gives us

$$n' \geq 43 \tag{3.91}$$

In other words, more than 43 elementary charges are needed per particle before electrostatic induced charge effects are expected to become nonnegligible compared to inertial impaction for typical tidal breathing flow rates (we used $Q = 300 \text{ cm}^3 \text{ s}^{-1}$ i.e., $18\,1\,\text{m}^{-1}$ in deriving Eq. (3.91), which is a typical tidal breathing flow rate). From Eq. (3.86) it is evident that the ratio $Inc/Stk$ varies as charge$^2$/(flow rate)$^2$, so doubling the flow rate to $36\,1\,\text{min}^{-1}$ means we need to double the charge to have the same $Inc/Stk$. As a result, we may need on the order of 100 or more charges per particle for electrostatic induced charge effects to be important compared to inertial impaction in single breath inhalers where flow rates are more than double typical tidal values.

Beyond the first dozen or so generations in the lung, inertial impaction plays a minor role in deposition compared to gravitational sedimentation, so that we should consider the ratio $Inc \times U_0/v_{\text{settling}}$ to decide if electrostatic induction is important there. Recalling the definition of settling velocity (Eq. (3.22))

$$v_{\text{settling}} = C_c \rho_{\text{particle}}\, gd^2/18\mu \tag{3.92}$$

then Eqs (3.83) and (3.92) give

$$Inc \times U_0/v_{\text{settling}} = \frac{3e^2}{8\pi^2 gd^3 \rho_{\text{particle}}\varepsilon_0 D^2}\frac{n'^2}{x'^2} \tag{3.93}$$

Again, we see that this parameter will be largest for the smallest diameter particle and for the smallest diameter lung dimension $D$. Thus, a conservatively large estimate of this parameter is obtained for inhaled pharmaceutical aerosols by using $d = 1\,\mu\text{m}$ and $D = 400\,\mu\text{m}$ (the latter is a typical diameter of an alveolus). Substituting these numbers into (3.93) and requiring $Inc \times U_0/v_{\text{settling}} \geq 0.1$ for electrostatic induced charge effects to be nonnegligible we then obtain

$$7 \times 10^{-8} n'^2/x'^2 \geq 0.1 \tag{3.94}$$

which simplifies to

$$n' \geq 1200\, x' \tag{3.95}$$

Using $x' = 0.025$ as above, we then find that we require $n' \geq 30$ for electrostatic induction to be nonnegligible compared to gravitational sedimentation in the lung. This result agrees well with experimental data on charged aerosols inhaled by human subjects. For example, Melandri et al. (1983) found that electrostatic effects began to appear for charged aerosols of diameters 0.6–1.0 μm approximately when

$$(10^{-10}\, C_c\, n^2/3\pi d\mu)^{1/3} \geq 10 \tag{3.96}$$

For a 1 μm diameter particle this implies $n \geq 26$, in good agreement with our rough estimate of $n \geq 30$.

From the above analysis, we conclude that for inhaled pharmaceutical aerosols we need on the order of 30 or more elementary charges on a particle before electrostatic induced charge effects need to be considered. Yu (1985) provides a review of electrostatic effects in the lung and gives a similar conclusion.

Typical values of charge on dry powder and metered dose inhaler aerosols are on the order of 1 $\mu$C kg$^{-1}$ (Peart *et al.* 1996, 1998), but can reach values as high as 4 mC kg$^{-1}$ for some powder aerosols such as budesonide (Byron *et al.* 1997). To determine whether or not such charge will result in electrostatic effects we need to estimate the charge per particle. For an aerosol consisting of particles of size $d$ and density $\rho$ with charge $\Theta$ per unit mass of particles, the charge per particle, $q$, is given by

$$q = \Theta \rho \pi d^3/6 \qquad (3.97)$$

which can be written in terms of elementary charges by dividing by the charge on an electron $e = 1.6 \times 10^{-19}$ C

$$n' = q/e \qquad (3.98)$$

to give

$$n' = \Theta \rho \pi d^3/(6e) \qquad (3.99)$$

Since we found above that we expect $n' \geq 30$ for electrostatic effects to significantly affect deposition, we can use Eq. (3.99) to decide what level of charge per unit particle mass, $\Theta$, is needed to make such effects important. Thus, we can write

$$n' = \Theta \rho \pi d^3/(6e) \geq 30 \qquad (3.100)$$

i.e.

$$\Theta \geq 30(6e/\rho \pi d^3) \qquad (3.101)$$

or

$$\Theta \geq 10^{-17}/\rho d^3 \qquad (3.102)$$

If we let $\rho = 1000$ kg m$^{-3}$ and $d = 5$ $\mu$m, then we obtain $\Theta \geq 0.1$ mC kg$^{-1}$ or so. Some pharmaceutical powder formulations do have charges above this amount, e.g. budesonide in the Turbuhaler$^R$ has been measured at 4 mC kg$^{-1}$ (Byron *et al.* 1997), so electrostatic effects may need to be included when predicting the fate of such aerosols in the lung. However, many pharmaceutical aerosols have charge levels well below this level, and for these aerosols we do not expect electrostatic effects to be important in determining deposition in the lung. However, it appears that charge levels vary widely between different pharmaceutical inhalation aerosol formulations, so it is necessary to examine each aerosol in order to justify neglecting electrostatic effects on deposition for that aerosol.

## 3.8 Space charge

In the preceding discussion we have considered an isolated charged particle. However, if a large number of charged particles are inhaled as an aerosol, the charge on nearby particles can affect the motion of a particle. For example, if the inhaled particles all have the same sign of charge (e.g. they are all positively charged), this can cause the particles to move away from each other due to Coulombic repulsion and thereby deposit on nearby walls. This is referred to as the 'space charge' effect, and can be considered as follows.

The force of Coulombic repulsion of two neighboring particles with charges $q_1$ and $q_2$ is given by

$$F = q_1 q_2 / (r^2 \, 4\pi\varepsilon_0) \tag{3.103}$$

where $r$ is the distance between the particles and recall $\varepsilon_0 = 8.85 \times 10^{-12} \, C^2 \, N^{-1} \, m^{-2}$. The actual force on a particle will be the sum of all the Coulombic forces of its neighboring particles. Thus, we cannot evaluate this force exactly without knowing the position of all particles. However, this force decays as $1/r^2$, so for uniformly distributed particles the force due to a particles' nearest neighbors will be four times stronger than due to the next to nearest neighbor. Thus, as a first approximation, we can evaluate this force by considering only a particle's nearest neighbors.

To estimate this force, let us assume the particle and its nearest neighbor have equal charge of magnitude $q$, and the spacing between the particles is $\Delta x$. Then the force between these two particles is

$$F = q^2 / (\Delta x^2 \, 4\pi\varepsilon_0) \tag{3.104}$$

An estimate for $\Delta x$ can be made by using the number of particles per unit volume:

$$\Delta x \approx N^{-1/3} \tag{3.105}$$

Thus, the force of repulsion can be estimated as

$$F \approx q^2 / (N^{-2/3} \, 4\pi\varepsilon_0) \tag{3.106}$$

Including this force in the particle equation of motion, the nondimensional equation of motion, Eq. (3.40), becomes

$$Stk \frac{dv'}{dt'} = \frac{v_{settle}}{U_0} \hat{g} - v'_{rel} + \left( \frac{C_c}{3\pi\mu U_0 d} \frac{e^2}{4\pi\varepsilon_0} \frac{n'^2}{N^{-2/3}} \right) \hat{x} \tag{3.107}$$

where $n' = q/e$ is the number of elementary charges on the particle, $e$ is the charge on an electron, and $\hat{x}$ is a unit vector pointing from away from the particle.

Defining the coefficient

$$Spc = \frac{C_c}{3\pi\mu U_0 d} \frac{e^2}{4\pi\varepsilon_0} \frac{n'^2}{N^{-2/3}} \tag{3.108}$$

then in an analogous manner to our earlier induced charge analysis, the ratio $Spc/Stk$ indicates the importance of the electrostatic term compared to the inertial term, e.g. if $Spc/Stk \ll 1$, space charge effects are negligible compared to inertial effects. Also, $Spc \times U_0/v_{settling}$ tells us the importance of the space charge term compared to gravity so if $Spc \times U_0/v_{settling} \ll 1$, space charge effects are negligible compared to gravitational settling.

Since inertial effects are most important in the larger airways, the ratio $Spc/Stk$ is the parameter of most importance in the larger airways. Deeper in the lung, gravitational settling is the most important deposition mechanism for inhaled pharmaceutical aerosols, so the parameter $Spc \times U_0/v_{settling}$ is the parameter of most importance in the smaller airways and alveolar region.

Proceeding as we did with the induced charge force, let us look at these two ratios separately. The ratio $Spc/Stk$ can be written as

$$Spc/Stk = \frac{3e^2 D}{2\pi^2 U_0^2 d^3 \rho_{particle}\varepsilon_0} \frac{n'^2}{N^{-2/3}} \tag{3.109}$$

In the conducting airways where this ratio is of most importance (i.e. where impaction is not negligible), the ratio $D/U_0^2$ varies slowly with generation number, but is largest for the largest diameter airways (since it can be shown that $D/U_0^2$ slowly decreases with increasing generation number in typical idealized lung models – see Chapter 5). Thus, in deciding if space charge might be important, we should evaluate Eq. (3.109) in the trachea to avoid underestimating this ratio. Using the value of $D = 0.018$ m from the trachea then, and requiring $Spc/Stk = 0.1$ for space charge to be nonnegligible, we must have

$$\frac{(8 \times 10^{-30})n'^2 N^{2/3}}{\rho_{particle} U_{0t}^2 d^3} \geq 0.1 \tag{3.110}$$

which simplifies to

$$n'N^{1/3} \geq 10^{14} \rho_{particle}^{0.5} U_{0t} \, d^{1.5} \tag{3.111}$$

where $U_{0t}$ is the tracheal airway velocity. Both Eqs (3.110) and (3.111) assume the use of SI units. Note also that Eq. (3.111) suggests that smaller particles at lower flow velocities can result in small charge levels and low number densities giving rise to significant space charge effects compared to inertial impaction. This makes sense, since inertial impaction decreases with particle size and airway velocity.

In the alveolar region, it is the ratio of space charge force to gravitational force that is important, so we consider the ratio

$$Spc \times U_0/v_{settling} = \frac{3e^2 n'^2}{2\pi^2 g d^3 \rho_{particle}\varepsilon_0 N^{-2/3}} \tag{3.112}$$

Setting a value of 0.1 on this ratio, then we must have

$$n'N^{1/3} \geq 4.7 \times 10^{13} \, \rho_p^{0.5} \, d^{3/2} \tag{3.113}$$

for space charge to be nonnegligible relative to gravitational settling.

Equations (3.111) and (3.113) can be used to estimate whether space charge is important relative to inertial and gravitational effects. The values of $n'$ obtained from Eqs (3.111) and (3.113) for space charge to be nonnegligible for various number densities $N$ are shown in Fig. 3.8 for 1 and 3 μm particles where a tracheal velocity of 1.4 m s$^{-1}$ (equivalent to a flow rate of 22 l min$^{-1}$ in a 1.8 cm diameter trachea) has been used in Eq. (3.111).

Figure 3.8 suggests that for a 1 μm particle with 100 elementary charges, a number density of approximately $10^{12}$ m$^{-3}$ is needed for space charge to become nonnegligible, which is similar to that suggested in a more detailed analysis by Yu (1985). It can also be seen in Fig. 3.8 that space charge becomes nonnegligible compared to gravity at lower number densities than for inertial impaction.

Also note in Fig. 3.8 that with 30 elementary charges (which we saw in Section 3.7 was the minimum charge needed for induced charge effects to become nonnegligible), a 3 μm diameter particle requires a number density $\geq 10^{16}$ particles m$^{-3}$ for space charge to

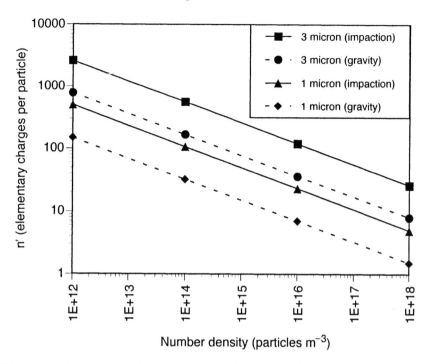

**Fig. 3.8** The number of electronic charges per particle predicted by Eqs (3.111) and (3.113) for space charge effects to become nonnegligible relative to inertial impaction or gravitational sedimentation is shown for various aerosol number densities $N$ and two particle diameters (1 μm and 3 μm). A tracheal velocity of 1.4 m s$^{-1}$ was used in Eq. (3.111).

become nonnegligible compared to gravity, while for a 1 μm diameter particle this number density is $10^{14}$ particles m$^{-3}$. If the particle density is 1000 kg m$^{-3}$, these number densities amount to 141 kg of particles m$^{-3}$ (141 g l$^{-1}$) for the 3 μm particles, and 0.052 kg m$^{-3}$ (52 mg l$^{-1}$) for the 1 μm particles. It is unlikely that such high masses of aerosol would be inhaled with inhaled pharmaceutical aerosols (since coughing would occur), so that induced charge is important when space charge is still negligible with such aerosols. Note, however, that for aerosol particles with 100 elementary charges per particle, space charge becomes important relative to gravity in the above analysis at a number density of $10^{12}$ particles m$^{-3}$ (which is 1.7 mg of aerosol per liter) for a 1 μm particle and 1.41 g l$^{-1}$ for a 3 μm particle. The former number is quite achievable with many pharmaceutical inhalation devices, so that space charge considerations could affect the deposition of such aerosols.

## 3.9 Effect of high humidity on electrostatic charge

The preceding sections indicate that electrostatic charge might affect the deposition of some pharmaceutical aerosols in the respiratory tract if enough charge is present on these aerosols. However, at high humidity, electrostatic charge is not usually present since it leaks away rapidly. In the lung the humidity is very high, typically 99.5%RH. Thus, it is possible that electrostatic charge on particles is not important in the lung since such charges may be neutralized by ions in the high humidity environment of the respiratory tract due to the presence of clusters of water molecules referred to as 'hydrated ion clusters' (Nguyen and Nieh 1989). These clusters contain a few tens of water molecules and have either a single positive or negative charge. The number of such hydrated ion clusters depends on the humidity – the higher the humidity the more there are. At high humidities these ion clusters rapidly adsorb onto any charged surfaces, thereby neutralizing any such surfaces.

Experiments with charged aerosols have shown that charge on particles does change the deposition pattern in the lung (Melandri *et al.* 1983). However these experiments are normally performed with aerosols having undergone 'field charging' whereby a large number of ions all of the same sign of charge are placed in the air that carries the particles. This is different from what happens with charged pharmaceutical aerosols, which are typically charged by triboelectric charging where the air containing the aerosol does not have a net charge of ions. This difference could cause a difference in what happens to charged pharmaceutical aerosols in the lung, since, if field charging results in a large number of free ions all of the same sign as the charge on a particle, the ion clusters of opposite sign charge to the particle (that would normally neutralize the particle's charge) are themselves neutralized by the free ions. This then could prevent the hydrated ion clusters from neutralizing the particle charge, and may give rise to experiments with charged particles in the lung that indicate electrostatic charge does play a role in deposition of inhaled particles, whereas with pharmaceutical aerosols electrostatic charge on particles may be neutralized in the high humidity environment in the lung, possibly making such charge unimportant. Research is needed to determine the importance of electrostatic charge on the *in vivo* deposition of inhaled pharmaceutical aerosols.

# References

Barton, I. E. (1995) Computation of particle tracks over a backward-facing step, *J. Aerosol Sci.* **26**:881.

Byron, P. R., Peart, J. and Staniforth, J. N. (1997) Aerosol electrostatics I: properties of fine powders before and after aerosolization by dry powder inhalers, *Pharm. Res.* **14**:698–705.

Clift, R., Grace, J. R. and Weber, M. E. (1978) *Bubbles, Drops and Particles*, Academic Press, New York.

Crowe, C., Sommerfeld, M. and Tsuji, Y. (1998) *Multiphase Flows with Droplets and Particles*, CRC Press, Boca Raton.

Di Felice, R. (1994) The voidage function for fluid-particle interation systems, *Int. J. Multiphase Flow*, **20**:153.

Fuchs, N. A. (1964) *The Mechanics of Aerosols*, Dover, New York.

Hadamard, J. S. (1911) *C. R. Acad. Sci.* **152**:1735.

Happel, J. and Brenner, H. (1983) *Low Reynolds Number Hydrodynamics*, Martinus, Nijhoff, The Hague.

Kim, I., Elghobahi, S. and Sirignano, W. A. (1998) On the equation for spherical-particle motion: effect of Reynolds and acceleration numbers, *J. Fluid Mech.* **367**:221–253.

Maxey, M. R. and Riley, J. J. (1983) Equation of motion for a small rigid sphere in a nonuniform flow, *Phys. Fluids* **26**:883–889.

Melandri, C., Tarroni, G., Prodi, V., Zaiacomo, T. De, Formignani, M. and Lombardi, C. C. (1983) Deposition of charged particles in the human airways, *J. Aerosol Sci.* **14**:657–669.

Nguyen, T. and Nieh, S. (1989) The role of water vapor in the charge elimination process for flowing powders, *J. Electrostatics* **22**:213–227.

Peart, J., Staniforth, J. N., Byron, P. R. and Meakin, B. J. (1996) Electrostatic charge interactions in pharmaceutical dry powder aerosols, in *Respiratory Drug Delivery V*, eds R. N. Dalby, P. R. Byron and S. J. Farr, Interpharm Press, Buffalo Grove, IL, pp. 85–93.

Peart, J., Magyar, C. and Byron, P. R. (1998) Aerosol electrostatics – metered dose inhalers (MDIs). reformulation and device design issues, in *Respiratory Drug Delivery VI*, eds R. N. Dalby, P. R. Byron and S. J. Farr, Interpharm, Buffalo Grove, IL, pp. 227–233.

Reitz, J. R., Milford, R. J. and Christy, R. W. (1979) *Foundations of Electromagnetic Theory*, Addison Wesley, Reading, MA.

Rybczynski, W. (1911) *Bull. Int. Acad. Sci. Cracov* **1911A**:40.

Stokes, G. G. (1851) *Trans. Cambridge Phil. Soc.* **9**:8.

Wallis, G. B. (1974) The terminal speed of single drops or bubbles in an infinite medium, *Int. J. Multiphase Flow* **1**:491–511.

Willeke, K. and Baron, P. (1993) *Aerosol Measurement: Principles, Techniques and Applications*, Van Nostrand Reinhold, New York.

Yu, C. P. (1985) Theories of electrostatic lung deposition of inhaled aerosols, *Ann. Occup. Hygiene* **29**:219–227.

# 4

# Particle Size Changes due to Evaporation or Condensation

## 4.1 Introduction

Particle size is an important property of an inhaled aerosol, since it strongly affects deposition of inhaled particles in the respiratory tract (as discussed in detail in Chapter 7). Thus, factors that cause a particle to change its size can be important. For inhaled pharmaceutical aerosols, particle size changes often occur because the particles absorb or lose mass from their surface due to evaporation or condensation. For example, evaporation or condensation readily occurs for water droplets produced by nebulizers delivering aqueous drug formulations. For propellant-driven metered dose inhalers, evaporation of propellant droplets is an important factor in determining their delivery, and for powder particles inhaled from dry powder inhalers, particle growth can occur as water condenses onto the particles from moist air in the respiratory tract. Understanding and predicting these size changes are important in optimizing the respiratory tract deposition of inhaled pharmaceutical aerosols.

When water is the substance being transferred at the surface of the particle, the accompanying size change is called 'hygroscopic' (*hygros* means moist in Greek, and scopic comes from the verb *skopeein*, meaning 'to watch' in Greek). For simplicity, unless otherwise stated, we will consider hygroscopic size changes (i.e. we will consider water to be the substance being transferred at the particle surface), but the concepts are easily generalized to other substances. Indeed, we generalize them to examine the evaporation of metered dose inhaler propellant droplets later in this chapter.

The mechanism responsible for hygroscopic effects can be understood in an introductory manner by considering the case of evaporation at an air–water interface. At such an interface, molecules are continually being exchanged back and forth between the air and water. If there are more water molecules per unit volume in the gas next to the interface than at locations away from the interface, diffusion of water molecules will occur away from the interface. This diffusion will cause a net motion of water out of the liquid phase into the gas phase, resulting in evaporation. Thus, hygroscopic effects ultimately arise because of gradients in water vapor concentration in the air next to a droplet surface.

## 4.2 Water vapor concentration at an air–water interface

Because it is the gradient in water vapor concentration near a droplet's surface that drives hygroscopic effects, an important factor in determining hygroscopicity is the water vapor concentration $c_s$ in the air next to an air–water surface, defined as the mass of water vapor per unit volume in the gas adjacent to the interface.

The value of $c_s$ depends on temperature as well as several other factors that we will consider in turn, and this dependence can be understood from a molecular viewpoint as follows. First, we know that a water molecule at the interface of a saturated liquid in air has a choice of two states, one being the liquid state (state 1) and the other a vapor state (state 2). A molecule in state 1 is surrounded closely by other water molecules, while a molecule in state 2 is separated by relatively large distances between it and other molecules (which are mostly air molecules under typical inhalation conditions). From equilibrium statistical mechanics, molecules that have a choice of two states will have a Boltzmann distribution so that the mole fraction $X$ of substance in each state will be related by

$$X_2/X_1 = \exp[-\Delta\mu/kT] \tag{4.1}$$

where $T$ is the temperature in Kelvin, $k = 1.38 \times 10^{-23}$ J K$^{-1}$ is Boltzmann's constant, $X_1$ and $X_2$ are the mole fractions of substance in each state, and $\Delta\mu = \mu_2 - \mu_1$ is the average difference in energy between molecules in the two states. In our case, $\Delta\mu$ is the amount of energy required to pluck a water molecule out of the liquid water and place it in air, i.e. it is the energy needed to overcome the intermolecular attraction that the liquid water molecules exert on each other (due to the attraction of the electrons and protons between the molecules).

For an air–water interface, Eq. (4.1) explains the well-known exponential temperature dependence of water vapor concentration at an air–water interface, since for pure water, the mole fraction $X_1 = 1$, and we obtain

$$\text{mole fraction of water vapor in air} = \exp[-\Delta\mu/kT] \tag{4.2}$$

But, since one mole of water vapor has a mass of $18.0 \times 10^{-3}$ kg, and one mole of air occupies a volume $R_u T/p$ (from the ideal gas law $pV = NR_u T$ where $N$ is the number of moles of gas) we obtain at atmospheric pressure

$$c_s = a \exp[-\Delta\mu/kT]/T \tag{4.3}$$

where $a$ is a constant.

Equation (4.3) describes the temperature dependence of vapor concentration (as well as the temperature dependence of the saturated vapor pressure $p_s = c_s R_u T/M$ where $M$ is the molar mass, or 'molecular weight', of the vapor molecules and $R_u$ is the universal gas constant). Unfortunately, the change in interaction energy $\Delta\mu$ cannot be derived from first principles. Instead, a continuum thermodynamic argument is normally used to derive this result, which gives the Clausius–Clapyeron equation governing the saturation pressure versus temperature ($dp_s/dT = Lp/RT^2$, see Saunders 1966), integration of which gives

$$p_s = F \exp[-L/(RT)] \tag{4.4}$$

where $L$ is the latent heat of vaporization (J kg$^{-1}$) and is temperature dependent. Here, $F$ is a constant and $R = R_u/M$, where $R_u = 8.314$ kg mol$^{-1}$ m$^2$ s$^{-2}$ K$^{-1}$ is the universal gas constant and $M$ is the molar mass (kg mol$^{-1}$).

Because the temperature dependence of the latent heat $L$ must be given empirically in Eq. (4.4), for many substances it is common to use instead the following empirical version of Eq. (4.4):

$$c_s = \frac{Me^{1-\frac{B}{T-C}}}{R_u T} \tag{4.5}$$

where $A$, $B$ and $C$ are constants determined experimentally (and given in the literature for many substances; see Reid *et al.* 1987). Here, $M$ is the molar mass (i.e. molecular weight) in kg mol$^{-1}$. For water with $T$ in kelvin, $A = 23.196$, $B = 3816.44$, $C = 46.13$. Another commonly used empirical equation for water is

$$c_s|_{\text{pure H}_2\text{O}} = (3.638 \times 10^5) \exp(-4943/T) \tag{4.6}$$

Equation (4.6) is an approximate fit to experimental data and does not have the $1/T$ dependence that is present in the Clausius–Clapyron equation (or Eq. (4.3)), i.e. Eq. (4.6) is truly an empirical equation. The subscript 'pure' in Eq. (4.6) is warranted since the value for $c_s$ is affected by the presence of dissolved solutes in the water, as we will see in the next section.

Note that it is important to use a reasonably accurate approximating equation for $c_s$, since errors in $c_s$ can cause significant errors in estimates of droplet temperatures and mass transfer rates because of the rapid, nonlinear variation of $c_s$ with $T$.

It should also be noted that the presence of surface active agents (surfactants) at the surface can modify the vapor pressure. Indeed, Otani and Yang (1984), among others, find significant reductions in droplet growth and evaporation when surfactants are present. The reader should bear these effects in mind when applying the principles developed in this chapter to pharmaceutical systems where surfactants are present (see Eq. (4.30) and the discussion there for further consideration of this effect).

Finally, for droplets smaller than approximately 1 μm, the Kelvin effect should be considered. This effect increases the vapor pressure at the surface of the droplet above the values given for flat surfaces in Eq. (4.6) and is discussed later in this chapter.

## 4.3 Effect of dissolved molecules on water vapor concentration at an air–water interface

If we dissolve salt or drug molecules in water, the amount of energy $\Delta\mu$ needed to pluck out a water molecule at the interface and place it in the gas phase in the air adjacent to the interface is different than it was in the absence of the solute, since the salt or drug molecules affect the intermolecular attractive force (due to forces between the electrons and protons) on the water molecule we are plucking out.

For example, consider dissolving NaCl in water. The Na$^+$ and Cl$^-$ ions exert strong attractive forces on the water molecules, and we need more energy than before to pull out a water molecule from the liquid phase (Israelachvili 1992). Thus, $\Delta\mu$ in Eq. (4.3) will be larger than before and $c_s$ in Eq. (4.3), which is the water vapor concentration at the surface, will be less than if we had pure water. This leads to the conclusion that ions dissolved in water can reduce the water vapor concentration at an air–water interface.

To see this in more detail, we can write the interaction energy $\Delta\mu$ as

$$\Delta\mu = \Delta\mu|_{\text{pure H}_2\text{O}} + \delta\mu \tag{4.7}$$

where $\Delta\mu|_{\text{pure H}_2\text{O}}$ is the energy needed to move a water molecule out of pure water into the vapor above the interface, and $\delta\mu$ is the extra energy associated with the interaction of the water molecule and the dissolved ions. From Eq. (4.3), the concentration of water vapor at the interface is given by

$$c_s = a \exp[(-\Delta\mu|_{pure\ H_2O} + \delta\mu)/kT] \tag{4.8}$$

But the first half of the right-hand side of Eq. (4.8) is what we would obtain if we had pure water, so we can write

$$c_s = c_s|_{pure\ H_2O} \exp(-\delta\mu/kT)$$

Introducing the factor $S = \exp(-\delta\mu/kT)$, we can write this result as

$$c_s = Sc_s|_{pure\ H_2O} \tag{4.9}$$

where $S \le 1$ and is determined experimentally. The factor $S$ is sometimes called the water activity coefficient.

The relative humidity at an air–water interface is defined as the actual water vapor concentration divided by that at an air–pure water interface, i.e.

$$RH = c_s/c_s|_{pure\ H_2O} \tag{4.10}$$

Comparing Eqs (4.9) and (4.10) we see that the relative humidity $RH$ at an air–water interface is equal to the factor $S = \exp(-\delta\mu/kT)$ associated with the change in interaction energy due to the dissolved solute. For example, the cellular fluid in our bodies usually has $S = 0.995$ and is referred to as being isotonic, so that an isotonic aqueous solution gives $RH = 0.995$ at an air interface.

For dilute solutions, $S$ is close to 1 and we can approximate the exponential expression for $S$, by using a Taylor series expansion, as

$$S = \exp(-\delta\mu/kT) \approx 1 - \delta\mu/kT \tag{4.11}$$

The change $\delta\mu$ in the transfer energy of a water molecule with neighboring ions is too complicated to determine from first principles; however, we would expect $\delta\mu$ in Eq. (4.11) to increase with the number of dissolved ions in the water. This leads to the empirical equation (sometimes called Raoult's law, after Raoult 1887):

$$S \approx 1 - \frac{ix_s}{x_w} \quad \text{(dilute solutions only)} \tag{4.12}$$

where $x_s$ is the molar concentration of solute (i.e. number of moles of solute molecules per unit volume of liquid), $x_w$ is the molar concentration of water (i.e. number of moles of water molecules per unit volume of liquid), and $i$ is sometimes called the van't Hoff factor and is determined experimentally. 'Ideal' solution behavior occurs if $i$ is equal to the number of ions that a molecule splits up into upon dissolution, e.g. $i = 2$ for NaCl (since one NaCl molecule dissolves into two ions). For dilute solutions, $i$ is usually not too far from the ideal solution value, e.g. for NaCl, the measured value of the van't Hoff factor is $i = 1.85$.

The van't Hoff factor $i$ or the activity coefficient $S$ can be measured by vapor pressure osmometry, or possibly freezing point osmometry if no phase transitions occur (Gonda et al. 1982). For concentrated solutions, $S$ varies in a nonlinear fashion with concentration of the dissolved substance (see Cinkotai 1971 for data on NaCl).

Note that as a droplet either grows and absorbs water or evaporates and loses water, the number of moles of water in the droplet relative to those of solute changes, so that the vapor concentration at the droplet surface changes as the droplet changes size. This is readily seen, for example, when Eq. (4.12) is used for the factor $S$ in Eq. (4.9), since the reduction in water vapor concentration is directly proportional to the molar concentration of solute in Eq. (4.12). Thus, when determining the hygroscopic size changes of a water droplet that contains dissolved drug or salt, we must keep track of how much water and how much drug is contained in the droplet and use this information when evaluating Eq. (4.12). Since only the water molecules undergo evaporation or condensation, the number of moles of drug or salt in a droplet is constant, so that this is not usually a difficult matter and is merely a matter of bookkeeping.

If more than one solute is present, it is reasonable to assume that each solute $j$ contributes an amount $\delta\mu_j$ to the transfer energy $\delta\mu$ in Eq. (4.11) independently of the other solutes[1], so that the water vapor concentration is then given by

$$S \approx 1 - \frac{\sum_j i_j x_{sj}}{x_w} \qquad \text{(multiple component dilute ionic solutions)} \qquad (4.13)$$

For multiple component solutions that are not dilute, more complex theories are available that allow prediction of $S$ using values of $S$ for solutions made up of the individual components (Robinson and Stokes 1959).

### Example 4.1

What is the relative humidity at the surface of a solution of Ventolin respiratory solution (2.5 mg ml$^{-1}$ salbutamol sulfate + 9.0 mg ml$^{-1}$ NaCl in water) at room temperature. The van't Hoff factor of NaCl is 1.85, while that of salbutamol sulfate is 2.5. The molecular weight is 18.0 g mol$^{-1}$ for water, 58.44 g mol$^{-1}$ for NaCl and 576.70 g mol$^{-1}$ for salbutamol sulfate (Budavari 1996).

### *Solution*

First we need to determine the molar concentrations of the dissolved substances in order to use Eqs (4.9) and (4.13). These can be obtained by taking the mass concentrations and dividing by the molecular weights. Thus, the molar concentration of NaCl is given by

$$x_{NaCl} = (9.0 \times 10^{-3} \text{ g ml}^{-1})/(58.44 \text{ g mol}^{-1}) = 1.54 \times 10^{-4} \text{ mol ml}^{-1}$$

and the molar concentration of salbutamol sulfate is given by

$$x_{ss} = (2.5 \times 10^{-3} \text{ g ml}^{-1})/(576.7 \text{ g mol}^{-1}) = 4.34 \times 10^{-6} \text{ mol ml}^{-1}$$

The molar concentration of water is given by

$$x_w = 1 \text{ g ml}^{-1}/18.0 \text{ g mol}^{-1} = 0.056 \text{ mol ml}^{-1}$$

where we have neglected the effect of the dissolved solutes on the molar concentration of water (the presence of the solutes reduces the amount of water per ml, but for the dilute solutions that we are dealing with this is negligible).

---

[1] This assumes the interaction energies of the different molecules are independent of each other, which will be true for dilute solutions if the interactions between the molecules are simply electrostatic forces like those occurring between ions and water molecules.

The effect of the solutes on the water vapor concentration at the surface is then obtained using Eq. (4.13):

$$S = 1 - (i_{NaCl}x_{NaCl} + i_{ss}x_{ss})/x_w$$
$$= 1 - (1.85 \times 1.54 \times 10^{-4} \text{ mol ml}^{-1} + 2.5 \times 4.34 \times 10^{-6})/(0.056 \text{ mol ml}^{-1})$$
$$= 1 - (2.85 \times 10^{-4} + 1.1 \times 10^{-5})/0.056$$
$$= 1 - 5.0 \times 10^{-3} - 0.02 \times 10^{-3}$$

We can see that the effect of the salbutamol sulfate on the water vapor concentration is essentially negligible compared with that due to the NaCl ($0.02 \times 10^{-3}$ compared to $5.0 \times 10^{-3}$), and we can neglect this effect. Indeed, for most nebulized inhaled pharmaceutical aerosols, water vapor reductions due to dissolved drugs or other solutes are often negligible compared with that due to NaCl present in the formulation.

Neglecting the salbutamol sulfate then, we have

$$S = 0.995$$

From Eqs (4.9) and (4.10), $RH = S$, so the water vapor concentration next to a solution of Ventolin respiratory solution is 99.5%, and this solution is thus isotonic.

The actual water vapor concentration can be determined by combining Eq. (4.6)

$$c_s|_{\text{pure H}_2\text{O}} = 3.638 \times 10^5 \exp(-4943/T) \tag{4.6}$$

and Eq. (4.9)

$$c_s = Sc_s|_{\text{pure H}_2\text{O}} \tag{4.9}$$

So that

$$c_s = 0.995 \times 3.638 \times 10^5 \exp[-4943/(273.15 + 20)]$$
$$= 0.0172 \text{ kg m}^{-3}$$

## 4.4 Assumptions needed to develop simplified hygroscopic theory

Although a basic understanding of hygroscopicity is possible with the above discussion, the general equations that govern droplet size changes are quite complicated and rather involved. However, we can often avoid using the general version of these equations and instead use a simplified version if a number of assumptions can be made. Let us now look at the assumptions that are commonly made to simplify the equations governing droplet growth and evaporation. As we make each assumption we will examine the conditions needed for each assumption to be valid.

(1) *Assume the mass transfer at the droplet surface does not cause any bulk motion in the air surrounding the droplet.*

This implies we neglect what is called Stefan flow (after Stefan 1881) in which, for example, vapor evaporates from a droplet at such a high rate that it sets up motion in the air surrounding the droplet. As we will see later in this chapter, neglecting Stefan flow requires $p_s \ll p$, where $p_s$ is the partial pressure of the vapor at the surface, and $p$ is the total gas pressure there. A similar derivation can be used to show that this assumption is also satisfied if $c_s/\rho_{gas} \ll 1$, where $\rho_{gas}$ is the density of the gas (vapor and air) next to the droplet surface and $c_s$ is the vapor concentration there.

### Example 4.2

How reasonable is assumption (1) for

(a) a water droplet in air at room temperature and at body temperature,
(b) an HFA 134a droplet that has been cooled by propellant evaporation to $-60°C$.

### *Solution*

(a) From Eq. (4.6) and the ideal gas law ($p_s = c_s RT$), the saturated vapor pressure of water in air at 101 kPa and 295 K is found to be 2617 Pa, so we have

$$p_s/p = 2617/101,000$$
$$= 0.026$$

while, at 310 K $p_s = 6221$ Pa and we obtain

$$p_s/p = 0.062$$

Since $p_s/p \ll 1$ in both cases, Stefan flow can be neglected for the evaporation/condensation of water droplets at the given temperatures.

(b) The vapor pressure of HFA 134a at –60°C is 15.9 kPA (ASHRAE 1997), so we obtain

$$p_s/p = 15.9 \text{ kPa}/101.32 \text{ kPa}$$
$$= 0.16$$

This is no longer much less than one (being greater than the usual cut-off of 0.1). Thus, assumption (1) is violated and we cannot use the simplified theory we are about to develop. Instead we must include Stefan flow and use the more complicated analysis discussed later in this chapter.

(2) *Assume the temperature inside the droplet does not vary spatially ('lumped capacitance' assumption).*

This assumption requires that the Biot number is small, where the Biot number, *Bi*, is defined as the ratio of the resistance to heat transfer within the droplet to the resistance to heat transfer at the droplet surface (Incropera and De Witt 1990). For liquid droplets surrounded by gas, *Bi* can be shown to be given by

$$Bi = Nu \, k_{gas}/k_{droplet} \tag{4.14}$$

where the Nusselt number, *Nu*, is a nondimensional measure of the heat transfer rate at the droplet surface (Incropera and De Witt 1990), while $k_{gas}$ is the thermal conductivity of the gas surrounding the droplet and $k_{droplet}$ is the thermal conductivity of the droplet. For a stationary, spherical droplet $Nu = 2$ if assumption 1 above is satisfied.

## Example 4.3

How reasonable is assumption (2) for a water droplet in air at room temperature and body temperature?

### *Solution*

For air, the thermal conductivity has a value $k \leq 0.03$ W m$^{-1}$ K$^{-1}$ for $T < 350$ K. If assumption (1) is satisfied, we can expect the water vapor to have little effect on the thermal conductivity of the gas surrounding the droplet (since if $p_s \ll p$ there is little water vapor compared to air in the gas), and we can use $k_{gas} = 0.03$ W m$^{-1}$ K$^{-1}$.

For liquid water at either room temperature or body temperature, $k_{droplet} = 0.6$ W m$^{-1}$ K$^{-1}$, so with $Nu = 2$ in Eq. (4.14) we obtain $Bi = 0.1$. Thus, assumption (2) is reasonable in this case.

(3) *Assume the concentration of water vapor and the temperature of the gas are functions only of distance* r *from the center of the drop.*

This assumption will be violated if the drop is moving with significant speed through the gas surrounding the drop, since then we have air flow past a liquid sphere, for which there can be azimuthal and polar variation of heat and mass transfer. Thus, only in the limit of zero particle velocity is this assumption rigorously valid, i.e. it is valid in the limit of zero Reynolds number, $Re$, and Peclet number, $Pe$:

$$Re = \frac{v_{rel}d}{v_{gas}} \to 0$$

$$Pe_h = \frac{v_{rel}d}{\alpha_{gas}} \to 0 \qquad (4.15)$$

$$Pe_m = \frac{v_{rel}d}{D} \to 0 \qquad (4.16)$$

Here $d$ is the droplet diameter, $v_{rel}$ is the velocity of the drop relative to the gas, $v_{gas}$ and $\alpha_{gas}$ are the kinematic viscosity and thermal diffusivity of the gas surrounding the droplet, respectively, while $D$ is the diffusion coefficient for mass diffusion of the given vapor in air.

For water vapor and HFA 134A propellant, the Schmidt number ($Sc = v/D$) and Prandtl number ($Pr = v/\alpha$) are of order one, so that the requirement of small Peclet number in Eqs (4.15) and (4.16) is automatically satisfied if the Reynolds number is small. In this case, then from experiments and computations of heat and mass transfer around droplets, it is known that as long as we have particle Reynolds numbers $Re \leq 0.1$ (Finlayson and Olson 1987, Taflin and Davis 1987, Zhang and Davis 1987), assumption (3) is reasonable.

We have seen in Chapter 3 that particle Reynolds numbers $Re \ll 1$ often occur for inhaled aerosols, except for metered dose inhalers. We may be on the verge of not being able to use this assumption for MDI aerosols close to the point where they exit the nozzle; however, these aerosols decelerate rapidly and may soon have $Re \ll 1$.

(4) *Assume particle radius* $\gg$ *mean free path.*

For very small particles, additional corrections are needed to properly predict the vapor concentration and temperature profiles near the droplet surface (Fuchs 1959, Ferron and

Soderholm 1990, Vesala *et al.* 1997), since the assumption of a continuum of gas molecules at the droplet surface is no longer valid. The corrections needed to account for noncontinuum effects are discussed later in this chapter.

*(5) Assume quasi-steadiness.*

This assumption requires that the droplet changes size slowly enough that at each point in time the heat and mass transfer rate at the droplet surface is the same as that occurring for the steady case of a droplet of this size but maintained at a fixed radius.

The quasi-steady solution to the more general transient problem can be viewed as the zeroth-order term in a perturbation series solution, so that for the quasi-steady solution to be accurate, the higher order terms in this series must be small (Duda and Vrentas 1971), which, using the fact that liquid densities are much higher than gas densities, can be shown to require the following two conditions:

(5a) the density of the vapor phase at the surface, $c_s$, must be much less than the density of the droplet, i.e. $c_s/\rho_{drop} \ll 1$, and

(5b) $D\tau/R_0^2 \gg 1$   and   $\alpha\tau/R_0^2 \gg 1$    (4.17)

where $D$ is the diffusion coefficient of the vapor in air, $R_0$ is the initial droplet radius, $\tau$ is a representative time scale over which hygroscopic effects occur and $\alpha$ is the thermal diffusivity of the gas surrounding the droplet.

These two conditions can be derived based on the perturbation analysis of Duda and Vrentas (1971), but can also be obtained by requiring that the distance, $x$, that mass or heat diffuses during the representative time $\tau$ be much greater than the droplet radius, where $x = (2D\tau)^{1/2}$ for mass transfer and $x = (2\alpha\tau)^{1/2}$ for heat transfer.

Actually, it can be shown that assumption (5 a) is all that is needed in order to have (5 b) satisfied, as long as assumptions (1), (3) and (4) are also satisfied. This can be done by first making the quasi-steady assumption, and then using the conservative estimate, $t_{L0}$, for the time scale $\tau$ (obtained using the simplified theory below – see Eq. (4.47)), i.e.

$$\tau = \frac{R_0^2 \rho_{drop}}{2Dc_s}$$    (4.18)

Putting Eq. (4.18) into Eq. (4.17) we then obtain for condition (5 b)

$$c_s/\rho_{drop} \ll 1 \quad \text{and} \quad (D/\alpha)c_s/\rho_{drop} \ll 1$$    (4.19)

If $D/\alpha \approx 1$, the second condition in Eq. (4.19) is the same as the first, and both are then the same as needed for assumption (5 a).

For a water droplet in air at room temperature, the diffusion coefficient of water vapor in air is $D = 2.5 \times 10^{-5}\,m^2\,s^{-1}$ while for HFA 134a in air $D = 7 \times 10^{-6}\,m^2\,s^{-1}$ (estimated using the formula 16.3-1 in Bird *et al.* 1960); the thermal diffusivity of air is approximately $\alpha = 0.2 \times 10^{-4}\,m^2\,s^{-1}$. Thus, $D/\alpha$ (which is called the Lewis number) is of order 1 and Eq. (4.19) reduces simply to $c_s/\rho_{drop} \ll 1$. Thus, we find condition (5 a) (i.e. $c_s/\rho_{drop} \ll 1$) is also the condition needed for (5 b) to be satisfied, as long as assumptions (1), (3), and (4) are satisfied (since these assumptions were also needed to derive the estimate in Eq. (4.47) of droplet lifetime using the simplified theory).

## Example 4.4

Is the quasi-steady assumption reasonable for water droplets in air at room temperature and body temperature at 1 atmosphere of pressure?

### Solution

We have already seen that water droplets in ambient room temperature air or body temperature air satisfy assumptions (1)–(4). Thus, (5 a) is enough to satisfy (5 b) as well, and so all we need for the quasi-steady assumption to be valid is

$$c_s/\rho_{drop} \ll 1$$

For a water surface at room temperature in 1 atm pressure, Eq. (4.6) gives $c_s = 0.02$ kg m$^{-3}$, while the density of water is 998 kg m$^{-3}$, so we have

$$c_s/\rho_{drop} = 0.02/998 = 0.0002 \; (\ll 1),$$

while at body temperature (310.65 K) $c_s = 0.04$ kg m$^{-3}$ and we obtain

$$c_s/\rho_{drop} = 0.04/998 = 0.0004 \; (\ll 1)$$

We thus see that the quasi-steady assumption is very reasonable for water droplets under normal ambient conditions or at body temperature.

Note that if any of the assumptions (1), (3) or (4) is not satisfied, then in general, we need to have an estimate for the droplet lifetime, or some other representative time scale for the droplet evaporation/condensation process, in order to decide if condition (5 b) is satisfied. This, however, requires us to have already solved the problem, and we run into difficulties. It may be possible to estimate $\tau$ if one has access to experimental data, as seen in the following example.

## Example 4.5

Are the requirements (5 a), i.e. $c_s/\rho_{drop} \ll 1$, and (5 b), i.e. $D\tau/R_0^2 \gg 1$ and $\alpha\tau/R_0^2 \gg 1$ reasonable for an HFA 134a droplet fired from a metered dose inhaler (MDI) and having a temperature of $-60°C$?

### Solution

For HFA 134a at $-60°C$ (213 K), the vapor pressure is 15.94 kPa (ASHRAE 1997), and we can obtain $c_s$ from the ideal gas law:

$$p_s = c_s R_u T/M \tag{4.20}$$

where $R_u = 8.314$ kg mol$^{-1}$ m$^2$ s$^{-2}$ K$^{-1}$, $T$ is in kelvin and $M$ is the molecular weight in kg mol$^{-1}$ where $p_s$ is the vapor pressure and is 1 atm in this case. Thus, we obtain

$$c_s = p_s M/(R_u T)$$
$$= \frac{15\,940 \text{ Pa } 102.03 \times 10^{-3} \text{ kg mol}^{-1}}{(8.314 \text{ kg mol}^{-1} \text{ m}^2 \text{ s}^{-2} \text{ K}^{-1} \; 223 \text{ K})}$$
$$= 0.9 \text{ kg m}^{-3}$$

The density of saturated liquid HFA 134a at this temperature is $1471\ kg\ m^{-3}$, and so we obtain

$$c_s/\rho_{drop} = 0.0006\ (\ll 1)$$

so that (5 a) is reasonable.

To look at (5 b), however, as we saw earlier, assumption 1 may not be satisfied for HFA 134a propellant droplets. Thus, the analysis carried out to show that (5 b) is equivalent to (5 a), which required assumption 1 to be satisfied since we used the simplified theory below to estimate $\tau$ from Eq. (4.47), is not valid. Thus, to see if (5 b) is valid, we must directly consider whether the inequalities in (5 b)

$$D\tau/R_0^2 \gg 1 \quad \text{and} \quad \alpha\tau/R_0^2 \gg 1$$

are satisfied.

From measurements with phase Doppler anemometry (PDA) or laser diffraction methods (Dunbar et al. 1997), we know that $R_0$ is not usually larger than 30 microns or so. Using this number as a conservatively large estimate of droplet size, along with a diffusion coefficient $D = 7 \times 10^{-6}\ m^2\ s^{-1}$ for 134a in air, and $\alpha = 2.5 \times 10^{-6}\ m^2\ s^{-1}$ (which is the value for air, but is also near the value for 134a), means that (5 b) requires

$$\tau \gg \frac{R_0^2}{D} = \frac{(30 \times 10^{-6})^2}{7 \times 10^{-6}} = 1 \times 10^{-4}\ s$$

$$\tau \gg \frac{R_0^2}{\alpha} = 4 \times 10^{-4}\ s$$

Droplet lifetimes from propellant metered dose inhalers are typically much longer than either of these times (Dunbar et al. 1997). In addition, propellant droplets smaller than 30 microns are likely (Dunbar et al. 1997), which would make the right-hand side of these inequalities even smaller and therefore easier to satisfy, so that in general the quasi-steady assumption appears reasonable.

## 4.5 Simplified theory of hygroscopic size changes for a single droplet: mass transfer rate

If assumptions (1)–(5) of Section 4.4 are satisfied, the classical theory for hygroscopic growth or shrinkage of a single droplet can be used. Let us first develop the equation that governs the rate of change of mass of the droplet in this theory, and then in the next section we will examine the temperature change that accompanies this change in mass.

The problem at hand is shown in Fig. 4.1 for the case of droplet evaporation. At the droplet surface there will be a certain vapor concentration, $c_s$, which will be given by Eq. (4.6) if the droplet consists of pure water, or Eq. (4.9) if the droplet consists of water with dissolved solutes. Far away from the droplet surface the concentration of vapor in the air has some value, $c_\infty$, that is in general different from $c_s$ (in fact $c_\infty$ can take on any value from 0 to $c_s|_{pure\ H_2O}$ given in Eq. (4.6)). If $c_s = c_\infty$ (which occurs for a pure water droplet immersed in 100% $RH$ air), the droplet will neither evaporate nor grow since there is no concentration gradient.

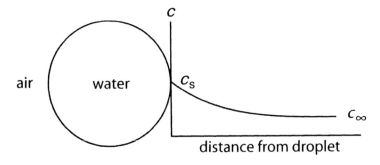

**Fig. 4.1** Water vapor concentration decreases with distance away from an evaporating droplet, as shown.

The mass flux of vapor at a point outside the droplet is given from Fick's first law of diffusion as

$$j = -D\nabla c \qquad (4.21)$$

where $j$ is the mass flux of vapor, $D$ is the diffusion coefficient for diffusion of the droplet vapor in air, and $c$ is the mass concentration (or density) of the vapor in the air. Equation (4.21) assumes that the total density of the gas phase ($\rho_{gas} = c + \rho_{air}$) around the droplet is independent of radial distance $r$, which is reasonable given that we must have $p_s \ll p$ for assumption (1) of Section 4.4 to be valid, so that there isn't much vapor around and most of the gas surrounding the droplet is air at atmospheric pressure (and therefore constant density). More complicated versions of Fick's law exist that include various subtle effects (Dufour and Soret effects – see Bird *et al.* 1960, Section 18.4, or references listed in Vesala *et al.* 1997), but we can neglect these effects here.

Because of assumption (3) of Section 4.4, we can replace the gradient operator $\nabla$ in Eq. (4.21) with $d/dr$, and Fick's law for our purposes simplifies to

$$j = -D\frac{dc}{dr} \qquad (4.22)$$

Now, the mass flux of vapor through a spherical surface at a distance $r$ from the center of the drop (where $r > d/2$ and $d$ is the droplet diameter) can be obtained by multiplying Eq. (4.22) by the surface area $4\pi r^2$. Thus, the mass flux $I$ through a sphere at radius $r$ is given by

$$I = -4\pi r^2 D\frac{dc}{dr} \qquad (4.23)$$

However, under steady state conditions, the mass flux $I$ must be independent of $r$, because of conservation of mass[2].

Thus, if we place a spherical shell just outside the droplet and another at any arbitrary radius $r$, the mass flux through them must be the same and equal to $I$ (assuming assumption (4) of Section 4.4 is valid, otherwise the mass transfer at the droplet surface must be modified by noncontinuum corrections as discussed above under assumption

---

[2]If this were not the case then the vapor mass flux through two concentric shells, as in Fig. 4.2, can be different, implying that vapor is appearing out of nowhere between the two shells, which defies conservation of mass.

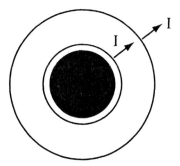

**Fig. 4.2** The mass flux of vapor, $I$, through concentric spherical shells centered around a droplet must be the same under steady-state conditions.

(4)). From this we can conclude that $I$ is equal to the mass flux of vapor from the droplet's surface, i.e.

$$I = -\frac{dm}{dt} \qquad (4.24)$$

where $m$ is the droplet's mass. The minus sign arises because $I$ is the mass transferred away from the droplet's surface.

From Eq. (4.23), the concentration outside the droplet satisfies

$$\frac{dc}{dr} = \frac{-I}{(D4\pi r^2)} \qquad (4.25)$$

where our quasi-steady assumption causes $I$ to be independent of $r$. Assuming a constant value of the diffusion coefficient $D$ (the validity of which is discussed below), then we can integrate Eq. (4.25) from the droplet surface to $r = \infty$ to obtain

$$c_s - c_\infty = \frac{I}{(2\pi dD)} \qquad (4.26)$$

which can be rewritten as

$$I = 2\pi dD(c_s - c_\infty) \qquad (4.27)$$

Equation (4.27) was derived in 1855 by Maxwell (Maxwell 1890) and is sometimes called Maxwell's equation.

Combining Eqs (4.24) and (4.27), we arrive at the equation that governs the rate of change of the mass of a droplet:

$$\frac{dm}{dt} = 2\pi dD(c_s - c_\infty) \qquad (4.28)$$

Since, for a sphere $m = \rho_{drop}\pi d^3/6$, where $d$ is the droplet diameter, we can also write this equation as

$$\frac{dd}{dt} = -\frac{4D(c_s - c_\infty)}{\rho_{drop}d} \qquad (4.29)$$

Equation (4.29) is the standard equation used to include hygroscopic effects on droplet size.

If the assumption of constant diffusion coefficient, $D$, is not reasonable (which would occur if large temperature gradients develop in the gas near the droplet because of its evaporation), Fuchs (1959) shows that Eq. (4.28) can be improved by using $\sqrt{D_s D_\infty}$ instead of $D$ in Eq. (4.28), where $D_s$ is the value of $D$ at the droplet surface, and $D_\infty$ is its value far from the droplet surface.

The ability of Eq. (4.29) to predict the rate of change of size of droplets has been extensively examined experimentally, and found to give excellent agreement with experiment as long as the assumptions (1)–(5) of Section 4.4 made in deriving this theory remain valid. Reviews of some of the work that validates this theory can be found in Fuchs (1959) and more recently in Miller *et al.* (1998).

It was mentioned earlier that the presence of surface active molecules (surfactants) can alter the rate of mass transfer at the surface of a droplet. In such cases, the surfactant monolayer may have little effect if it does not entirely coat the exterior surface of the droplet (Derjaguin *et al.* 1966), but when the surface of the drop is saturated with a surfactant monolayer, the evaporation kinetics are altered to such an extent that the equation governing droplet growth/evaporation (Eq. 4.29) is invalid, and can be replaced with

$$\frac{\mathrm{d}d}{\mathrm{d}t} = \frac{-av(c_s - c_\infty)}{8\rho_{\mathrm{drop}}} \tag{4.30}$$

where $a$ is an experimentally determined coefficient of condensation and $v$ is the mean speed of water vapor molecules (Derjaguin *et al.* 1966). Otani and Wang (1984) propose an equation that merges Eq. (4.30) and Eq. (4.29) to cover the range from no surfactant monolayer to saturated monolayer, finding agreement with their experimental data.

## 4.6 Simplified theory of hygroscopic size changes for a single droplet: heat transfer rate

We can develop an equation that governs the temperature of the droplet in a manner analogous to the manner in which we derived Eq. (4.28) for mass transfer. We begin by writing the heat flux at any point, using Fourier's law, as

$$q = -k\nabla T \tag{4.31}$$

where we must invoke assumptions (1) and (3) of Section 4.4 so that we have heat transfer only by conduction[3]. Assumption (3) also implies we can write the gradient operator $\nabla$ in Eq. (4.31) as $\mathrm{d}/\mathrm{d}r$, so we have

$$q = -k\frac{\mathrm{d}T}{\mathrm{d}r}$$

Here, $k$ is the thermal conductivity of the gas surrounding the droplet, which we assume is constant (since from assumption (1) there isn't much vapor around to cause a change in physical properties, so we can use $k_{\mathrm{air}}$, which varies little with temperature in the range of temperatures we expect). The energy flux into a spherical surface at radius $r$ will then be

---

[3] That is convection is negligible; we also assume radiative heat transfer is negligible, which is reasonable – see Fuchs (1959) for a calculation on the relative importance of radiative heat transfer.

$$Q = 4\pi r^2 k \frac{dT}{dr} \tag{4.32}$$

But, assuming quasi-steadiness (assumption (5) of Section 4.4), there can be no energy accumulation within any volume outside the drop, so that $Q$ must be independent of $r$.[4] Thus, we can integrate Eq. (4.32) with respect to $r$ from the droplet surface to $r = \infty$ to obtain

$$Q = -2\pi d k_{air} (T_s - T_\infty) \tag{4.33}$$

This is the radially inward heat flux through any sphere of radius $r > d/2$ (where $d$ is the droplet diameter), and is due solely to heat conduction.

Now if we consider an energy balance at any instant for the droplet itself, the rate of change of energy due to heat conduction plus the rate of change of energy due to energy being carried away by evaporating vapor must add up to the rate of change of thermal energy of the droplet. More specifically, considering a control volume $V$ surrounding a droplet with surface $S$, we have

$$\int_S q \, dS - \int_S \rho h_g \mathbf{v} \cdot \hat{n} \, dS = \frac{d}{dt} \int_V \rho \hat{u} \, dV \tag{4.34}$$

where $h_g$ is the enthalpy per unit mass of the gas vapor leaving the droplet, and $\hat{u}$ is the internal energy of the liquid in the droplet per unit mass. This equation can be rewritten using $\hat{u} = h_l - (p/\rho_l)$, where the subscript l indicates liquid, and using the fact that it is reasonable to assume the pressure and density of the liquid inside the drop are constant in space and independent of time. Thus, Eq. (4.34) can be rewritten as

$$-2\pi d k_{air}(T_s - T_\infty) + h_g \frac{dm}{dt} = \frac{d}{dt}(mh_l) - \frac{p}{\rho_l}\frac{dm}{dt} \tag{4.35}$$

But the first term on the right-hand side of this equation can be expanded as

$$\frac{d}{dt}(mh_l) = h_l \frac{dm}{dt} + m \frac{d}{dt}(h_l) \tag{4.36}$$

Substituting Eq. (4.36) into Eq. (4.35) and rearranging we then have

$$m \frac{d}{dt}(h_l) - \frac{p}{\rho_l}\frac{dm}{dt} = L\frac{dm}{dt} + Q \tag{4.37}$$

where $L = h_g - h_l$ is the enthalpy change associated with the phase change from the liquid to gas state and is called the latent heat of evaporation (or latent heat of vaporization). Substituting a linear temperature dependence, $h_l = c_p T + \text{constant}$, into Eq. (4.37) we obtain

$$mc_p \frac{dT}{dt} = \left(L + \frac{p}{\rho_l}\right)\frac{dm}{dt} + Q \tag{4.38}$$

Usually $p/\rho_l$ can be neglected compared to $L$ for most liquids of interest to us, so this equation becomes

---

[4]Otherwise we could have a volume enclosed by two spheres where more energy comes into the inside sphere than leaves the outside sphere, resulting in time-dependent accumulation of energy.

$$mc_p \frac{dT}{dt} = L \frac{dm}{dt} + Q \tag{4.39}$$

If the droplet volume is $V$ and its density $\rho_{drop}$, Eq. (4.39) can be rewritten as

$$L \frac{dm}{dt} + Q = \rho_{drop} c_p V \frac{dT}{dt} \tag{4.40}$$

Substituting $V = \pi d^3/6$, $Q$ from Eq. (4.33), $dm/dt$ from Eq. (4.28), realizing that the droplet surface temperature is $T_s = T$ and simplifying, we obtain

$$-LD(c_s - c_\infty) - k_{air}(T - T_\infty) = \frac{dT}{dt} \rho_{drop} c_p \frac{d^2}{12} \tag{4.41}$$

Equation (4.41) is the equation that governs the temperature of an evaporating or growing droplet under assumptions (1)–(5) of Section 4.4. In general it must be solved in conjuction with Eq. (4.29) and an equation giving $c_s(T)$ (Eq. (4.5), (4.6) or (4.9)) for the vapor concentration at the interface. However, it is common to assume that the right-hand side of this equation is negligible, which makes the analysis easier, as described in the next section. Note that if the droplet temperature is constant, then $dT/dt = 0$ and our equation for the droplet temperature Eq. (4.40) reduces to

$$L \frac{dm}{dt} + Q = 0 \tag{4.42}$$

In words, Eq. (4.42) states that the energy lost or gained by the droplet due to evaporation or condensation must be balanced by conduction of heat to or from the droplet surface.

## 4.7 Simplified theory of droplet growth or evaporation of a single droplet whose temperature is constant

Equation (4.41) shows that, in general, an evaporating or growing droplet will undergo a temperature change, with the temperature of the droplet governed by Eq. (4.41). To solve Eq. (4.41), we must also solve Eq. (4.29) in combination with an equation for the exponential temperature dependence of the vapor concentration $c_s(T)$ like that given in Eqs (4.6) or (4.9). However, if somehow the droplet temperature remains at a constant value, then we can solve the equation governing the droplet's diameter analytically, as follows.

We begin with Eq. (4.41) with $dT/dt$ set to zero on the right-hand side, which gives

$$LD(c_s - c_\infty) + k_{air}(T - T_\infty) = 0 \tag{4.43}$$

The ambient conditions $c_\infty$, $T_\infty$ and the thermophysical parameters $k_{air}$ and $D$ are considered known. The vapor concentration at the droplet surface, $c_s$, is a known function of temperature $T$, i.e. $c_s(T)$. Thus, Eq. (4.43) is an algebraic equation with temperature as the only variable. This equation is readily solved iteratively using standard numerical methods, from which we obtain the droplet temperature, $T$. For pure liquid droplets with no Kelvin effects, the resulting temperature is often called the 'wet bulb temperature', since it is the temperature that a wetted thermometer bulb would read (and is used in measuring humidity with a traditional wet/dry bulb apparatus).

Once the droplet temperature is known, we can obtain $c_s(T)$ from Eq. (4.6) and substitute this into the equation governing the droplet diameter (Eq. (4.29))

$$\frac{\mathrm{d}d}{\mathrm{d}t} = -\frac{4D(c_s - c_\infty)}{\rho_{\mathrm{drop}}d} \tag{4.29}$$

The right-hand side of Eq. (4.29) is a constant for a single droplet in ambient air, since the ambient vapor concentration far from the droplet, $c_\infty$, can be considered to be constant[5]. Also, with the droplet temperature being constant, the diffusion coefficient $D$ and the vapor concentration at the surface are also constant, since they depend only on temperature (neglecting dissolved solute effects, as well as the Kelvin effect). Thus, the only time-dependent variable in Eq. (4.29) is the droplet diameter, $d$, and we can integrate Eq. (4.29) with respect to time to obtain

$$d_0^2 - d^2 = \frac{8D(c_s - c_\infty)}{\rho_{\mathrm{drop}}}t \tag{4.44}$$

where $d_0 = d(t = 0)$ is the droplet's initial diameter.

The lifetime of an evaporating droplet, $t_L$, is obtained by setting $d = 0$ in Eq. (4.44) to obtain

$$t_L = \rho_{\mathrm{drop}}d_0^2/[8D(c_s - c_\infty)] \tag{4.45}$$

or in terms of the droplet's initial radius $R_0 = d_0/2$,

$$t_L = \rho_{\mathrm{drop}}R_0^2/[2D(c_s - c_\infty)] \tag{4.46}$$

For 0% $RH$ ($c_\infty = 0$), the droplet lifetime is

$$t_{L0} = \rho_{\mathrm{drop}}R_0^2/[2Dc_s] \tag{4.47}$$

Equation (4.47) for $t_{L0}$ was used earlier as a conservative (lower bound) representative time scale over which hygroscopic effects occur, since it is both the time over which a droplet will shrink to nothing when the humidity is 0%, as well as the time for a droplet to grow to $\sqrt{2}$ times its initial radius when the humidity at the droplet surface $c_s/c_s|_{\mathrm{pure\ H_2O}}$ is 50% (perhaps due to dissolved solids). In general, droplet lifetimes or the time for droplets to grow by $\sqrt{2}$ will be larger than these values, since 0% ambient air or an $RH$ of 50% at the droplet surface are extreme cases.

Remember that in addition to the droplet temperature being constant, both Eqs (4.46) and (4.47) require assumptions (1)–(5) of Section 4.4 to be valid.

## 4.8 Use of the constant temperature equation for variable temperature conditions and a single droplet

The temperature of a single hygroscopic droplet is obtained from

$$-LD(c_s - c_\infty) - k_{\mathrm{air}}(T - T_\infty) = \frac{\mathrm{d}T}{\mathrm{d}t}\rho_{\mathrm{drop}}c_p\frac{d^2}{12} \tag{4.41}$$

---

[5]This will, of course, not be true if we have many droplets evaporating, as we will discuss later in this chapter, nor will it be true if the droplet is traveling through regions of varying humidity, e.g. traveling through the lung.

If the right-hand side of this equation can be neglected, then we obtain simply the algebraic equation (Eq. (4.43)) which is easier to solve as we saw above. Also, if the ambient conditions are constant, neglecting the right-hand side leads to a constant value of the temperature obtained by solving Eq. (4.43) so that a single evaporating or growing droplet of pure $H_2O$ placed in constant ambient conditions will have a constant temperature (except for an initial transient which we will discuss later).

To address the question as to when the right-hand side of Eq. (4.41) can be neglected, we introduce the following nondimensional variables, which are all expected to be O(1):

$T' = (T - T_\infty)/\Delta T$ where $\Delta T$ is some characteristic temperature difference
$c' = (c_s - c_\infty)/\Delta c$ where $\Delta c$ is some characteristic vapor concentration difference
$d' = d/d_0$ where $d_0$ is the droplet's initial diameter
$t' = t/t_L$ where $t_L$ is the droplet's lifetime

Introducing these nondimensional variables into Eq. (4.41), we then obtain

$$(LD\Delta c)c' + (k_{air}\Delta T)T' = -\left(\frac{\rho_{drop}c_p\Delta T d_0^2}{12t_L}\right)d'^2\frac{dT'}{dt'} \qquad (4.48)$$

The terms in brackets are constant coefficients, while the terms outside the brackets are expected to be O(1) because of our nondimensionalization. Thus, the right-hand side will be negligible if the coefficient in brackets on the right-hand side is much less than either of the coefficients on the left-hand side. Comparing with the second coefficient on the left-hand side, the right-hand side can be neglected if

$$\frac{\rho_{drop}c_p\Delta T d_0^2}{12t_L} \ll k_{air}\Delta T \qquad (4.49)$$

Simplifying, this gives

$$\frac{\rho_{drop}c_p d_0^2}{12t_L k_{air}} \ll 1 \qquad (4.50)$$

To decide if this inequality is satisfied we need an estimate of the droplet lifetime. For this purpose, we first assume the inequality is satisfied so that we can use a quasi-steady estimate from Eq. (4.47) for the droplet lifetime $t_L$. We then calculate the left-hand side of the inequality and if it is still satisfied, a constant temperature assumption is valid. (Although this seems like a circular argument, it is logically sound.)

Thus, from Eq. (4.47) we estimate the droplet lifetime as $t_L = \rho_{drop}d_0^2/[8D(c_s - c_\infty)]$, and substitute this into Eq. (4.50), to obtain

$$8D(c_s - c_\infty)c_p/12k_{air} \ll 1 \qquad (4.51)$$

For water droplets in air, $D = 2.5 \times 10^{-5}$ m$^2$ s$^{-1}$, the specific heat of liquid water is $c_p = 4.2 \times 10^3$ J kg$^{-1}$ K$^{-1}$, and the thermal conductivity of air is 0.026 W m$^{-1}$ K$^{-1}$. The inequality will be hardest to satisfy if we set $c_\infty = 0$, so let us stay on the conservative side and assume $c_\infty = 0$. A typical value of the water vapor concentration $c_s$ in air at temperatures between 285 and 310 K is 0.01–0.04 kg m$^{-3}$ (from Eq. (4.6)). Putting these numbers into Eq. (4.51), we find $8D(c_s - c_\infty)c_p/12k_{air} \leq 0.1$, and indeed our inequality is satisfied.

Thus, for water droplets in room temperature or body temperature air we can neglect the right-hand side of Eq. (4.41), which means the droplet temperature is approximately constant, and we can use Eq. (4.43) and the simplified analysis in Section 4.6 that this equation implies. For droplets made of other substances than water, we would need to check inequality Eq. (4.51) to see if we can use Eq. (4.43) and the constant-temperature simplified analysis or if we must instead solve the more complicated Eq. (4.41) for the temperature of the droplets.

For water droplets with dissolved solutes, it is common to neglect the right-hand side of Eq. (4.41) to obtain the temperature for a given diameter $d$ and use this temperature to advance Eq. (4.29) over a short interval of time $\Delta t$. At the new diameter $d(t + \Delta t)$ there will be a new value of $c_s$ since the solute concentration in the drop is different now, but by again using Eq. (4.43) we can find the new droplet temperature $T$ at the new droplet size and continue iterating explicitly in this fashion to obtain $d(t)$ for all $t$.

## Example 4.6

Estimate the temperature of a 3 micron isotonic saline droplet evaporating in 50% *RH* room temperature air (i.e. $23°C$).

## Solution

As we have just seen, for water droplets we can use the constant temperature solution embodied by Eq. (4.43):

$$LD(c_s - c_\infty) + k_{air}(T - T_\infty) = 0 \qquad (4.43)$$

Solving for $T$ we obtain

$$T = T_\infty - \frac{LD(c_s - c_\infty)}{k_{air}} \qquad (4.52)$$

To evaluate the second term on the right-hand side, let us assume the droplet temperature is not too far from room temperature and use the values of the physical properties $L$, $D$ and $k_{air}$ for room temperatures. If we find a temperature that is far from room temperature, then we may want to correct these values and iterate.

At room temperature, the latent heat of vaporization for water is $2.44 \times 10^6\ J\ kg^{-1}$, the diffusion coefficient of water vapor in air $D = 2.5 \times 10^{-5}\ m^2\ s^{-1}$, and the thermal conductivity of air is $k_{air} = 0.026\ W\ m^{-1}\ K^{-1}$.

The vapor concentration at the droplet surface is given by Eq. (4.9) with $S = 0.995$ (since this is the definition of an isotonic solution), so we have

$$c_s = Sc_s|_{pure\ H_2O} \qquad (4.9)$$

Thus, we have

$$c_s = 0.995 \times c_s|_{pure\ H_2O}(T)$$

Since the humidity is 50% *RH*, we also have

$$c_\infty = 0.5 \times c_s|_{pure\ H_2O}(T = 296.15\ K)$$

Using Eq. (4.6)

$$c_s|_{pure\ H_2O}(T) = 3.638 \times 10^5 \times \exp[-4943/T]$$

then we obtain $c_\infty = 0.0102584$ kg m$^{-3}$. Putting these numbers into Eq. (4.52), we have

$$T = 296.15 - 2346.15(0.995 \, c_s|_{pure \, H_2O}(T) - 0.0102584)$$

But since we don't know $T$ yet, we don't know $c_s|_{pure \, H_2O}(T)$, and we must iterate to solve this algebraic equation. Simplifying, we have

$$T = 296.15 - 8.49263 \times 10^8 \exp[-4943/T] + 24.0679$$

Using fixed point iteration (which solves the equation $x = g(x)$ using the iterative equation $x^{n+1} = g(x^n)$), we obtain

$$T = 288.8 \text{ K (i.e. } 15.7°\text{C)}$$

This is approximately 7°C below room temperature; $L$, $D$ and $k_{air}$ are little affected by such a small temperature difference, so that this answer will change little if we adjust the values of $L$, $D$ and $k_{air}$ to 288.8 K.

Notice that the temperature would appear to be independent of the droplet size. As the droplet evaporates, however, the ratio of salt to water in the droplet increases. This causes $S$ to decrease (see Eq. (4.12)), so that this temperature will in fact change as the droplet evaporates.

## 4.8.1 Inapplicability of constant temperature assumption during transients

In deciding whether we can ignore the right-hand side of Eq. (4.41) we have implicitly assumed that $dT/dt$ is not large. However, if we place a room temperature isotonic saline droplet in room temperature air, using Eq. (4.43) we find that the temperature of this droplet must instantly drop to its wet bulb temperature. We have a contradiction here – an instantaneous drop in $T$ means an infinite $dT/dt$, whereas Eq. (4.43) was derived assuming finite (and moderate) $dT/dt$. This paradox is, of course, caused by the fact that there is an initial transient change in temperature of the droplet, which is not described using Eq. (4.43) and must instead be obtained using Eq. (4.41).

We can estimate the transient time during which Eq. (4.43) is not valid by assuming that during this transient time the conduction heat transfer is balanced by a change in thermal energy of the droplet with no mass transfer. Thus, during the transient period, Eq. (4.41) can be approximated by

$$-k_{air}(T - T_\infty) = \frac{dT}{dt} \rho_{drop} c_p \frac{d^2}{12} \tag{4.53}$$

To reach a temperature drop of $\Delta T = T - T_\infty$ degrees in time $\Delta t_{transient}$, this equation implies that

$$k_{air} \Delta T \approx \frac{\Delta T}{\Delta t_{transient}} \rho_{drop} c_p \frac{d^2}{12} \tag{4.54}$$

Simplifying, we obtain

$$\Delta t_{transient} \approx \rho_{drop} c_p d^2 / 12 k_{air} \tag{4.55}$$

Comparing this time to our conservative estimate of the droplet lifetime $t_{L0} = \rho_{drop} R_0^2 / [2Dc_s]$ from Eq. (4.47), we see that

$$\frac{\Delta t_{transient}}{t_{L0}} = \frac{8Dc_p c_s}{12k_{air}} \tag{4.56}$$

For a water droplet in air, substituting in the various physical properties which we have encountered before ($k_{air} = 0.03$ W m$^{-1}$ K$^{-1}$, $c_p = 4.2 \times 10^3$ J kg$^{-1}$ K$^{-1}$, $c_s = 0.01 - 0.04$ kg m$^{-3}$, $D = 2.5 \times 10^{-5}$ m$^2$ s$^{-1}$), we obtain $\Delta t_{transient}/t_{L0} \leq 0.1$, and we see that the transient portion of a water droplet's lifetime in room or body temperature air occupies only a short portion of its life, so that the droplet responds rapidly to temperature changes. Thus, except for a very short transient period after the droplet is exposed to a new environment, the temperature of a single water droplet under typical inhalation conditions is obtained by using Eq. (4.43) rather than Eq. (4.41).

## 4.9 Modifications to simplified theory for multiple droplets: two-way coupled effects

The simplified theory developed in the preceding sections has an inherent flaw if we are to consider what happens when many droplets are present in a volume of air. In particular, if many water droplets are evaporating, the concentration of water vapor in the air will increase. This in turn will slow the evaporation rate. As a result, if many droplets undergo hygroscopic size changes in a given volume, the mass transferred to or from the air surrounding them causes a change in the ambient conditions, which causes the droplets to change size at a different rate than the case of a single droplet considered in the preceding sections. Thus, when many droplets are present in a given volume, the droplets are affected by the ambient air, but the ambient air is affected by the evaporation or condensation of the droplets. Hygroscopic size changes occurring under such conditions are referred to as 'two-way coupled' hygroscopic effects. The single droplet case described above is sometimes called a one-way coupled hygroscopic treatment, since the droplets are affected by the ambient air, but the ambient conditions are unaffected by the droplets.

To account for two-way coupling it is necessary to treat the vapor concentration and temperature of the air as additional unknowns that we must solve for in addition to the temperature and mass or size of the droplets, as described in Finlay and Stapleton (1995). The analysis is somewhat involved, but to illustrate the basic idea, consider two-way coupled effects for a monodisperse aerosol contained by adiabatic walls with no mass transfer at the walls. In this case a simple mass balance implies that the rate of change of vapor concentration of ambient air is equal to the rate of mass transferred to air per unit volume, i.e.

$$\frac{dc_\infty}{dt} = -N\frac{dm}{dt} \qquad (4.57)$$

where $N$ is the number of particles per unit volume. The minus sign arises since droplets increasing in mass will reduce the vapor concentration in the ambient air.

Similarly, an energy balance implies that the rate of change of thermal energy of ambient air is equal to the rate of heat transfer from droplets, i.e.

$$\rho_{air} c_{p_{air}} \frac{dT_\infty}{dt} = -NQ \qquad (4.58)$$

where we have neglected the thermal energy of the vapor (since from assumption (1) of Section 4.4 there shouldn't be much vapor compared to air), and the minus sign arises because $Q$ was defined previously as the rate of heat transfer *to* the droplets from the air, and is as given earlier from Eq. (4.33), i.e.

$$Q = -2\pi dk_{air}(T - T_\infty)$$

where $T$ is the temperature of the droplets. Equations (4.57) and (4.58) must be solved in combination with Eq. (4.28) for the mass of a droplet

$$\frac{dm}{dt} = -2\pi dD(c_s - c_\infty) \tag{4.28}$$

equation (4.41) for the droplet temperature

$$-LD(c_s - c_\infty) - k_{air}(T - T_\infty) = \frac{dT}{dt}\rho_{drop}c_p\frac{d^2}{12} \tag{4.41}$$

and Eqs (4.6) or (4.9) for the temperature dependence of the vapor concentration $c_s$.

Equations (4.28), (4.41), (4.57) and (4.58) are four coupled, nonlinear, first-order ordinary differential equations that can be solved numerically using a standard numerical ordinary differential equation (ODE) solver.

Inhaled pharmaceutical aerosols are not monodisperse, but are instead polydisperse. In this case, we can divide the size distribution into a number, $K$, of discrete sizes and treat each size using the equations we have written here. However, we then have an equation similar to Eqs (4.28) and (4.41) for each particle size, with the right-hand side of Eqs (4.57) and (4.58) modified to be summations over all $K$ particle sizes (see Finlay and Stapleton 1995 for details). If we divide our particle size distribution into $K$ particle sizes, we then have a total of $K$ equations for the mass of each particle size, $K$ equations for the temperature of each particle size, plus an equation like Eq. (4.57) for the vapor concentration in the ambient air and one like Eq. (4.58) for the temperature of the ambient air. Thus, we now have $2K + 2$ coupled, nonlinear, ODEs that must be solved. This is much more work than the simple analytical result obtained earlier for a single droplet. Additionally, the different particle sizes will respond at different rates, resulting in the possibility of a stiff system of ODEs and the need for an implicit ODE solver (Finlay and Stapleton 1995), which complicates matters.

## 4.10 When are hygrosopic size changes negligible?

The previous section shows that two-way coupled hygroscopic effects can add considerably to the complexity of the fate of a polydisperse aerosol. Because the size of inhaled droplets is a primary factor that determines where these droplets will deposit in the lung, and because hygroscopic droplets can change size with time, hygroscopic effects may need to be included if we are to obtain reasonable predictions of where inhaled droplets deposit in the lung. But, because of the complexity of hygroscopic effects, we would rather not include such effects if we do not have to. For this reason, it is useful to be able to estimate whether hygroscopic effects are small enough that they can be neglected. Finlay (1998) addresses this question in some detail, and what follows is based on this work.

To answer the question as to whether hygroscopic size changes are negligible, we must realize that two-way coupled effects will always act to reduce the magnitude of hygroscopic size changes, since two-way coupling causes the ambient air to respond to the particles and, in a sense, meet the droplets halfway in their efforts to undergo hygroscopic size changes. Thus, if a one-way coupled hygroscopic treatment predicts negligible size changes, then we can safely say that a two-way coupled treatment would give even smaller size changes, so that an assumption of negligible hygroscopic size changes is reasonable.

Thus, we can use Eq. (4.28) (which is based on a one-way coupled treatment) to decide if large changes in droplet mass can be expected over a typical time that we are interested in. Recall Eq. (4.28) is

$$\frac{dm}{dt} = -2\pi dD(c_s - c_\infty) \tag{4.28}$$

If we define $\Delta m$ as the change in mass of a hygroscopic droplet over a representative time scale $\Delta t$ (e.g. $t_L$ from Eq. (4.45)), then Eq. (4.28) gives the value of $\Delta m$:

$$\Delta m = \Delta t 2\pi D d \overline{\Delta c} \tag{4.59}$$

where $\overline{d\Delta c}$ is the mean value of $d|c_s - c_\infty|$ over the time interval $\Delta t$.

If hygroscopic effects are to be small, then we must have $\Delta m$ much less than the mass of the droplet; i.e. hygroscopic size changes are small if $\zeta \ll 1$ where $\zeta = \Delta m/m$. Using Eq. (4.59) for $\Delta m$, with $m = \rho_{drop}\pi d^3/6$, $\overline{d} = d$ and using $\overline{d\Delta c} = d\overline{\Delta c}$ (since $d$ doesn't change much if hygroscopic size changes are small), we have

$$\text{hygroscopic size changes are small if } \zeta \ll 1 \tag{4.60}$$

where

$$\zeta = 12\Delta t D \overline{\Delta c}/\rho_{drop}d^2$$

Here, $\overline{\Delta c}$ is the mean value of $|c_s - c_\infty|$ over the time interval $\Delta t$.

Note that Eq. (4.60) does not mean that hygroscopic effects are necessarily important if $\zeta \ll 1$ does not hold. For example, if we have enough droplets per unit volume, then if all of these droplets evaporate only a little bit, they can humidify the ambient air and halt hygroscopic size changes with only small changes in the droplets size. Thus, we must realize that two-way coupled effects can come into play and make hygroscopic size changes much smaller than predicted by the one-way coupled treatment used to derive Eq. (4.60). For this reason, if $\zeta \ll 1$ does not hold, we must look further to see whether hygroscopic size changes might be important.

Consider, for example, an aerosol that needs to change its mass only a small amount in order to reduce $c_s - c_\infty$ to zero. Such an aerosol will not undergo large hygroscopic size changes. Thus, we see that if $\zeta \ll 1$ does not hold, we should examine the parameter

$$\gamma = \text{mass of droplets per unit volume}/\Delta c^* \tag{4.61}$$

where $\Delta c^*$ is the amount of water vapor per unit volume that needs to be exchanged between the droplets and the surrounding air in order to reach equilibrium. If $\gamma \gg 1$, the droplets will hardly change size at all before they reduce $c_s - c_\infty$ to zero and halt further hygroscopic effects. On the other hand, if $\gamma \ll 1$, then the mass contained in droplets is much less than the amount of mass transfer needed to reduce $c_s - c_\infty$ to zero, and hygroscopic size changes will be large, so that a one-way coupled treatment can be used.

In summary, we have the following conditions:

$\zeta \ll 1$: hygroscopic size changes are small
$\zeta$ not $\ll 1$ and $\gamma \gg 1$: hygroscopic size changes are small
$\zeta$ not $\ll 1$ and $\gamma \ll 1$: hygroscopic size changes are not small, but can be treated as one-way coupled

$\zeta$ not $\ll 1$ and $\gamma = O(1)$: hygroscopic size changes are not small and two-way coupled treatment is needed

A parameter sometimes used instead of the mass of droplets per unit volume is the volume of the aerosol droplets per unit volume, $\beta$, which is simply the volume fraction occupied by droplets. If the volume fraction, $\beta$, is known, the mass of droplets per unit volume is then $\beta \rho_{drop}$ and $\gamma$ can be calculated from Eq. (4.61) to obtain

$$\gamma = \beta \rho_{drop} / \Delta c^* \tag{4.62}$$

where $\rho_{drop}$ is the mass density of the droplet.

## Example 4.7

Are hygroscopic size changes important for a monodisperse aerosol consisting of 5 μm, isotonic saline droplets with number density $5 \times 10^5$ droplets cm$^{-3}$ inhaled into the lung with 50% *RH*, room temperature air? If yes, then is a two-way coupled or one-way coupled treatment needed?

## *Solution*

We must calculate values of $\zeta$ and $\gamma$ in Eqs (4.60) and (4.61).
First let us calculate

$$\zeta = 12 \Delta t D \overline{\Delta c} / \rho_{drop} d^2$$

in Eq. (4.60). For this purpose, we must estimate the representative time scale $\Delta t$. In the present case, a representative time scale over which hygroscopic size changes occur is the droplet time scale $t_L$ given by Eq. (4.45):

$$t_L = \rho_{drop} d_0^2 / [8 D(c_s - c_\infty)] \tag{4.45}$$

Substituting $\rho_{drop} = 1000$ kg m$^{-3}$, $d_0 = 5 \times 10^{-6}$ m, $D = 2.5 \times 10^{-5}$ m$^2$ s$^{-1}$, $c_s = 0.02$ kg m$^{-3}$, $c_\infty = 0.5 c_s$ (since the ambient humidity is 50% *RH*), we obtain $t_L = 0.01$ s.
We must also estimate a value of $\overline{\Delta c}$, where $\overline{\Delta c}$ is the mean value of $|c_s - c_\infty|$ over the time $\Delta t$, i.e. the mean difference in vapor concentration at the particle surface and in ambient air. This is difficult to estimate, since $c_\infty$ changes as we go from the mouth, where the *RH* is 50% and *T* is room temperature, to its value deep in the lung where *RH* is 99.5%. Also, if the droplet evaporates, the concentration of salt in the droplet will increase, causing $c_s$ to change with time. However, we can supply a rough estimate of $\overline{\Delta c}$ by knowing that the particle will work to reduce the value of $c_s - c_\infty$ from its initial value at the mouth. Thus, $\overline{\Delta c}$ should be somewhere between its lowest value (which is zero) and its initial value at the mouth.
At the mouth we have an isotonic droplet for which we have,

$$c_s = 0.995 \, c_s|_{pure \, H_2O}$$

and at room temperature $c_s|_{pure \, H_2O} \approx 0.02$ kg m$^{-3}$.
Thus, we have

$$c_s \approx 0.02 \text{ kg m}^{-3}$$

Also, at the mouth we have 50% $RH$ and room temperature, so

$$c_\infty \approx 0.50 \; c_s|_{\text{pure H}_2\text{O}}$$

i.e.

$$c_\infty \approx 0.01 \text{ kg m}^{-3}$$

Thus, at the mouth we have

$$c_s - c_\infty \approx 0.01 \text{ kg m}^{-3}$$

A rough estimate for $\overline{\Delta c}$ is halfway between this value and zero, i.e.

$$\overline{\Delta c} \approx 0.005 \text{ kg m}^{-3}$$

Putting this value into the equation for $\zeta$ we have

$$\zeta = (12 \times 0.01 \text{ s} \times 2.5 \times 10^{-5} \text{ m}^2 \text{ s}^{-1} \times 0.005 \text{ kg m}^{-3})/(998 \text{ kg m}^{-3}(5 \times 10^{-6} \text{ m})^2)$$

$$\zeta = 0.6$$

Thus $\zeta \ll 1$ does not hold, and we cannot yet say whether hygroscopic size changes are important. Instead, we must look at the value of $\gamma$.

For this purpose, we can use Eq. (4.62) where the volume fraction of the aerosol is given by taking the number of particles per cm$^3$ and multiplying by the volume of a single particle ($\pi d^3/6$), i.e.

$$\beta = (5 \times 10^5 \text{particle cm}^{-3}) \times \pi(5 \times 10^{-4})^3/6 \text{ cm}^3 \text{ per particle}$$

which gives

$$\beta = 3.2 \times 10^{-5}$$

Putting this into Eq. (4.62) with $\rho_{\text{drop}} = 1000 \text{ kg m}^{-3}$, and estimating $\Delta c^*$ as[6] $\Delta c^* = \overline{\Delta c}/2 = 0.0025 \text{ kg m}^{-3}$, we have

$$\gamma = \beta \times \rho_{\text{drop}}/\Delta c^*$$
$$= 3.2 \times 10^{-5} \times 1000 \text{ kg m}^{-3}/0.0025 \text{ kg m}^{-3}$$
$$\gamma = 13$$

Thus, we see that $\zeta \ll 1$ does not hold and $\gamma \gg 1$, so that hygroscopic effects are probably not important for this aerosol.

## 4.11 Effect of aerodynamic pressure and temperature changes on hygroscopic effects

To this point, our examination of the rate at which droplets change size has been done without any consideration of the fact that the motion of the air carrying the droplets (i.e. the fluid dynamics) can cause pressure and temperature changes that may affect the conditions that surround the droplets. Such effects are not expected in the respiratory

---

[6]Recall that $\Delta c^*$ is the amount of water vapor that needs to be exchanged between the droplets and the surrounding air to come into equilibrium – this should be less than $\overline{\Delta c}$ in the present case since water vapor transfer at the airway walls will also humidify the air, so that the droplets don't have to supply all of the water vapor needed to reach equilibrium.

tract, but can be quite pronounced in laboratory settings where the aerosol is carried through large pressure drops that occur due to fluid dynamics, such as in some cascade impactors. There are two principal effects that occur because of this:

(a) A drop in air pressure due to fluid motion at constant temperature (where the pressure drops to a fraction $x$ of the upstream pressure) causes a directly proportional drop in $c_v$ (to a value $x$ of its upstream value) because of the ideal gas law ($p = cR_uT/M$), which in turn affects hygroscopic effects through Eq. (4.28) (Fang *et al.* 1991).

(b) The temperature of the air can be reduced through isentropic compressible flow effects if large increases in air velocity occur (which usually occur as a result of large drops in pressure). These temperature changes in turn result in reductions in the temperature of the droplets through conductive heat transfer. This affects $c_s$ in Eq. (4.28) through the temperature dependence of $c_s$ seen in Eq. (4.5), and results in hygroscopic effects (Biswas *et al.* 1987).

These two effects have opposite effects on droplet evaporation rates, with (a) tending to increase droplet evaporation rates, while (b) reduces evaporation rates. Since pressure drops usually go along with flow acceleration, the two effects usually occur together, and detailed calculations are necessary to decide whether the end result is reduced or accelerated droplet size changes. However, both these effects are due to flow compressibility, which normally can be neglected if Mach numbers are below approximately 0.3. Thus, if the aerosol remains at air velocities below approximately 100 m s$^{-1}$, such flow-induced effects can usually be neglected. Although many impactors do satisfy this condition (e.g. the Anderson Mark II impactor) such effects may need to be included for impactors with high speed jets (e.g. the last stage of the MOUDI impactor, see Fang *et al.* 1991).

## 4.12 Corrections to simplified theory for small droplets

For droplets of the order 1 μm in diameter or smaller, there are two effects that must be accounted for if we are to obtain accurate results from the simplified theory above, the first being the effect of radius of curvature on vapor pressure (the Kelvin effect), the second being due to the noncontinuum nature of heat and mass transfer at very small length scales (the so-called Fuchs corrections or Knudsen number corrections). Let us examine each of these in turn.

### 4.12.1 Kelvin effect

It has long been known that the vapor concentration at surfaces with very rapid curvature is higher than for that next to flat surfaces. A simple explanation of this effect can be obtained by considering what a highly curved surface does to the energy required to take a water molecule from the liquid phase to the vapor phase given in Eq. (4.3). Recall that Eq. (4.3) stated that the concentration of vapor next to an air–water surface is given by

$$c_s = a \exp[-\Delta\mu/kT] \tag{4.3}$$

where $\Delta\mu$ is the energy needed to take a molecule out of the droplet (where it is surrounded by many nearby water molecules) and put it in vapor state (where the molecules are far apart). For a flat surface, we can think of the nearest molecules to a surface molecule as lying within a hemispherical shell of thickness equal to the average distance between water molecules in the liquid state. However, if the surface is highly curved, then from simple geometric considerations, the nearest molecules to this surface molecule will lie in a shell that is now only a partial hemisphere. As a result, there are fewer molecules immediately next to a surface molecule, and we must break fewer intermolecular bonds to pull a surface molecule into the vapor phase and away from its neighboring water molecules in the droplet.

Arguing on these grounds, then surface curvature should reduce the value of $\Delta\mu$, and we can write

$$\Delta\mu = \Delta\mu|_{flat} - \delta\mu$$

where $\delta\mu$ is the reduction in energy associated with the reduction in the interaction of a water molecule on the surface and the other molecules nearby. From Eq. (4.3), the concentration of water vapor at the interface is thus given by

$$c_s = a \exp[(-\Delta\mu|_{flat} + \delta\mu)/kT] \tag{4.63}$$

But the first half of the right-hand side is just what we would obtain if we had a flat surface, so we can write

$$c_s = c_s|_{flat\ surface}\exp(\delta\mu/kT) \tag{4.64}$$

From Eq. (4.64) we see that the vapor concentration at the surface will be increased by the presence of surface curvature, since $\delta\mu/kT \geq 0$ so that $\exp(\delta\mu/kT)$ is always $\geq 1$.

Introducing the factor $K = \exp(\delta\mu/kT)$, we can write Eq. (4.64) as

$$c_s = Kc_s|_{flat\ surface} \tag{4.65}$$

where $K \geq 1$ can be determined from continuum thermodynamic and mechanical considerations (Adamson 1990) and is given by

$$K = \exp[4\sigma M/(N_0\rho dkT)] \tag{4.66}$$

or

$$K = \exp[4\sigma M/(R_u\rho dT)] \tag{4.67}$$

where $R_u = 8.314$ kg mol$^{-1}$ m$^2$ s$^{-2}$ K$^{-1}$ $= N_0k$ is the universal gas constant, $N_0$ is Avogadro's number ($N_0 = 6.023 \times 10^{-23}$ mol$^{-1}$), $k$ is Boltzmann's constant ($k = 1.3807 \times 10^{-23}$ kg m$^2$ s$^{-2}$), $M$ is the molar mass of the vapor molecules (kg mol$^{-1}$) and $\sigma$ is the surface tension at the droplet surface (which is 0.073 N m$^{-1}$ for water at room temperature and is 0.069 N m$^{-1}$ for water at body temperature, i.e. 37°C).

## Example 4.8

What is the correction due to the Kelvin effect for the vapor concentration at the surface of a 1 micron droplet of isotonic saline at room temperature and body temperature?

## Solution

From Eq. (4.67), we have

$$K = \exp[4\sigma M/(R_u \rho dT)]$$
$$= \exp[4 \times 0.073 \times 18.0 \times 10^{-3}/(8.314 \times 998 \times 1 \times 10^{-6} \times 293.15)]$$
$$= 1.002$$

Because the surface tension of water is only slightly different at body temperature, we also obtain $K = 1.002$ at 37°C.

As seen in the previous example, the Kelvin effect is typically quite small for most inhaled pharmaceutical aerosol droplets, since they are not usually intended to have diameters much smaller than 1 μm. A more detailed examination as to when we need to include the Kelvin effect can be made based on Heidenriech and Büttner (1995), as follows. To examine the size of the error incurred if we neglect the Kelvin effect, we can use Eq. (4.29), which shows that the rate of change of droplet size is given by

$$\frac{dd}{dt} = -\frac{4D(c_s - c_\infty)}{\rho_{drop} d} \tag{4.29}$$

Thus, at any instant in time, the relative error we make in the rate of change of droplet size by not including the Kelvin effect is given by

$$\varepsilon = \frac{dd/dt|_{\text{without Kelvin}} - dd/dt|_{\text{with Kelvin}}}{dd/dt|_{\text{with Kelvin}}}$$

Using Eq. (4.29), this error can be written as

$$\varepsilon = \frac{[(c_s|_{\text{without Kelvin}} - c_s|_{\text{with Kelvin}})/c_\infty]}{[(c_s - c_\infty)|_{\text{with Kelvin}}/c_\infty]}$$

For the Kelvin effect to be small it is necessary that $|\varepsilon| \ll 1$. As seen in the previous example, the difference between the vapor concentration when including or neglecting the Kelvin effect is quite small for typical sizes of inhaled pharmaceutical aerosols, so that the numerator here is, in general, a small number. Thus, in order for the Kelvin effect to not be small, we must have the denominator, $(c_s - c_\infty)/c_\infty$, also being small. However, if $(c_s - c_\infty)/c_\infty$ is small then the rate of change of size of a particle will also be small, since $(c_s - c_\infty)$ is what drives our hygroscopic size changes, as seen in Eq. (4.29). Thus, for our purposes, the Kelvin effect only comes into play when we have small rates of change of particle diameter. But this is exactly when hygroscopic size changes are not important. As a result, from the point of view of predicting where inhaled droplets will deposit in the lung, the Kelvin effect can normally be neglected for inhaled pharmaceutical aerosols.

However, there is one effect that the Kelvin effect does cause, that should be realized and which might be important for some polydisperse aerosols. In particular, the Kelvin effect can cause a drug or dissolved salt to be preferentially contained in smaller droplets. This can be seen by considering what happens when a droplet is in equilibrium with its surrounding environment, so that $dd/dt = 0$ in Eq. (4.29). In this case, then we have

$$c_s = c_\infty \quad \text{(droplet in equilibrium with its environment)} \tag{4.68}$$

Including the Kelvin effect, we can write, from Eqs (4.9) and (4.67)

$$c_s = SKc_s|_{\text{pure flat H}_2\text{O surface}} \tag{4.69}$$

where $S$ is the vapor concentration reduction due to dissolved salts, and is given for dilute solutions from Eq. (4.12) as

$$S = 1 - i\frac{x_s}{x_w} \tag{4.12}$$

Here, $x_s$ gives the moles of salt or drug in the droplet, while $x_w$ gives the moles of water in the droplet. Combining Eqs (4.68), (4.69) and (4.12) we can write

$$\left(1 - i\frac{x_s}{x_w}\right)K = \frac{c_\infty}{c_s|_{\text{pure flat H}_2\text{O surface}}}$$

However, the right-hand side of this equation is simply the relative humidity ($RH$). Thus, we have

$$\left(1 - i\frac{x_s}{x_w}\right)K = RH \tag{4.70}$$

Now, if we neglect the Kelvin effect, $K = 1$, and we have

$$\left(1 - i\frac{x_s}{x_w}\right)\bigg|_{\text{without Kelvin}} = RH \tag{4.71}$$

For a small droplet, if we include the Kelvin effect we have instead

$$\left(1 - i\frac{x_s}{x_w}\right)\bigg|_{\text{with Kelvin}}K = RH \tag{4.72}$$

Dividing Eq. (4.72) by Eq. (4.71), we have

$$\left(1 - i\frac{x_s}{x_w}\right)\bigg|_{\text{with Kelvin}}K = \left(1 - i\frac{x_s}{x_w}\right)\bigg|_{\text{without Kelvin}} \tag{4.73}$$

which can be rewritten as

$$\left(\frac{x_s}{x_w}\right)\bigg|_{\text{with Kelvin}}\bigg/\left(\frac{x_s}{x_w}\right)\bigg|_{\text{without Kelvin}} = \left(K - 1 + i\frac{x_s}{x_w}\bigg|_{\text{without Kelvin}}\right)\bigg/\left(Ki\frac{x_s}{x_w}\bigg|_{\text{without Kelvin}}\right) \tag{4.74}$$

For dilute solutions, the value of $x_s/x_w$ is approximately equal to the concentration of drug or salt given as moles of drug per mole of water, i.e. $x_s/x_w$ is equal to the mole fraction. Thus, defining $X$ as the mole fraction of drug in the droplet including the Kelvin effect and $X_0$ as the mole fraction of drug in droplet obtained by neglecting the Kelvin effect, we can rewrite Eq. (4.74) as

$$X/X_0 = (K - 1 + iX_0)/(KiX_0) \tag{4.75}$$

For large droplets, $K = 1$ and Eq. (4.75) implies $X = X_0$. However, for smaller droplets, $K \geq 1$ so that Eq. (4.75) implies $X \geq X_0$ and the drug concentration in smaller droplets is larger. Indeed, since $K$ increases with decreasing droplet diameter, for a polydisperse aerosol, the drug concentration is largest in the smallest droplets, as is seen in the following example. Note that this preferential concentration of drug in small droplets comes into play when time-dependent hygroscopic size changes are expected to be small from considerations of $\zeta$ and $\gamma$.

## Example 4.9

What is the difference in concentration of drug between a 1 µm diameter droplet and a 10 µm diameter droplet for a polydisperse saline aerosol in equilibrium at 99.5% *RH* and body temperature?

### *Solution*

From the definition of relative humidity, we know that if the ambient humidity is 99.5%, then

$$c_\infty = 0.995 \ c_s|_{\text{pure flat H}_2\text{O}}$$

For a droplet to be in equilibrium we must have $c_s = c_\infty$, so the vapor pressure at the droplet surfaces must satisfy

$$c_s = 0.995 \ c_s|_{\text{pure flat H}_2\text{O}}$$

From the definition of isotonicity, we know that a bulk isotonic saline solution (9 mg ml$^{-1}$ NaCl in water) will have this vapor concentration next to it. Thus, if the Kelvin effect was not present, all the droplets would have the same concentration of saline, and would have a concentration of 9 mg ml$^{-1}$; i.e. we have

$$S|_{\text{without Kelvin}} = 0.995 \tag{4.76}$$

But, from Eq. (4.12) we know that

$$S = 1 - i \frac{X_s}{X_w}$$

so that

$$S|_{\text{without Kelvin}} = 1 - iX_0 \tag{4.77}$$

where

$$X_0 = \frac{X_s}{X_w}\bigg|_{\text{without Kelvin}}$$

is the mole fraction for isotonic saline.

Combining Eqs (4.76) and (4.77), we have

$$0.995 = 1 - iX_0$$

Thus, we must have $iX_0 = 0.005$. Using a van't Hoff factor for NaCl of $i = 1.85$, we then find

$$X_0 = 0.0027$$

We are now in a position to use Eq. (4.75)

$$X/X_0 = (K - 1 + iX_0)/(KiX_0)$$

All we need now is the value of the Kelvin correction $K$, where, recall from Eq. (4.67),

$$K = \exp[4\sigma M/(R_u \rho dT)]$$

For a 1 µm droplet, we obtain

$$K = \exp[4 \times 0.069 \times 18 \times 10^{-3}/(8.314 \times 998 \times 1 \times 10^{-6} \times 310)] = 1.00193$$

Thus, we obtain

$$X/X_0|_{\text{1micron drop}} = (1.00193 - 1 + 0.005)/(1.00193 \times 0.005) = 1.38$$

and we see that the 1 µm saline droplet will have 38% higher concentration than a bulk solution of isotonic saline.

For a 10 µm droplet, we obtain

$$K = \exp[4 \times 0.069 \times 18 \times 10^{-3}/(8.314 \times 998 \times 10 \times 10^{-6} \times 310)]$$
$$= 1.0002$$

and Eq. (4.75) gives

$$X/X_0|_{\text{10 micron drop}} = (1.0002 - 1 + 0.005)/(1.0002 \times 0.005) = 1.04$$

and there is only a 4% difference between the concentration in a 10 µm drop and a bulk isotonic saline solution. Thus, the 1 µm droplet consists of a 34% more concentrated solution than the 10 µm droplet.

For inhaled pharmaceutical aerosols, the above example demonstrates how the Kelvin effect can cause the smaller droplets to contain more drug than the larger droplets. Whether this preferential concentration effect has a significant effect on where drug is deposited in the lung depends on the aerosol being inhaled. Aerosols with most of the mass contained in larger droplets ($\geq 4$ µm or so) would not be expected to be significantly affected. However, aerosols with small MMDs ($\leq 2$ µm) contain significant mass in droplets for which this effect may come into play.

Note, however, that this preferential concentration effect decreases rapidly for hypertonic aerosols (i.e. ones more concentrated than isotonic saline), since, for example, if we recalculate the previous example but instead use a saline solution twice as concentrated, i.e. 18 mg ml$^{-1}$ (1.8% by weight), we find negligible difference in concentration between the 1 µm and 10 µm droplets. This is because for low solute concentrations (isotonic or hypotonic solutions) the Kelvin correction has nearly as large an effect on the vapor concentration as that due to the dissolved solutes, and the Kelvin effect causes a nonuniform distribution of solute in the different size droplets. However, for more hypertonic solutions, the effect of the dissolved solutes on equilibrium vapor concentrations is much larger than the Kelvin effect, and the nonuniform distribution effect is not nearly as large.

### 4.12.2 Fuchs (or Knudsen number) corrections

For droplets with radii approaching the mean free path of the molecules in the gas surrounding the droplet (which is 0.07 µm for air), the rate at which heat and mass is transferred by diffusion at the surface of small droplets is not accurately predicted by the equations we have given, since the continuum assumption (i.e. the assumption that matter is continuous, rather than molecular in nature) breaks down for such small droplets. For such droplets, we must correct the simple Fick's diffusion law used above. The corrections cause extra factors to appear in the heat and mass transfer rates given by our simplified theory. Including these corrections in our derivation of the mass transfer rate, Eq. (4.28) is replaced by

$$\frac{dm}{dt} = -2\pi d C_m D(c_s - c_\infty) \tag{4.78}$$

where

$$C_m = \frac{1 + Kn}{1 + \left(\dfrac{4}{3\alpha_m} + 0.377\right) Kn + \dfrac{4}{3\alpha_m} Kn^2} \tag{4.79}$$

(Fuchs and Sutugin 1970, Vesala *et al.* 1997). Here, $Kn$ is the Knudsen number, defined as $Kn = 2\lambda/d$ and $\lambda$ is the mean free path of the gas surrounding the droplet.

A similar correction is necessary for the heat transfer equation, causing Eq. (4.41) to become

$$-LD C_m(c_s - c_\infty) - k_{air} C_T(T - T_\infty) = \frac{dT}{dt} \rho_{drop} c_p \frac{d^2}{12} \tag{4.80}$$

where $C_T$ is given by the same expression as (4.79) but with $\alpha_m$ replaced with $\alpha_T$ and $Kn$ replaced by a thermal Knudsen number (Wagner 1982). The symbols $\alpha_m$ and $\alpha_T$ are referred to as the mass accommodation coefficient and thermal accommodation coefficient, respectively. Their values are dependent on the composition of the droplet and surrounding gas, and there is some disagreement on what values to use for aqueous droplets in air. Values of $\alpha_m = \alpha_T = 1$ have been used (Ferron *et al.* 1988, Hinds 1993), but values of $\alpha_m = 0.04$ have also been used (Ferron *et al.* 1988). If we use $\alpha_m = 1$, for a 1 μm droplet in air, we obtain

$$Kn = (2 \times 0.07/1) = 0.14$$

and

$$C_m = 0.9$$

We thus see that neglecting the Knudsen correction for a 1 μm droplet results in a 10% overestimate of the rate of change of mass of the droplet. Note that this is a larger correction to the mass transfer rate than the Kelvin correction. However, since most pharmaceutical aerosols are greater than 1 μm in diameter, and because of the uncertainty in the values of $\alpha_m$ and $\alpha_T$ that should be used for aqueous droplets containing salts and drugs, neglecting the Knudsen correction may be a reasonable approximation for many inhaled pharmaceutical aerosol applications. If not, this correction can be included using the above corrections. Hinds (1982) gives droplet lifetimes for pure water at 50% *RH* and 20°C both with and without the Fuchs corrections (using values of 1 for the accommodation coefficients) and finds for example that a 1.0 μm droplet has a lifetime of 1.4 ms without the corrections, and 1.7 ms with the corrections.

An alternative formula to Eq. (4.79) for the Knudsen number correction is suggested by Miller *et al.* (1998) and found to agree reasonably well with experiment even when convection (i.e. nonzero droplet Reynolds numbers) and Stefan flow are present.

Note, however, that only the *rate* of mass transfer is affected by these corrections, and not the equilibrium size of the droplets. Thus, only when $\zeta$ and $\gamma$ suggest hygroscopic size changes are significant will there be potential for the Knudsen corrections to become significant (unlike the Kelvin effect which only had a chance of being important for typical inhaled pharmaceutical aerosols when the hygroscopic aerosols were near equilibrium).

## 4.13 Corrections to account for Stefan flow

In deriving the classical, simplified theory for droplet growth and evaporation considered thus far in this chapter, it was necessary to assume that evaporation or condensation did not result in any bulk motion of air surrounding the droplet (assumption (1) of Section 4.4). However, when a droplet evaporates or condenses rapidly, vapor moves away from or toward the droplet at considerable velocities. For example, for a rapidly evaporating droplet, vapor is ejected into the volume surrounding the droplet at such a rate that significant velocities occur in the gas surrounding the droplet. In this case, we must modify Fick's law that we wrote down earlier.

The version of Fick's law that we wrote down earlier was as follows:

$$j = -D\nabla c \tag{4.21}$$

This equation gives us the mass flux, $j$, of vapor *relative to the velocity of the bulk mixture of vapor and air*, i.e.

$$j = \text{diffusion mass flux relative to the velocity of the gas}$$

If there is no bulk motion of the gas phase, then this does indeed tell us the mass flux due to diffusion. However, when there is motion of the gas, the vapor is carried by the gas at the same time the vapor is diffusing, so that we must add in the velocity of the gas. Doing so gives us the following equation:

vapor mass flux relative to stationary coordinate system

$$= \text{mass flux due to velocity of gas mixture}$$

$$+ \text{diffusion mass flux relative to the velocity of the gas} \tag{4.81}$$

This might seem like a small correction, however, it causes considerable complications because the velocity of the gas mixture depends on the mass transfer rate at the droplet surface, which is an unknown that we are trying to solve for. To rigorously describe the effect requires some analysis. However, Stefan flow is important for rapidly evaporating droplets (as occur in propellant metered dose inhalers).

Since the effect is not important for water droplets at normal room or body temperatures, let us generalize our notation so that we are instead considering droplets made of a pure substance A (which might be HFA 134a for example). We will use the label B to refer to the air that surrounds the droplets. The gas outside the droplet is then a mixture of A and B. The problem at hand is then a binary diffusion problem of substance A diffusing through substance B, which is a standard problem in transport theory of multicomponent systems, and is described in various textbooks (see, for example, Chapter 18 of Bird *et al.* 1960).

To write down the equations that we must solve, let us define some notation, as follows.

$n_A$ = mass flux of species A
$n_B$ = mass flux of species B

$v$ = velocity of gas mixture (consisting of A and B; strictly speaking $v$ is the 'mass average' velocity of the mixture)

$\rho_A$ = mass concentration of species A (e.g. in kg m$^{-3}$)

$\rho_B$ = mass concentration of species B

$\rho = \rho_A + \rho_B$ = mass concentration of mixture (the mass density of the mixture, e.g. in kg m$^{-3}$)

$Y_A = \rho_A/\rho$ = mass fraction of species A

$Y_B = \rho_B/\rho$ = mass fraction of species B

Assuming that assumption (3) of Section 4.4 is still in place (which as we saw earlier requires that the droplet Reynolds number $\ll 1$), then these variables are functions only of radial location $r$. In addition, vector quantities (like the velocity or the mass flux) will have only a radial component, so that when we refer to such quantities we are referring only to their radial component, i.e. $v$ refers to the radial velocity, $n_A$ refers to the radial mass flux of species A, etc.

With the above notation, Eq. (4.81) gives us Fick's law for the diffusion of species A:

$$n_A = \rho_A v - \rho D \nabla Y_A \qquad (4.82)$$

In words, this equation implies that at any point outside the droplet, the mass flux of species A is equal to the mass flux of A due to the motion of the gas mixture at velocity $v$, plus the diffusion of A relative to the gas mixture.

Similarly, we have the mass flux of species B given by

$$n_B = \rho_B v - \rho D \nabla Y_B \qquad (4.83)$$

If we assume both the quasi-steady assumption (5) and the one-dimensional assumption (4) of Section 4.4 are still valid, which is reasonable as we saw earlier, then the amount of mass and energy in any spherical shell outside the droplet must be constant. As a result, the flux of mass and energy through any spherical surface must be constant (i.e. independent of $r$ and $t$). Considering first the mass flux of species A or B through a surface of radius $r$ and surface area $4\pi r^2$ then we must have

$$4\pi r^2 n_A = \text{constant} = I \qquad (4.84)$$

where $I$ is $-1$ times the rate of change of droplet mass. Similarly for species B,

$$4\pi r^2 n_B = \text{constant} \qquad (4.85)$$

Assuming air cannot dissolve into the droplet, the mass flux of air is zero, i.e. $n_B = 0$. Thus, Eq. (4.85) can instead be written

$$4\pi r^2 n_B = 0 \qquad (4.86)$$

Using Eq. (4.82) to eliminate $v$ from Eq. (4.83), using the definitions of $Y_A$ and $Y_B$ plus the fact that $Y_A + Y_B = 1$, it can be shown that Eq. (4.86) can be rewritten as

$$(1 - Y_A)n_A 4\pi r^2 + 4\pi r^2 \rho D \frac{dY_A}{dr} = 0 \qquad (4.87)$$

This is the equation governing the vapor mass fraction $Y_A$. We cannot solve this equation directly since the gas mixture density $\rho$ is an unknown function of $r$, in addition to the fact that $n_A$ is also unknown. Thus, additional considerations are needed.

In particular, we must consider energy conservation. For this purpose, we note that the flux of energy through a spherical surface of radius $r$ consists of conduction of heat plus convection due to mass flux of internal energy. But, since $n_B = 0$, we need consider convection of energy of species A only. We thus have the energy equation:

$$4\pi r^2(-k\nabla T + n_A c_{pA} T) = \text{constant} \tag{4.88}$$

where $-k\nabla T$ appears due to conduction and $n_A c_{pA} T$ appears due to convection (associated with Stefan flow). Differentiating with respect to $r$ and replacing the gradient operator using the one-dimensional assumption (3) of Section 4.4, we then have

$$\frac{d}{dr}\left[4\pi r^2\left(-k\frac{dT}{dr} + n_A c_{pA} T\right)\right] = 0 \tag{4.89}$$

where we have assumed $v^2 \ll c_{pA} T$, and have neglected viscous heating and Dufour effects (Bird et al. 1960). Here, $k$ is the thermal conductivity of the mixture and can be obtained from the thermal conductivities of A and B using an equation such as that given by Bird et al. (1960):

$$k = \frac{X_A k_A}{X_A + X_B \Phi_{AB}} + \frac{X_B k_B}{X_A \Phi_{BA} + X_B} \tag{4.90}$$

where $X$ indicates mole fraction, i.e.

$$X_A = \frac{\dfrac{\rho_A}{M_A}}{\dfrac{\rho_A}{M_A} + \dfrac{\rho_B}{M_B}} = \frac{\dfrac{Y_A}{M_A}}{\dfrac{Y_A}{M_A} + \dfrac{(1-Y_A)}{M_B}} \tag{4.91}$$

$X_B$ is obtained by interchanging A and B in Eq. (4.91). Also,

$$\Phi_{AB} = \frac{1}{\sqrt{8}}\left(1 + \frac{M_A}{M_B}\right)^{-1/2}\left[1 + \left(\frac{\mu_A}{\mu_B}\right)^{1/2}\left(\frac{M_B}{M_A}\right)^{1/4}\right]^2 \tag{4.92}$$

where $\Phi_{BA}$ is obtained by interchanging A and B in Eq. (4.92). Here $\mu_A$ and $\mu_B$ are the viscosities of pure A and B, respectively.

By considering Newton's second law, the velocity must also satisfy the Navier–Stokes equation

$$\rho\frac{Dv}{Dt} = -\nabla p - \nabla \times (\mu\nabla \times v) + \nabla(2\mu\nabla \cdot v) \tag{4.93}$$

where $D/Dt$ is the so-called total, substantial or material derivative. One can show that for droplets with diameters typical of inhaled pharmaceutical aerosol droplets (i.e. not much smaller than 1 μm) this equation reduces, with our one-dimensional assumption (assumption (3) of Section 4.4) to

$$\frac{dp}{dr} = 0 \tag{4.94}$$

since standard nondimensional analysis of this equation shows that the coefficients in front of all the other terms are small compared to the coefficients in front of the pressure gradient term. Thus, this equation simply reduces to the fact that the pressure of the gas mixture is a constant, i.e.

$$p = p_\infty = \text{constant} \tag{4.95}$$

Recall that our principal goal here is to obtain the rate of change of mass of the droplet

$$\frac{dm}{dt} = -I = 4\pi R^2 n_A \qquad (4.96)$$

where $R$ is the radius of the droplet. However, in order to find $I$ we must solve Eqs (4.87) and (4.89). Examining these equations we see that they involve the two unknowns $Y_A$, and $T$. Note that $\rho$ is not considered an unknown since it can be obtained by combining an assumption of ideal gas behavior for species A and B with Dalton's law for the gas mixture, so that we have

$$p = p_A + p_B = \frac{\rho_A R_u T}{M_A} + \frac{\rho_B R_u T}{M_B} \qquad (4.97)$$

where, from Eq. (4.95), $p = p_\infty = $ constant and is considered known. Equation (4.97) can be rewritten by dividing through by $\rho$, using the definitions of the mass fractions, and rearranging to give

$$\rho = \frac{p}{\dfrac{Y_A R_u T}{M_A} + \dfrac{(1 - Y_A) R_u T}{M_B}} \qquad (4.98)$$

where $R_u$ is the universal gas constant. Thus, $\rho$ is a function of $Y_A$ and $T$ and need not be considered an unknown.

Thus, Eqs (4.87) and (4.89) give two equations for the two unknowns $Y_A$ and $T$. If we look at these equations, we see that Eq. (4.89) is a second-order ODE, which requires two boundary conditions, while Eq. (4.87) is a first-order ODE, which requires one boundary condition. Thus, we need three boundary conditions for these equations. Far from the droplet surface, we must specify $T$ and $Y_A$ (and of course $p_\infty$ so that Eq. (4.94) is not considered as one of our equations), which gives us two boundary conditions. Also, at the droplet surface, we must have $p_A$ and $T$ related by the usual vapor–pressure relation, which gives us a third final boundary condition.

In addition to these three boundary conditions, the mass flow rate $4\pi r^2 n_A$ is an unknown constant that requires specification – an equation governing its value is obtained by use of the boundary condition obtained by evaluating Eq. (4.88) at the droplet surface so that the conductive and convective heat flux at the droplet surface are equal to the latent heat times the mass flux there, i.e.

$$\text{at the droplet surface } n_A L = -k \frac{dT}{dr} + n_A c_{pA} T \qquad (4.99)$$

where $L$ is the latent heat of evaporation. This condition, in addition to the above three straightforward boundary conditions, completes the problem specification.

## 4.14 Exact solution for Stefan flow

Solving the above equations would allow us to determine the value of $I = 4\pi r^2 n_A = -dm/dt$ and thereby know the rate at which a droplet will change size. However, these equations as written cannot be solved exactly, forcing us to discretize them and solve the above boundary value problem numerically (for example with a finite difference method). However, with two additional assumptions we can use a well-known solution in the literature on droplets (Godsave 1953; see also Sirignano 1993) that gives us an exact, analytical solution to Eqs (4.87) and (4.89). In particular, we can transform Eqs (4.87) and (4.89) into equations that are easily solved, by replacing $r$ with the new independent variable $\beta$, where

$$\beta(r) = \int_r^\infty \frac{n_A c_{pA}}{k} \, dr' \tag{4.100}$$

Then we can replace $d/dr$ in Eqs (4.87) and (4.89) using the result that

$$\frac{d}{dr} = -\frac{n_A c_{pA}}{k} \frac{d}{d\beta} \tag{4.101}$$

to convert Eq. (4.87) into

$$\frac{1}{Le} \frac{dY_A}{d\beta} = -(Y_A - 1) \tag{4.102}$$

where

$$Le = k/(\rho c_{pA} D)$$

is the Lewis number. With this transformation, Eq. (4.89) is also converted to give

$$\frac{d^2 T}{d\beta^2} + T = 0 \tag{4.103}$$

If we assume the Lewis number $Le$ is a constant, then Eqs (4.102) and (4.103) can be solved exactly using standard ODE methods with the boundary conditions discussed earlier. The result is

$$T(\beta) = T_\infty - e^B (1 - e^{-\beta}) L/c_{pA} \tag{4.104}$$

$$Y(\beta) = 1 - (1 - Y_\infty) e^{(-\beta Le)} \tag{4.105}$$

where the subscript $\infty$ indicates values at $r = \infty$. The parameter $B$ in these equations is the value of $\beta$ at the droplet surface, i.e. at $r = R$. By setting $r = R$ in Eqs (4.104) and (4.105) we obtain two different expressions for $B$. From Eq. (4.104), we have

$$B = \ln\left[ c_{pA} \frac{T_\infty - T_s}{L} + 1 \right] \tag{4.106}$$

while from Eq. (4.105) we obtain

$$B = \frac{-\ln[(1 - Y_{As})/(1 - Y_{A\infty})]}{Le} \tag{4.107}$$

where the subscript s indicates a value at the droplet surface. Equating these two expressions for the value of $B$, then we must have

$$\ln\left[ c_{pA} \frac{(T_\infty - T_s)}{L} + 1 \right] = -\ln[(1 - Y_{As})/(1 - Y_{A\infty})]/Le \tag{4.108}$$

where $Y_{As}$ is a known function of droplet surface temperature obtained from the saturated vapor pressure $p_{As}(T)$ by combining ideal gas behavior (so that $X_{As} = p_{As}/p$)

with the definitions of mole and mass fractions (so that $Y_{As} = X_{As}M_A/(X_{As}M_A + X_{Bs}M_B)$ and $X_{Bs} = 1 - X_{As}$) to give

$$Y_{As}(T_s) = \frac{p_{As}M_A}{p_{As}M_A + (p_\infty - p_{As})M_B} \tag{4.109}$$

Equation (4.109) can be written instead using the saturated vapor concentration at the droplet surface, $c_s = \rho_{As}$, that we saw earlier in the chapter by combining a Clausius–Clapeyron vapor–pressure relation for $\rho_{As}(T)$, with Dalton's law of partial pressures ($p = p_\infty = p_A + p_B$) and assuming ideal gas behavior for A and B, to give

$$Y_{As}(T_s) = c_s/[c_s + (p/R_u T_s - c_s/M_A)M_B] \tag{4.110}$$

Equations (4.108) and (4.109) (or Eq. (4.110)) are an algebraic set of equations that we can solve by iteration (using for example Newton–Raphson iteration) to obtain the droplet temperature $T_s$. Putting this value of $T_s$ into Eq. (4.106), we then know $B$. To obtain the mass flux at the droplet surface we take this value of $B$ and use the definition that $B = \beta|_{r=R}$ in Eq. (4.100), where $R$ is the droplet radius, to obtain

$$B = \int_R^\infty \frac{n_A}{k/c_{pA}}\,dr'$$

Multiplying the numerator and denominator in this integrand by $4\pi r^2$ and using our previous result (Eq. (4.84)) that

$$4\pi r^2 n_A = \text{constant} = I$$

where $I = -dm/dt$, we have

$$B = I \int_R^\infty \frac{1}{4\pi r^2(k/c_{pA})}\,dr' \tag{4.111}$$

Using our previous assumption of a constant Lewis number, i.e. $Le = k/(\rho c_{pA}D) = $ constant, we can rewrite this equation as

$$\text{constant } Le: \ B = \frac{I}{Le}\int_R^\infty \frac{1}{4\pi\rho Dr^2}\,dr' \tag{4.112}$$

Making one final additional assumption that $\rho D = $ constant, we can evaluate the integral to finally obtain

$$I = \frac{4\pi R k B}{c_{pA}} \tag{4.113}$$

or in terms of droplet diameter $d = 2R$

$$I = \frac{2\pi d k B}{c_{pA}} \tag{4.114}$$

Using $I = -dm/dt$, we can thus write

$$\frac{dm}{dt} = \frac{-2\pi d k B}{c_{pA}} \tag{4.115}$$

which can also be written using $m = \rho_{drop}\pi d^3/6$ as

$$\frac{dd}{dt} = \frac{-4\pi k\,B}{(\rho_{drop}c_{pA}d)} \tag{4.116}$$

Equation (4.116) is the result we have been after. It gives us the droplet rate of change of size after we have obtained $T_s$ by solving Eqs (4.108) and (4.109) iteratively, just as we had to solve Eq. (4.43) before we could obtain the droplet's rate of change of size from Eq. (4.28) or Eq. (4.29).

One question to ask at this point is how accurate the constant $Le$ and constant $\rho D$ assumptions are. This issue can be addressed by solving the Stefan flow problem for HFA 134a and CFC 12 droplets without the constant $Le$ and $\rho D$ assumptions. One finds that Eq. (4.115) gives values of $dm/dt$ that are within a few percent of the full (nonconstant $Le$, nonconstant $\rho D$) Stefan flow solution over a wide range of ambient conditions if the geometric mean values of the transport properties are used in these equations (i.e. $Le$, $c_{pA}$, etc. are evaluated as, for example, $Le = \sqrt{Le_s Le_\infty}$, etc.).

Note that Eqs (4.115) and (4.116) can be reduced to Eqs (4.28) and (4.29) in the limit of small vapor mass fractions, since if we perform a Taylor series expansion on $B$ in Eq. (4.107) and keep only the first term in this expansion, Eq. (4.115) reduces to Eq. (4.28) and Eq. (4.116) reduces to Eq. (4.29). This brings us to the following question.

## 4.15 When can Stefan flow be neglected?

It should be noted that even without solving the above equations, we can examine the importance of Stefan flow by obtaining a result for the mass transfer rate at the droplet surface, $n_A$, as follows.

From Eq. (4.83), we have Fick's law for species B given by

$$n_B = \rho_B v - \rho D\nabla Y_B \tag{4.83}$$

However, we have already said that for air as species B, $n_B = 0$ otherwise we would have air diffusing into the liquid droplet. Thus, with $n_B = 0$, we must have

$$\rho_B v = \rho D\nabla Y_B \tag{4.117}$$

But we know that the mass fractions must add to one, i.e.

$$Y_B + Y_A = 1 \tag{4.118}$$

Taking the gradient of this equation gives

$$\nabla Y_B = -\nabla Y_A \tag{4.119}$$

Combining (4.117) and (4.119) and the fact that $\rho_B = \rho - \rho_A$, then we can write the velocity $v$ in terms of the other variables as

$$v = -\frac{1}{1 - \dfrac{\rho_A}{\rho}}D\nabla Y_A \tag{4.120}$$

Substituting Eq. (4.120) into Eq. (4.82), which is Fick's law for the vapor component of the droplet species (species A):

$$n_A = \rho_A v - \rho D \nabla Y_A \tag{4.82}$$

we obtain

$$n_A = -\left(\frac{1}{1 - \dfrac{\rho_A}{\rho}}\right) \rho D \nabla Y_A \tag{4.121}$$

If we assume $\rho_A/\rho \ll 1$, the term in brackets is approximately equal to one. In this case, there is not much vapor in the ambient air and then it also makes sense that the density of the air + vapor mixture around the droplet is essentially constant, i.e. $\rho = $ constant and we can take $\rho$ inside the $\nabla$ operator to obtain the earlier result, Eq. (4.21), when we neglected Stefan flow:

$$\text{Stefan flow neglected: } n_A = -D\nabla\rho_A = -\rho D\nabla Y_A \tag{4.122}$$

This equation is valid for $\rho_A/\rho \ll 1$, so that the mass transfer rate at the droplet surface obtained earlier as Eq. (4.21) neglecting Stefan flow will be accurate when $\rho_A/\rho \ll 1$. This is the condition stated earlier under assumption (1) of Section 4.4.

By using Fick's law written instead in terms of molar concentrations $x_A$ and $x_B$ (moles liter$^{-1}$), and mole fractions $X_A$ and $X_B$, one can derive a similar result to show that Stefan flow has a negligible effect on the mass transfer rate at the droplet surface if $x_A/x \ll 1$ at the droplet surface. Using the ideal gas law, we know that

$$p = x R_u T \tag{4.123}$$

$$p_A = x_A R_u T \tag{4.124}$$

so that the condition $x_A/x \ll 1$ reduces to

$$p_A/p \ll 1 \tag{4.125}$$

i.e. the vapor pressure at the droplet surface must be much less than the total pressure there, which is the alternative condition given when discussing assumption (1) of Section 4.4 in the simplified theory of hygroscopic size changes.

Note that since $\rho_A/\rho = 1 - Y_A$ must always be $\leq 1$, the coefficient $1/(1 - (\rho_A/\rho))$ in Eq. (4.121) will always be $\geq 1$. As a result, the mass flow rate at the droplets's surface with Stefan flow included will always be larger than that given by Eq. (4.122) when Stefan flow is neglected.

## Example 4.10

(a) Calculate the droplet 'wet bulb' temperature of an HFA 134a droplet evaporating in room air at 20°C (i.e. zero ambient propellant mass fraction) including Stefan flow and assuming ideal gas behavior. Use a molecular weight of 102 g mol$^{-1}$ for HFA 134a and 28.97 g mol$^{-1}$ for air. In order to maintain reasonable accuracy, the constant value of the Lewis number in Eq. (4.108) should be evaluated using

$$Le = (Le_s Le_\infty)^{1/2} \tag{4.126}$$

where $Le_s$ is the Lewis number evaluated at the droplet surface conditions, and $Le_\infty$ is the Lewis number evaluated at ambient conditions (far from the droplet surface). In addition, the use of at least a linear variation of transport properties with temperature is necessary for reasonable accuracy, so use the following linear functions for the variation of transport properties of HFA 134a with temperatures between $-70°C$ and $20°C$:

vapor thermal conductivity
$$k_{134} = 1000(-13.44168 + 0.0921486T) \text{ W m}^{-1} \text{ K}^{-1} \qquad (4.127)$$

diffusion coefficient for HFA 134a in air:
$$D = -5.725646 \times 10^{-6} + 5.265307 \times 10^{-8}T \quad \text{m}^2 \text{ s}^{-1} \qquad (4.128)$$

vapor specific heat $c_{p134} = 1000(-0.06682556 + 0.003577778T)$  J kg$^{-1}$ K    (4.129)

gas constant $R_{134} = 81.56$  J kg$^{-1}$ K$^{-1}$    (4.130)

vapor dynamic viscosity
$$\mu_{134} = -9.4602778 \times 10^{-7} + 4.389 \times 10^{-8}T \quad \text{kg m}^{-1}\text{s}^{-1} \qquad (4.131)$$

latent heat of vaporization $L = 1000 \times (388.3988 - 0.7025714T)$  J kg$^{-1}$    (4.132)

For the saturated vapor pressure use the following empirical relation, whose form comes from the Clasius–Clapyeron equation:

$$L = \frac{RT^2 \, dp_s}{p_s \ dT}$$

i.e. $p_s(T) = 6.021795 \times 10^6 \exp[-2714.4749/T]$    (4.133)

For air between $-70°C$ and $20°C$, use the following values for the transport properties:

thermal conductivity $k_{air} = 0.0017 + 0.000082T$  W m$^{-1}$ K$^{-1}$    (4.134)

dynamic viscosity $\mu_{air} = 2.83 \times 10^{-6} + 5.21 \times 10^{-8}T$  kg m$^{-1}$ s$^{-1}$    (4.135)

and for air, the gas constant is $R_{air} = 287$ J kg$^{-1}$ K$^{-1}$.

(b) What wet bulb temperature is predicted if Stefan flow is neglected (i.e. using Eq. (4.43)) and the above equations for transport properties are used?

### Solution

(a) To find the droplet temperature we must solve Eqs (4.108) and (4.109) as a coupled pair of equations, where the subscript A in these equations represents HFA134a and B represents air. Equation (4.109) gives the mass fraction $Y_{As}$ as a function only of the droplet temperature

$$Y_{As}(T_s) = \frac{p_{As}(T_s)M_A}{p_{As}(T_s)M_A + [p_\infty - p_{As}(T_s)]M_B} \qquad (4.109)$$

where we are given the molecular weights $M_A = 0.102$ kg mol$^{-1}$ and $M_B = 0.02897$ kg mol$^{-1}$, while Eq. (4.133) gives us the function $p_{As}(T_s)$, so that the right-hand side of Eq. (4.109) is a known function of droplet temperature.
Equation (4.108) is given by

$$\ln\left[c_{pA}\frac{(T_\infty - T_s)}{L} + 1\right] = -\ln[(1 - Y_{As})/(1 - Y_{A\infty})]/Le \qquad (4.108)$$

where we are given $T_r = 293$ K and there is no propellant in the ambient air, so that $Y_{A_r} = 0$. The specific heat of the propellant, $c_{pA}$, and its latent heat of vaporization, $L$, are known functions of temperature, given by Eqs (4.129) and (4.132).

The most tedious part of solving Eq. (4.108) is evaluating the Lewis number

$$Le = k/(\rho c_{pA} D) \tag{4.136}$$

since for accuracy we need to use the geometrically averaged Lewis number given by Eq. (4.126)

$$Le = (Le_s Le_\infty)^{1/2} \tag{4.137}$$

A difficulty arises here because $Le_s$ depends on the surface properties, which we don't know yet since they depend on the droplet temperature that we are to find. Before addressing this difficulty, note that we can evaluate $Le_r$ easily from the given ambient conditions where only room air is present, so that

$$Le_\infty = \frac{k_{air}}{\rho_{air} c_{p134} D} \bigg|_{293\,K} \tag{4.138}$$

where Eq. (4.134) gives

$$k_{air}(293\ K) = 0.026\ W\ m^{-1}\ K^{-1}$$

while the ideal gas law gives

$$\rho_{air} = p/R_{air} T = 1.2\ kg\ m^{-3}$$

for $p = 101\ 320$ Pa and $T = 293$ K. From Eq. (4.128), the diffusion coefficient is

$$D(293\ K) = 9.7 \times 10^{-6}\ m^2\ s^{-1}$$

Putting in these values, Eq. (4.138) gives us $Le_\infty = 2.27$.

We must now determine an expression for $Le_s$ in Eq. (4.137), which is more difficult, since $Le_s$ is a function of the droplet surface temperature $T_s$. However, this can be done as follows. First, we must realize that the density $\rho$ appearing in the Lewis number $Le_s$ is the density of the air–134a gas mixture at the droplet surface. The density of an air–134a mixture is given in general by

$$\rho = \rho_{air} + \rho_{134} \tag{4.139}$$

At the droplet surface, the densities on the right-hand side of Eq. (4.139) can be calculated from the pressures at the surface. For the propellant, we know the pressure of the propellant at the droplet surface is simply the saturated vapor pressure $p_s(T_s)$ given by Eq. (4.133), so that the ideal gas law gives us

$$\rho_{134s}(T_s) = \frac{p_s}{R_{134} T_s} \tag{4.140}$$

From Dalton's law of partial pressures for ideal gases, the pressure of the air at the droplet surface must make up the remainder of the pressure, i.e.

$$p_{air_s} = p_\infty - p_s(T_s) \tag{4.141}$$

where, as discussed earlier, the total pressure is constant throughout the entire gas and must have the value $p_{\infty} = 101\,320$ Pa for the given ambient conditions. From the ideal gas law, the density of the air at the droplet surface must be $p_{air}/RT$, i.e.

$$\rho_{air_s}(T_s) = \frac{p_\infty - p_s}{R_{air}T_s} \tag{4.142}$$

Combining Eqs (4.139), (4.140) and (4.142) allows us to write the density at the droplet surface, $\rho_s$, as a known function depending only on the droplet temperature $T_s$.

From Eq. (4.136), we see that to evaluate the Lewis number at the droplet surface also requires knowing the value of the thermal conductivity of the air–134a gas mixture at the droplet surface. For this purpose, we use Eqs (4.90)–(4.92) evaluated at the droplet surface, from which we obtain the thermal conductivity at the droplet surface, $k_s$, as a function only of droplet temperature (making use of the Eqs (4.131) and (4.135) for the viscosities $\mu$).

Putting all of the above together, we substitute Eq. (4.109) into Eq. (4.108) to obtain an equation of the form $f(T_s) = 0$, where $f$ is now a known nonlinear function. Use of a numerical root finding method (or even simple trial and error) allows us to find the droplet temperature $T_s$ that satisfies this equation. The result is

$$T_s = 211 \text{ K, i.e. } -62°\text{C}$$

It is interesting to note that this value is within 1% of the value obtained using a more complex procedure that includes quadratic functions for the variation of transport properties with temperature, a Martin–Hou equation of state for HFA 134a rather than ideal gas behavior, and solution of the governing Eqs (4.87) and (4.89) numerically by finite differencing (rather than by using the analytical, constant Lewis number approximation of Eqs (4.108) and (4.109)).

(b)  If we instead neglect Stefan flow and use Eq. (4.43)

$$LD(c_s - c_\infty) + k_{air}(T_s - T_\infty) = 0 \tag{4.43}$$

where $c_\infty = 0$, while $L$, $D$ and $k_{air}$ are known functions of temperature $T_s$ given by Eqs (4.132), (4.128) and (4.134) and $c_s = \rho_{134}$ is given by Eq. (4.140), we find the only solution to this equation occurs at absolute zero, i.e. $T_s = 0$. This is a nonphysical result. Thus, we see that inclusion of Stefan flow is essential in calculating the evaporation of a HFA 134a droplet, which agrees with the *a posteriori* realization based on part (a) that the requirement $p_s/p \ll 1$, which is needed in order to neglect Stefan flow, is not satisfied.

# References

Adamson, A. W. (1990) *Physical Chemistry of Surfaces*, 5th edn, Wiley, New York.
ASHRAE (1997) *1997 American Society of Heating, Refrigerating and Air-Conditioning Engineers Handbook: Fundamentals*, ASHRAE, Atlanta, GA.
Bird, R. B., Stewart, W. E. and Lightfoot, E. N. (1960) *Transport Phenomena*, Wiley, New York.
Biswas, P., Jones, C. L. and Flagan, R. C. (1987) Distortion of size distributions by condensation and evaporation in aerosol instruments, *Aerosol Sci. Technol.* 7:231–246.
Budavari, S. (1996) *The Merck Index: An Encyclopedia of Chemicals, Drugs, and Biologicals*, 12th edn, Merck, Whitehouse, NJ.
Cinkotai, F. F. (1971) The behaviour of sodium chloride particles in moist air, *J. Aerosol Sci.* 2:325–329.
Derjaguin, B. V., Fedoseyev, V. A. and Rosenzweig, L. A. (1966) Investigation of the adsorption

of cetyl alcohol vapor and the effect of this phenomenon on the evaporation of water drops, *J. Colloid Interface Sci.* **22**:45–50.

Duda, J. L. and Vrentas, J. S. (1971) Heat or mass transfer-controlled dissolution of an isolated sphere, *Int. J. Heat Mass Transfer* **14**:395–408.

Dunbar, C. A., Watkins, A. P. and Miller, J. F. (1997) Theoretical investigation of the spray from a pressurized metered-dose inhaler, *Atomization and Sprays* **7**:417–436.

Fang, C. P., McMurry, P. H., Marple, V. A. and Rubow, K. L. (1991) Effect of flow-induced relative humidity changes on size cuts for sulfuric acid droplets in the microorifice uniform deposit impactor (MOUDI), *Aerosol Sci. Technol.* **14**:266–277.

Ferron, G. A. and Soderholm, S. C. (1990) Estimation of the times for evaporation of pure water droplets and for stabilization of salt solution particles, *J. Aerosol Sci.* **21**:415–429.

Ferron, G. A., Kreyling, W. G. and Hauder, B. (1988) Inhalation of salt aerosol particles – II. Growth and deposition in the human respiratory tract, *J. Aerosol Sci.* **19**:611–631.

Finlay, W. H. (1998) Estimating the type of hygroscopic behaviour exhibited by aqueous droplets, *J. Aerosol Med.* **11**:221–229.

Finlay, W. H. and Stapleton, K. W. (1995) The effect on regional lung deposition of coupled heat and mass transfer between hygroscopic droplets and their surrounding phase, *J. Aerosol Sci.* **26**:655–670.

Finlayson, B. A. and Olson, J. W. (1987) Heat transfer to spheres at low to intermediate Reynolds numbers, *Chem. Eng. Commun.* **58**:431–447.

Fuchs, N. A. (1959) *Evaporation and Droplet Growth in Gaseous Media*, Pergamon Press, London.

Fuchs, N. A. and Sutugin, A. G. (1970) *Highly Dispersed Aerosols*, Ann Arbor Science Publ., Ann Arbor, MI.

Godsave, G. A. E. (1953) Studies of the combustion of drops in a fuel spray: the burning of single drops of fuel, in *Proceedings of the Fourth Symposium (International) on Cumbustion*, Combustion Institute, Baltimore, MD, pp. 818–830.

Gonda, I., Kayes, J. B., Grrom, C. V. and Fildes, F. J. T. (1982) In *Particle Size Analysis*, eds N. G. Stanley-Wood and T. Allen, Wiley, pp. 31–43.

Hinds, W. C. (1982) *Aerosol Technology: Properties, Behaviour and Measurement of Airborne Particles*, Wiley, New York.

Hinds, W. C. (1993) Physical and chemical changes in the particulate phase, in K. Willeke and P. A. Baron (eds), *Aerosol Measurement, Principles, Techniques and Applications*, Van Nostrand Reinhold, New York.

Heidenreich, S. and Büttner, H. (1995) Investigations about the influence of the Kelvin effect on droplet growth rates, *J. Aerosol Sci.* **26**:335–339.

Incropera, F. P. and De Witt, D. P. (1990) *Introduction to Heat Transfer*, Wiley, New York.

Israelachvili, J. (1992) *Intermolecular and Surface Forces*, 2nd edn, Academic Press, New York.

Maxwell, J. C. (1890) *The Scientific Papers of Clerk Maxwell* (W. D. Niven, ed.), Cambridge University Press, London.

Miller, R. S., Harstad, K. and Bellan, J. (1998) Evaluation of equilibrium and non-equilibrium evaporation models for many-droplet gas-liquid flow simulations, *Int. J. Multiphase Flow* **24**:1025–1055.

Otani, Y. and Wang, C. S. (1984) Growth and deposition of saline droplets covered with a monolayer of surfactant, *Aerosol Sci. Technol.* **3**:155–166.

Raoult, F.-M. (1887) Loi générale des tensions de vapeur des dissolvants, *C. R Hebd. Seanes Aca. Sci.* **104**:1430–1433.

Reid, R. C., Prausnitz, J. M. and Poling, B. E. (1987) *The Properties of Gases and Liquids*, 4th edn, McGraw-Hill, New York.

Robinson, R. A. and Stokes, R. H. (1959) *Electrolyte Solutions*. Butterworth, London.

Saunders, L. (1966) *Principles of Chemistry for Biology and Pharmacy*, Oxford University Press, London.

Sirignano, W. A. (1993) Fluid dynamics of sprays – 1992 Freeman Scholar Lecture, *J. Fluids Eng.* **115**:345–378.

Stefan, J. (1881) *Wien. Ber.* **81**:943.

Taflin, D. C. and Davis, E. J. (1987) Mass transfer from an aerosol droplet at intermediate Peclet numbers, *Chem. Eng. Commun.* **55**:199–210.

Vesala, T., Kulmala, M., Rudolf, R. Vrtala, A. and Wagner, P. (1997) Models for condensational growth and evaporation of binary aerosol particles, *J. Aerosol Sci.* **28**:565.

Wagner, P. E. (1982) Aerosol growth by condensation, in *Aerosol Microphysics II*, ed. W. H. Marlow, pp. 129–178, Springer, Berlin.

Zhang, S. II. and Davis, E. J. (1987) Mass transfer from a single micro-droplet to a gas flowing at low Reynolds number, *Chem. Eng. Commun.* **50**:51–67.

# 5
# Introduction to the Respiratory Tract

In order to understand the deposition of aerosols in the respiratory tract, to which we turn in Chapter 7, it is useful to define some basic aspects of lung geometry and breathing.

## 5.1 Basic aspects of respiratory tract geometry

From an engineering point of view, the geometry of the respiratory tract is not well known. This lack of knowledge exists for several reasons. First, the geometry contains so much fine detail in the lung (there are millions of alveoli with diameters of the order 300 μm, each with a slightly different shape) that it is not possible to specify this detailed information to any great extent. Second, the geometry of the respiratory tract is time-dependent because of the very nature of breathing in which the fine structures (the alveoli) fill and empty with breathing (and the time-dependence to the shape of the alveoli is not known). Finally, the respiratory tract geometry varies considerably in its details from individual to individual. Thus, realistic three-dimensional characterization of the entire airspaces are not available for any human lung, and it is probably unrealistic to expect such information soon.

This lack of information is rather disheartening from the point of view of modeling the fate of aerosols in the lung. However, despite this lack of knowledge, there is enough known about the basic aspects of the respiratory tract to develop some simple models of respiratory tract geometry. Details of the known geometrical features of the respiratory tract are given in many texts, so that only a brief overview will be given here. The reader is referred to ICRP (1994), Morén *et al.* (1993), or a basic anatomy text (e.g. O'Rahilly 1983) for more detailed descriptions. From a topological point of view, the lungs essentially consist of a series of bifurcating pipes (each bifurcation leading to a 'generation' of the lung) with the pipes becoming progressively smaller and smaller. The pipes of course lead to the alveoli. The basic features of the respiratory tract are shown in Fig. 5.1.

Three basic regions of the respiratory tract can be defined: the extrathoracic region, the tracheo-bronchial region and the alveolar region. The extrathoracic region, shown in more detail in Fig. 5.2, is also referred to as the 'upper airways', or the nose, mouth and throat.

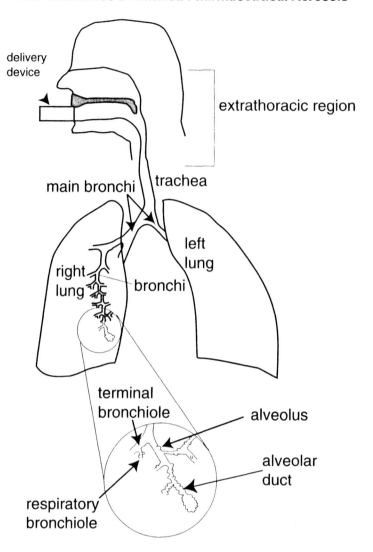

**Fig. 5.1** Diagram of the respiratory system. Adapted from ICRP (1994).

The extrathoracic region is the respiratory tract region proximal to the trachea (where proximal here means closer to the origin of the respiratory tract, i.e. closer to the face). The extrathoracic region includes the following components:

- the oral cavity (i.e. the mouth, sometimes also called the buccal cavity);
- the nasal cavity (i.e. the nose);
- the larynx, which is the constriction at the entrance to the trachea that contains the vocal cords; during swallowing a 'trap door' flap called the epiglottis swings down and covers the opening into the larynx to avoid aspiration of food or liquids into the lung;
- the pharynx, which is the throat region between the larynx and either the mouth or nose. The pharynx itself can be divided into parts that include the pathway from the larynx to the mouth (oropharynx) and the nose (nasopharynx). The term 'throat' usually means the pharynx and larynx.

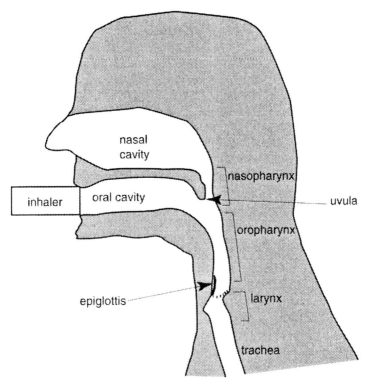

**Fig. 5.2** Schematic of the airways in the extrathoracic region. When swallowing, the epiglottis closes to cover the opening to the larynx, its closed position being shown schematically by the dashed line.

The extrathoracic region is a complicated geometry (see Stapleton *et al.* 2000 for a few characteristic dimensions), with considerable variation from individual to individual. Within an individual there can be considerable variation in the shape of the oral cavity due to changes in the position of the tongue and jaws. It should also be noted that the laryngeal opening into the trachea (called the 'glottis') changes shape with flow rate (opening with increasing flow rate – see Brancatisano *et al.* 1983), so that this is a time-dependent geometry. A typical cross-sectional area of the larynx is approximately $1 \text{ cm}^2$, and a typical dimension of this opening is 0.5–1.0 cm (Cheng *et al.* 1997, Stapleton *et al.* 2000).

Immediately distal to the extrathoracic region is the tracheo-bronchial region, sometimes also called the 'lower airways'. This region consists of the airways that conduct air from the larynx to the gas exchange regions of the lung, starting with the trachea, passing through the bronchi and stopping at the end of the so-called 'terminal bronchioles'. The bronchi (the singular of bronchi is bronchus) are the first three generations of branched airways after the trachea, and all have names. The two airways branching off the trachea are called the main bronchi. These branch into the lobar bronchi (of which there are two in the left lung and three in the right lung), which subsequently branch into the segmental bronchi. The parts of the lungs that are ventilated by the lobar bronchi are called lobes (so there are two lobes in the left lung and three in the right lung), and each lobe is broken up into the bronchopulmonary segments that are ventilated by each segmental bronchi.

Taken together, the extrathoracic and tracheo-bronchial airways are called the 'conducting airways' since they conduct air to the gas-exchange regions of the lung. The term 'central airways' is sometimes used to indicate the upper regions of the tracheo-bronchial airways.

The tracheo-bronchial airways are covered in a mucus layer that overlays fine hairs (called cilia) that are attached to the airway walls. The cilia continually wave in synch and act to clear the mucus layer to the throat, where it is swallowed or expectorated. The motion of the mucus caused by the cilia results in fairly rapid clearance of particles depositing on the mucus layer in the tracheo-bronchial region (i.e. often within 24 hours). However, some smaller particles (below a few microns in diameter) may burrow below the mucus layer due to a number of possible effects and may then not be cleared immediately by mucociliary clearance, but instead may remain in the conducting airways for 24 hours or longer. Thus, the strict division of the respiratory tract into regions where depositing particles are cleared rapidly (i.e. within 24 hours) and slowly (i.e. not cleared within 24 hours) is probably not appropriate for smaller particle sizes – see Chapter 7 for further discussion of this issue.

Cartilaginous rings are present on the trachea and main bronchi of the conducting airways, which cause these airways to have a corrugated inner surface which may be important in affecting the fluid dynamics and particle transport in these generations (Martonen *et al.* 1994).

Distal (meaning deeper into the lung) to the tracheo-bronchial airways is the alveolar region, sometimes also called the parenchyma or pulmonary region. Taken together, the tracheo-bronchial and alveolar regions are sometimes called the lung. The alveolar region includes all parts of the lung that contain alveoli, starting at the so-called 'respiratory bronchioles', the first generation of which are the daughter tubes branching off of the terminal bronchioles. The respiratory bronchioles are the most proximal tubes to have alveoli on them – see Fig. 5.1. The most proximal respiratory bronchioles have relatively few alveoli exiting off of them. Proceeding deeper into the lung, each generation of respiratory bronchioles has increasingly more alveoli on them, until reaching the 'alveolar ducts', which are entirely covered with alveoli. There are several generations of respiratory bronchioles as well as alveolar ducts, the latter of which end in alveolar sacs. The term acinus is sometimes used to mean all the daughter generations of a single terminal bronchiole.

Although there have been a number of studies that have measured various dimensions of the parts of the respiratory tract mentioned above, these structures are three-dimensional in nature and detailed dimensions of this three-dimensional structure are not generally known. However, many of the airways are approximately cylindrical in shape over much of their length, so that a diameter and length can be used to characterize them. A number of authors have used measurements of casts of normal lungs to suggest approximate models for the lengths, diameters and number of airways for each generation in the human lung (Weibel 1963, Horsfield and Cumming 1968, Hansen and Ampaya 1975, Phalen *et al.* 1978, Yeh and Schum 1980, Haefeli-Bleuer and Weibel 1988, Weibel 1991, among others). These models are of course quite drastic simplifications of the actual lung geometry, but the information they supply is instructive.

One of the most well known of these lung models is the symmetric model of Weibel (1963) (usually referred to as the Weibel A model), which, despite its flaws, has been used extensively in modeling airflow in the lung. The Weibel A model assumes that each generation of the lung branches symmetrically into two identical daughter branches; generations 0–16 are the tracheo-bronchial region, while generations 17–23 make up the alveolar region. Generations 17–19 are the respiratory bronchioles (with the number of alveoli on a bronchiole being 5, 8 and 12 for generations 17, 18 and 19, respectively), while generations 20–23 are alveolar ducts (with 20 alveoli per duct in this model).

The assumption of symmetric branching simplifies analysis dramatically, but is not entirely accurate since the diameters and lengths of daughter airways can be quite different from each other in actual human lungs. Several of the lung models mentioned above include some asymmetry in the branching. Note however, that the frequency distribution of bifurcation asymmetry is a monotonic function that decays rapidly with increasing asymmetry, with symmetric bifurcations being the most common (Phillips and Kaye 1997), so that the assumption of symmetrical branching can be quite reasonable for some purposes.

The diameters and lengths of the alveolated airways in the alveolar region of the Weibel A model are known to be too small and also to start at too high a generation number. A revised model which accounts for these facts is given by Haefeli-Bleuer and Weibel (1988). In addition, the original Weibel A model corresponds to a lung volume of 4.8 l, while an average adult male has a lung volume of approximately FRC = 3 l (FRC stands for functional residual capacity, which is the lung volume at the end of a tidal breath). It is usual to scale the lengths and diameters of the Weibel A model by a factor of $(FRC/4.8\ l)^{1/3}$ to account for this fact. More complicated scaling that accounts for the fact that the different conducting airway generations inflate by different fractions (e.g. the trachea changes little in size with lung inflation while the terminal bronchioles change more significantly with inflation) have been proposed (Lambert et al. 1982) but not widely adopted.

The Weibel A model is also known to underpredict the diameters of the tracheo-bronchial airways. Phillips et al. (1994) analyze the measurements from casts taken by Raabe et al. (1976) and suggest more realistic values for airway diameters. A symmetric lung geometry based on their airway data and the alveolar data of Haefeli-Bleuer and Weibel (1988) is given by Finlay et al. (2000) and shown in Table 5.1. Note that the conducting airways end at generation 14 in the Finlay et al. (2000) model, as opposed to generation 16 in the Weibel A model. For comparison purposes, a Weibel A model scaled to an FRC of 3 l is also shown in Table 5.1.

In Table 5.1, notice that most of the lung volume is contained in the alveolar region. For reference, the extrathoracic airways in an adult have a volume of approximately 50 ml, while the tracheo-bronchial region has a volume of approximately 100 ml. (A useful rule of thumb gives the conducting airway volume (in ml) as being approximately equal to body weight in pounds (West 1974)). The remainder of the lung volume (which is usually between 2000 and 4000 ml during tidal breathing, but is approximately 6000 ml in an adult male when fully inflated) is occupied by the 300 million or so alveoli.

Lung volumes are smaller for children, and various authors present idealized lung models for pediatric ages (e.g. Hofmann 1982, Phalen et al. 1985, Hofmann et al. 1989, ICRP 1994, Finlay et al. 2000).

Note that the above descriptions of lung geometry are all based on measurements made with normal lungs. Subjects with lung disease, such as asthma, cystic fibrosis, emphysema etc., may have parts of their lungs that differ quite drastically from the normal geometry. If one considers the normal lung geometry as being not particularly well characterized, then our knowledge of the effect of disease on the detailed geometry

**Table 5.1**   Dimensions of the Weibel A lung geometry (Weibel 1963) scaled to a 3 l lung volume, and using a volume of $10^{-5}$ ml per alveoli, is compared to the symmetric lung geometry used by Finlay *et al.* (2000). The thick lines in the table indicate the border between the alveolar and tracheo-bronchial regions in the models. The mouth–throat volume has not been included in the cumulative volume

| Generation | Finlay *et al.* model length (cm) | Scaled Weibel A length (cm) | Finlay *et al.* model diameter (cm) | Scaled Weibel A diameter (cm) | Finlay *et al.* model cumulative volume (cc) | Scaled Weibel A cumulative volume (cc) |
|---|---|---|---|---|---|---|
| 0 (trachea) | 12.456 | 10.26 | 1.81 | 1.539 | 32.05 | 19.07 |
| 1 | 3.614 | 4.07 | 1.414 | 1.043 | 43.401 | 25.64 |
| 2 | 2.862 | 1.624 | 1.115 | 0.71 | 54.572 | 28.64 |
| 3 | 2.281 | 0.65 | 0.885 | 0.479 | 65.786 | 29.5 |
| 4 | 1.78 | 1.086 | 0.706 | 0.385 | 76.918 | 31.7 |
| 5 | 1.126 | 0.915 | 0.565 | 0.299 | 85.948 | 33.76 |
| 6 | 0.897 | 0.769 | 0.454 | 0.239 | 95.237 | 35.95 |
| 7 | 0.828 | 0.65 | 0.364 | 0.197 | 106.236 | 38.39 |
| 8 | 0.745 | 0.547 | 0.286 | 0.159 | 118.458 | 41.14 |
| 9 | 0.653 | 0.462 | 0.218 | 0.132 | 130.922 | 44.39 |
| 10 | 0.555 | 0.393 | 0.162 | 0.111 | 142.711 | 48.26 |
| 11 | 0.454 | 0.333 | 0.121 | 0.093 | 153.381 | 53.01 |
| 12 | 0.357 | 0.282 | 0.092 | 0.081 | 163.119 | 59.14 |
| 13 | 0.277 | 0.231 | 0.073 | 0.07 | 172.644 | 66.26 |
| 14 | <u>0.219</u> | 0.197 | <u>0.061</u> | 0.063 | <u>183.13</u> | 77.14 |
| 15 | 0.134 | 0.171 | 0.049 | 0.056 | 204.967 | 90.7 |
| 16 | 0.109 | <u>0.141</u> | 0.048 | <u>0.051</u> | 239.898 | <u>190.26</u> |
| 17 | 0.091 | 0.121 | 0.039 | 0.046 | 284.101 | 139.32 |
| 18 | 0.081 | 0.1 | 0.037 | 0.043 | 357.893 | 190.61 |
| 19 | 0.068 | 0.085 | 0.035 | 0.04 | 474.046 | 288.17 |
| 20 | 0.068 | 0.071 | 0.033 | 0.038 | 689.872 | 512.95 |
| 21 | 0.068 | 0.06 | 0.03 | 0.037 | 1067.707 | 925.25 |
| 22 | 0.065 | 0.05 | 0.028 | 0.035 | 1742.742 | 1694.17 |
| 23 | 0.073 | 0.043 | 0.024 | 0.035 | 3000 | 3000 |

of the air passages in the lung would have to be considered as poor, and this is a topic for future work.

## 5.2 Breath volumes and flow rates

There is a large amount of literature available on normal breathing dynamics, and the reader is referred to any standard text on respiratory physiology (e.g. Chang and Paiva 1989) for more detail on the physiology of breathing. However, for our purposes we are interested only in the flow rates and volumes that can be expected during inhalation from a delivery device. It must be remembered that most of the work on this subject has been done on normal subjects without the presence of an aerosol delivery device. Neither the effect of an aerosol device at the face, nor the effect of disease on breathing patterns have been well characterized. A few basic definitions are as follows:

- tidal volume $V_t$: average volume inhaled and exhaled during periodic (i.e tidal) breathing (needed to satisfy metabolic requirements);

- breathing frequency $f$: the number of tidal breaths per minute – typically around 12 for adults;
- total lung capacity (*TLC*): the total volume of the airspaces in the lung when maximally inflated (by as large a breath as one can take) – typically around 6 l in adults;
- functional residual capacity (*FRC*): the volume of the airspaces during tidal breathing at the start of a tidal inhalation – typically around 3 l in adults;
- residual volume (*RV*): the volume of the airspaces when the lung is minimally inflated (by exhaling as much air as one can from the lung);
- vital capacity (*VC*): the biggest possible volume one can inhale (taken by exhaling to residual volume and then inhaling to *TLC*) – typically just over 4 l in adults;
- $FEV_1$ (forced expiratory volume in one second): the maximum volume that can be exhaled within one second starting from *TLC*.

Several of these lung volumes are shown schematically in Fig. 5.3.

Typical values of these volumes in adults are a function of age, height, weight and race (Quanjer *et al.* 1993, ICRP 1994) as well as disease. Because various lung volumes are measured in clinical lung function tests in order to diagnosis or assess lung disease, standard reference values of several lung volumes have been agreed upon (Quanjer *et al.* 1993). For healthy adult caucasians, some of these values are shown in Table 5.2.

Notice the large standard deviation in the lung volumes in Table 5.2. A single standard deviation is approximately one half of each lung volume, indicating there are large ($\pm 50\%$) variations in these values between individuals. Values for these volumes for various pediatric ages (5–18 years) are given in Stocks and Quanjer (1995).

For single breath devices such as metered dose inhalers and dry powder inhalers, inhaled volumes would be expected to normally be between vital capacity and inspiratory capacity (*IC*), where $IC = TLC - FRC$ and *FRC* and *TLC* are as given in

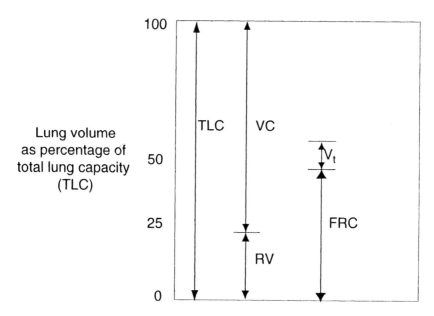

**Fig. 5.3** Lung volume definitions shown as approximate percentage of total lung capacity.

**Table 5.2**   Reference values and relative standard deviation for vital capacity, functional residual capacity and total lung capacity for healthy adult caucasians (from Quanjer *et al.* 1993)

| Volume | Reference value in liters (age $A$ in years, height $H$ in m) | Relative standard deviation |
|---|---|---|
| $VC$ | male:   $5.76H - 0.026A - 4.34$ | 0.61 |
|  | female: $4.43H - 0.026A - 2.89$ | 0.43 |
| $FRC$ | male:   $2.34H + 0.0009A - 1.09$ | 0.60 |
|  | female: $2.24H + 0.001A - 1$ | 0.50 |
| $TLC$ | male:   $7.99H - 7.08$ | 0.70 |
|  | female: $6.60H - 5.79$ | 0.50 |

Table 5.2. However, for these devices, the flow rate during inhalation of this volume is an important parameter since it can affect the uptake and deaggregation of powder from powder devices (see Chapter 9) and affects where the aerosol deposits in the respiratory tract (see Chapter 7). Because the flow rate through a particular inhaler depends on the resistance of the particular device (which is quite variable with dry powder inhalers, but typically averages 20–100 l min$^{-1}$ — see Clark and Hollingworth 1993), there is no 'reference' flow rate for inhalers. The use of a 'reference' pressure drop of 4 kPa across a device (United States Pharmacopeia) is a step towards giving a standardized breathing pattern for an inhaler. However, the use of a step function for the time-dependence of the pressure drop does not simulate the *in vivo* situation. It has been suggested that this may be a concern with some dry powder inhalers in which the character of the inhaled powder depends on the actual time dependent shape of the inhalation curve (Clark and Bailey 1996). However, Finlay and Gehmlich (2000) find that the use of square wave profiles is adequate for the inhalers they tested, as long as the flow rates are representative of those expected during powder entrainment when actual patients use the inhaler. However, independent of this issue, 'reference' curves for use with breath simulation of dry powder inhalers do not exist and presently must be measured for each device since they are different for different devices.

For nebulizers, tidal volume is the lung volume of most interest since tidal breathing is the normal breathing pattern used with these devices. Because tidal volume is not normally used clinically in lung function testing, standard reference values for $V_t$ do not exist. Like the other lung volumes, $V_t$ varies with age, height, weight, gender, race, but $V_t$ also varies with activity level, increasing with activity level. Unlike the other lung volumes mentioned above, tidal volumes are slightly different for natural unencumbered breathing when compared to breathing through a mouthpiece or facemask (Askanazi *et al.* 1980, Perez and Tobin 1985). This effect is not likely related to any added resistance of mouthpieces or facemasks, since such resistance is typically very small with such tidal breathing apparatuses (although certainly adding mouthpieces or facemasks with significant resistance would be expected to alter tidal breathing patterns). Several reasons for this effect have been proposed, including psychological load, facial sensory stimuli, and change in respiratory route (from nasal to oral). Whatever the reason, tidal breathing patterns through nebulizers (which have either mouthpieces or facemasks) is not identical with natural tidal breathing at rest. Although there are relatively few studies available in which tidal breathing has been measured with nebulizer mouthpieces or facemasks in place (Howite *et al.* 1987, Phipps *et al.* 1989, 1994, Chan *et al.* 1994, Diot

**Table 5.3** Approximate values of breathing parameters for tidal breathing expected with attached tidal breathing aerosol delivery devices at various ages. Considerable intersubject variation about these values should be expected

| Age | Tidal volume (l) | Flow rate ($l\ min^{-1}$) | Frequency (breaths $min^{-1}$) |
|---|---|---|---|
| 6 months | | | |
|   sleeping | 0.075 | 4.8 | 32 |
|   low activity | 0.175 | 6.0 | 17 |
| 2 years | 0.19 | 8.2 | 21.6 |
| 4 years | 0.23 | 11.1 | 24 |
| 8 years | 0.325 | 13 | 20 |
| Adult male (with mouthpiece) | 0.750 | 18 | 12 |

*et al.* 1997), those that have done so indicate tidal volumes that are somewhat higher than occur without mouthpieces or facemasks, resulting in values that are similar to natural tidal breathing during low activity. Note that breathing patterns with facemasks may be different from breathing patterns with mouthpieces (Askanazi *et al.* 1980 found $V_t$ increased by 15.5% during the use of a mouthpiece with noseclips, but by 32.5% with a face mask).

In tidal breathing, exhalation times are usually longer than inhalation times, so that inhalation flow rates are slightly higher than exhalation flow rates. The term 'duty cycle' is used to refer to the ratio of the inhalation time divided by the breathing period. The Task Group on Lung Dynamics (1966) suggests a breathing cycle where inhalation occupies 43.5% of the breathing period, while exhalation occupies 51.5% of a cycle, with a pause of 5% before exhalation. Duty cycles near this value are seen in the above mentioned studies where breathing has been measured with nebulizer mouthpieces or facemasks in place, as well as in pediatrics (ATS/ERS 1993), although duty cycles in children with chronic airway disease differs from those seen in normal subjects, with inhalation times occupying values as low as 25% of the breathing cycle (ATS/ERS 1993).

Typical values of tidal volumes and flow rates for various age groups are shown in Table 5.3 based on Hofmann *et al.* (1989) and Taussig *et al.* (1977) assuming low activity levels for the pediatric ages, and using the studies listed above for nebulizers with mouthpieces for the adult values.

Regarding the shape of tidal breathing waveforms, again there is considerable intersubject variability. However, tidal waveforms are quiet reproducible for a given individual (Benchetrit *et al.* 1989). Typical flow rate waveforms vary considerably but unpublished examinations in our laboratory with 12 subjects suggest they can usually be broadly classed as approximately sinusoidal, triangular or rectangular.

# References

Askanazi, J., Silverberg, P. A., Foster, R. J., Hyman, A. I., Milic-Emili, J. and Kinney, J. M. (1980) Effects of respiratory apparatus on breathing pattern, *J. Appl. Physiol.* **48**:577 580.
ATS/ERS (American Thoracic Society/European Respiratory Society) (1993) Respiratory mechanics in infants: physiologic evaluation in health and disease, *Am. Rev. Respir. Dis.* **147**:474–496.

Benchetrit, G., Shea, S. A., Dinh, T. P., Bodocco, S., Baconnier, P. and Guz, A. (1989) Individuality of breathing patterns in adults assessed over time, *Resp. Physiol.* **75**:199–210.

Brancatisano, T., Collett, P. W. and Engel, L. A. (1983) Respiratory movements of the vocal cords, *J. Appl. Physiol.* **54**:1269–1276.

Chan, H.-K., Phipps, P. R., Gonda, I., Cook, P., Fulton, R., Young, I. and Bautovich, G. (1994) Regional deposition of nebulized hypodense nonisotonic solutions in the human respiratory tract, *Eur. Respir. J.* **7**:1483–1489.

Chang, H. H. and Paiva, M. (1989) *Respiratory Physiology: An Analytical Approach*, Marcel Dekker.

Cheng, K.-H., Cheng, Y.-S., Yeh, H.-C. and Swift, D. L. (1997) Measurements of airway dimensions and calculations of mass transfer characteristics of human oral passages, *J. Biomech. Eng., Trans. ASME* **119**:476–482.

Clark, A. R. and Bailey, R. (1996) Inspiratory flow profiles in disease and their effects on the delivery characteristics of dry powder inhalers, in *Resp. Drug Delivery V*, Interpharm Press, Buffalo Grove, IL.

Clark, A. R. and Hollingworth, A. M. (1993) The relationship between powder inhaler resistance and peak inspiratory conditions in healthy volunteers – implications for in vitro testing, *J. Aerosol Med.* **6**:99–110.

Diot, P., Palmer, L. B., Smaldone, A., DeCelie-Germana, J., Grimson, R. and Smaldone, G. C. (1997) RhDNase I Aerosol deposition and related factors in cystic fibrosis, *Am. J. Respir. Crit. Care Med.* **156**:1662–1668.

Finlay, W. H. and Gehmlich, M. G. (2000) Inertial sizing of aerosol inhaled from two dry powder inhalers with realistic breath patterns vs. constant flow rates, *Int. J. Pharm.* **210**:83–95.

Finlay, W. H., Lange, C. F., King, M. and Speert, D. (2000) Lung delivery of aerosolized dextran, *Am. J. Resp. Crit. Care Med.* **161**:91–97.

Haefeli-Bleuer, B. and Weibel, E. R. (1988) Morphometry of the human pulmonary acinus, *Anatom. Rec.* **220**:401–424.

Hansen, J. E. and Ampaya, E. P. (1975) Human air space shapes, sizes, areas and volumes, *J. Appl. Physiol.* **38**:990–995.

Hofmann, W. (1982) Mathematical model for the postnatal growth of the human lung, *Respir. Physiol.* **49**:115–367.

Hofmann, W., Martonen, T. B. and Graham, R. C. (1989) Predicted deposition of nonhygroscopic aerosols in the human lung as a function of subject age, *J. Aerosol Med.* **2**:49–68.

Horsfield, K. and Cumming, G. (1968) Morphology of the bronchial tree in man, *J. Appl. Physiol.* **24**:373–383.

ICRP (1994) Publication 66. *Annals of the ICRP*, **24**, Nos. 1–3, Pergamon/Elsevier, Tarrytown NY.

Ilowite, J. S., Gorvoy, J. D. and Smaldone, G. C. (1987) Quantitative deposition of aerosolized gentamicin in cystic fibrosis, *Am. Rev. Respir. Dis.* **136**:1445–1449.

Lambert, R. K., Wilson, T. A., Hyatt, R. E. and Rodarte, J. R. (1982) A computation methodology for expiratory flow, *J. Appl. Physiol.*, **52**:44–56.

Martonen, T. B., Yang, Y. and Xue, Z. Q. (1994) Influences of cartilaginous rings on tracheobronchial fluid dynamics, *Inhal. Tox.* **6**:185–203.

Morén, F., Dolovish, M. B., Newhouse, M. T. and Newman, S. P. (1993) *Aerosols in Medicine: Principles, Diagnosis and Therapy*, Elsevier, New York.

O'Rahilly, R. (1983) *Basic Human Anatomy*, W. B. Saunders, Philadelphia, PA.

Perez, W. and Tobin, M. J. (1985) Separation of factors responsible for change in breathing pattern induced by instrumentation, *J. Appl. Physiol.* **59**:1515–1520.

Phalen, R. F., Yeh, H. C., Schum, G. M. and Raabe, O. G. (1978) Application of an idealized model to morphometry of the mammalian tracheobronchial tree, *Anatom. Rec.* **190**:167–176.

Phalen, R. F., Oldham, M. J., Beaucage, C. B., Crocker, T. T. and Mortensen, J. D. (1985) Postnatal enlargement of the human tracheobronchial airways and implications for particle deposition, *Anatom. Rec.* **242**:368–380.

Phillips, C. G. and Kaye, S. R. (1997) On the asymmetry of bifurcations in the bronchial tree, *Resp. Physiol.* **107**:85–98.

Phillips, C. G., Kaye, S. R. and Schroter, R. C. (1994) A diameter-based reconstruction of the branching pattern of the human bronchial tree, *Resp. Physiol.* **98**:193–217.

Phipps, P. R., Gonda, I., Anderson, S. A., Bailey, D., Borham, P., Bautovich, G. and Anderson, S. D. (1989) Comparisons of planar and tomographc gamma scintigraphy to measure the penetration of inhaled aerosols, *Am. Rev. Respir. Dis.* **139**:1516–1523.

Phipps, P. R., Gonda, I., Anderson, S. A., Bailey, D. and Bautovich, G. (1994) Regional deposition of saline aerosols of different tonicities in normal and asthmatic subjects, *Eur. Respir. J.* **7**:1474–1482.

Quanjer, Ph. H., Tammeling, G. J., Cotes, J. E., Pederson, O. F., Peslin, R. and Yernault, J.-C. (1993) Lung volumes and forced ventilatory flows, *Eur. Respir. J.* **6**, Suppl. 16:5–40.

Raabe, O.-G., Yeh, H. C., Schum, G. M. and Phalen, F. F. (1976) *Tracheobronchial Geometry: Human, Dog, Rat, Hamster.* LF-53. Albuquerque, NM: Lovelace Foundation for Medical Education and Research.

Stapleton, K. W., Guentsch, E., Hoskinson, M. K. and Finlay, W. H. (2000) On the suitability of k-ε turbulence modelling for aerosol deposition in the mouth and throat: a comparison with experiment, *J. Aerosol Sci.* **31**:739–749.

Stocks, J. and Quanjer, Ph. H. (1995) Reference values for residual volume, functional residual capacity and total lung capacity, *Eur. Respir. J.* **8**:492–506.

Task Group on Lung Dynamics (1966) Deposition and retention models for internal dosimetry of the human respiratory tract, *Health Physics* **12**:173–207.

Taussig, L. M., Harris, T. R. and Lebowitz, M. D. (1977) Lung function in infants and young children, *Am. Rev. Respir. Dis.* **116**:233–239.

Weibel, E. R. (1963) *Morphometry of the Human Lung,* Academic Press, New York.

Weibel, E. R. (1991) Design of airways and blood vessels considered as branching trees, in *The Lung: Scientific Foundations,* eds R. G. Crystall, J. B. West *et al.,* Raven Press, New York.

West, J. B. (1974) *Respiration Physiology – the Essentials,* Williams & Wilkins, Baltimore.

Yeh, H. C. and Schum, G. M. (1980) Models of human lung airways and their application to inhaled particle deposition, *Bull. Math. Biol.* **42**:461–480.

# 6

# Fluid Dynamics in the Respiratory Tract

In order to answer many questions about the fate of inhaled aerosols in the respiratory tract, it is necessary to first understand the fluid motion that occurs in the respiratory tract. In a sense, fluid dynamics in the respiratory tract is the cornerstone upon which particle deposition is built. However, in order to solve a problem in fluid mechanics it is necessary to specify the detailed geometry that the fluid flow is occurring in. As we have seen in Chapter 5, the geometry of the respiratory tract is not known in detail, is quite complex, and varies greatly from individual to individual. As a result, we cannot yet specify the detailed fluid dynamics in the entire respiratory tract with any great accuracy, particularly for any one individual. Despite this though, we can make a number of informative statements about the general nature of the fluid dynamics in the respiratory tract, as follows.

## 6.1 Incompressibility

The flow of a pure fluid can normally be considered incompressible if the Mach number $< 0.3$, and temperature differences $\Delta T$ in the fluid are small relative to a reference temperature $T_0$ (Panton 1996). For typical inhalation conditions this requires velocities less than about $100 \text{ m s}^{-1}$ and temperature differences of less than about 30 K, both conditions normally being satisfied in pharmaceutical aerosol applications. It is conceivable that the temperature condition is violated for inhalation of metered dose inhaler sprays (which may cool the inhaled gas considerably), in which case Rayleigh line effects (White 1999) will accelerate the fluid in the upper and central airways. However, this effect is probably small since even air at $-25°C$ when heated to $37°C$ (body temperature) would undergo a velocity increase of only approximately 25% at an inhalation flow rate of $60 \text{ l min}^{-1}$ in a constant area duct, which is not likely to cause a large effect on deposition, but may be worth including if detailed numerical simulations of the flow and deposition of inhaled metered dose inhaler sprays are being done.

Because gas in the lung is not a pure substance, but instead contains varying amounts of oxygen and carbon dioxide, the assumption of constant density associated with incompressibility could be violated even if Mach number or temperature changes are small. However, nitrogen makes up more than three-quarters of the density of air and is not exchanged across the lung epithelium under ambient conditions, so that variations in the content of oxygen or carbon dioxide of gases in the lung would not be expected to significantly alter the density of air in the lung.

Thus, when inhaling pharmaceutical aerosols, the fluid dynamics of the bulk gas motion in the lung can probably be reasonably approximated as an incompressible flow of air under most circumstances.

## 6.2 Nondimensional analysis of the fluid equations

For incompressible flow of air, consideration of Newton's second law for the fluid results in the Navier–Stokes equations:

$$\frac{\partial \mathbf{v}}{\partial t} + \mathbf{v} \cdot \nabla \mathbf{v} = -\frac{1}{\rho}\nabla p + \frac{\mu}{\rho}\nabla^2 \mathbf{v} \tag{6.1}$$

where $\mu$ is the dynamic viscosity, $\rho$ is the density, $p$ is the pressure and $\mathbf{v}$ is the fluid velocity. If we nondimensionalize this equation (as we did in Chapter 3 for the equation of motion of a particle) using a characteristic velocity $U$, length $D$ and (if the flow is unsteady) a time scale $\tau$, we obtain the following nondimensional equation

$$\frac{1}{St}\frac{\partial \mathbf{v}'}{\partial t'} + \mathbf{v}' \cdot \nabla \mathbf{v}' = -\nabla p' + \frac{1}{Re}\nabla^2 \mathbf{v}' \tag{6.2}$$

where $v' = v/U$, $p' = p/(\rho U^2)$, $t' = t/\tau$, $x' = x/D$ are dimensionless versions of their dimensional counterparts, and $St = \tau U/D$ and $Re = \rho UD/\mu$ are dimensionless parameters that determine the importance of the unsteady term ($\frac{1}{St}\frac{\partial \mathbf{v}'}{\partial t'}$) and the viscous term ($\frac{1}{Re}\nabla^2 \mathbf{v}'$), relative to the convective term ($\mathbf{v}' \cdot \nabla \mathbf{v}'$). In particular, the Reynolds number tells us how important the viscous term is relative to the convective term, while the Strouhal number tells us how important the unsteady term is relative to the convective term. Thus, for very high $Re$ we may be able to neglect the viscous term, while for very low $Re$ we may be able to neglect the convective term. Similarly, for very high $St$, we may be able to neglect the unsteady term relative to the convective term. Thus, the values of these two nondimensional parameters are important quantities in determining what effects need to be included if we are to model the fluid dynamics in the airways.

By considering the simplified geometrical models of the respiratory tract discussed in Chapter 5, we can estimate the Reynolds number and Strouhal number in the various generations of these model respiratory tracts. For example, Fig. 6.1 shows $Re$ and $St$ for the idealized lung model geometry given in Chapter 5 for tidal breathing (frequency $f = 12$, tidal volume $V_t = 0.75\,\text{l}$, flow rate $Q = 300\,\text{cm}^3\,\text{s}^{-1}$) and a typical single inhalation pattern for an MDI or DPI (inhalation time 5 s, flow rate 60 l min$^{-1}$).[1]

Several results follow from this data. First, it can be seen that the Reynolds number is quite high in the larynx and is very low when deep in the lung. Internal flows become turbulent at high Reynolds numbers and are laminar at low Reynolds numbers. Thus, we must examine the possibility that turbulence is present in the upper and central airways, but we do not expect turbulence in the deep lung. In fact, experimental

[1] Note that the use of the inhalation time $\tau = 5$ s in obtaining Strouhal number values in Fig. 6.1 means that we are examining the importance of externally imposed unsteadiness at time scales associated with inhalation flow rate unsteadiness, and we are not examining unsteadiness at time scales intrinsic to the fluid motion (for the latter we would need to use $\tau$ associated with a characteristic flow phenomena time, such as a vortex shedding frequency of turbulent eddy time scale. However, except in the oropharynx, such intrinsic unsteady flow phenomena are not expected at the low Reynolds numbers seen and the smoothly branching pipe flow geometry of the lung.)

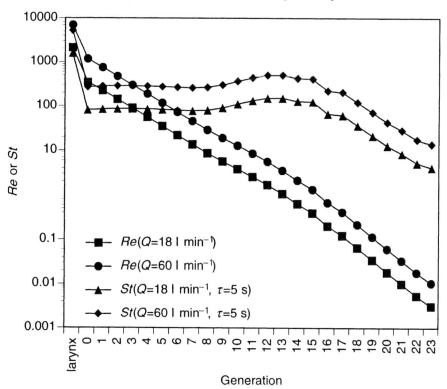

**Fig. 6.1** Reynolds number (*Re*) and Strouhal number (*St*) plotted against generation number of the idealized lung geometry from Chapter 5 for two flow rates (18 and 60 l min$^{-1}$).

observations do indicate the presence of turbulence in the upper airways and trachea (the laryngeal jet produces much of this turbulence), but the turbulence produced in the upper airways decays rapidly as it is convected into the lung (Simone and Ultmann 1982, Ultmann 1985). Even if turbulent production occurred distal to the larynx by shear in the boundary layers of the first few generations, such turbulence would not exist long enough to be convected significantly into the next generation (based on an analysis of turbulence time scales that we have done using concepts explained in Tennekes and Lumley 1972). Thus, it is reasonable to expect that turbulence is produced in the extrathoracic airways and may be convected into the first few generations of the lung. However, distal to these regions, the flow can probably be considered as laminar for the purposes of predicting typical pharmaceutical aerosol deposition.

By examining the Strouhal numbers in Fig. 6.1, we see that in the tracheo-bronchial region, the Strouhal number is quite high and unsteadiness in inhalation flow rate is not expected to play a large role in the fluid dynamics. In the distal parts of the lung the Reynolds number is small, so that the convective terms are small here. Thus, even though the Strouhal number is O(1), this does not mean that unsteadiness is important, since the Strouhal number compares the unsteady term to the convective term and if the convective terms are small, then a Stouhal number O(1) would indicate that the unsteady terms are small as well. However, if the Reynolds number is small, it makes more sense to compare the unsteady terms directly to the viscous terms, since the viscous terms are then the dominant terms in the Navier–Stokes equations. Thus, if we multiply Eq. (6.2) through by $Re$, we obtain

$$\frac{Re}{St}\frac{\partial \mathbf{v}'}{\partial t'} + Re\,\mathbf{v}' \cdot \nabla \mathbf{v}' = -Re\nabla p' + \nabla^2 \mathbf{v}' \qquad (6.3)$$

The ratio $Re/St$, where

$$\frac{Re}{St} = \frac{D^2}{\tau v}$$

is thus a measure of the importance of the unsteady term compared to the viscous term, where $v = \mu/\rho$ is the kinematic viscosity ($v = 1.5 \times 10^{-5}$ for air). For large values of $Re/St$ we expect unsteadiness to be important relative to viscous forces, while for small values we expect unsteadiness to be unimportant. Values of $Re/St$ for the idealized lung geometry given in Chapter 5 are shown in Fig. 6.2.

Figure 6.2 shows that for the alveolar region and much of the distal portions of the tracheo-bronchial region, the unsteady terms are expected to be small compared to the viscous terms.

Instead of considering the parameter $Re/St$, the parameter

$$\alpha = (Re/St)^{1/2} = D\,(f/v)^{1/2} \qquad (6.4)$$

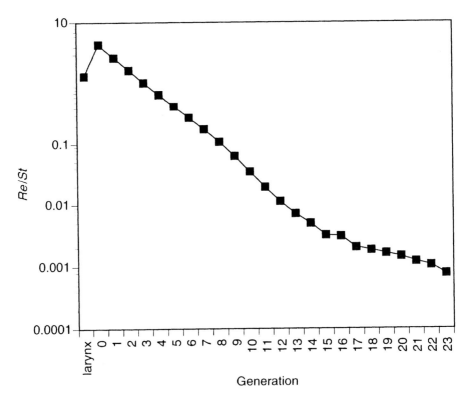

**Fig. 6.2** The ratio of Reynolds number to Strouhal number (which gives the relative importance of unsteady to viscous forces) plotted against lung generation for the idealized lung geometry given in Chapter 5.

is sometimes used; $\alpha$ is called the Womersley number and is also a measure of the importance of the unsteady term compared to the viscous term. The Womersley number appears directly in the solution for laminar sinusoidally oscillating flow in a circular pipe (Womersley 1955).

Combining the results from Figs 6.1 and 6.2, we see that the unsteady terms are small in comparison to either the convective terms (which are important in the proximal tracheo-bronchial airways based on $Re$) or in comparison to the viscous terms (which are important in the distal tracheo-bronchial and alveolar regions, again based on $Re$). Thus, the fluid dynamics associated with inhalation of pharmaceutical aerosols can typically be considered by neglecting the unsteady term in the equations (except of course in the oropharynx where turbulence occurs, which is inherently unsteady). Note that in making this assertion, we have used an average value of the unsteady term to determine its magnitude. In actual fact, this term may be much larger than its average value at certain times in a breath (e.g. at the start of a single inhalation or between inhalation and exhalation in tidal breathing), so that it does not make sense to use the average value when nondimensionalizing Eq. (6.1) at these times in the breath. Instead, if we say $dU/dt$ has a characteristic value $U'$ and we use this to nondimensionalize the unsteady term in Eq. (6.1) for times near when $dU/dt$ has the value $U'$, then we obtain

$$\varepsilon \frac{\partial \mathbf{v}'}{\partial t'} + \mathbf{v}' \cdot \nabla \mathbf{v}' = -\nabla p' + \frac{1}{Re}\nabla^2 \mathbf{v}' \qquad (6.5)$$

where

$$\varepsilon = DU'/U^2 \qquad (6.6)$$

(Pedley 1976). In Eq. (6.6) we must use the value of $U$ that is characteristic of the flow field at the time when $dU/dt$ takes on the value $U'$. The value of $\varepsilon$ gives an indication of the importance of the unsteady term relative to the convective term that can be used at different times in a breath.

Since we know the convective term is important only in the extrathoracic region and upper tracheo-bronchial airways, the parameter $\varepsilon$ is most meaningful in these regions. In the distal tracheo-bronchial regions and the alveolar regions, we can multiply Eq. (6.5) through by $Re$ and then the parameter $\varepsilon Re$ will give us an indication of the importance of the unsteady terms relative to the viscous terms (where, when calculating $Re$, we must use the value of $U$ that occurs at the same time in the breath that our chosen value of $U'$ occurs).

If we examine the flow at a time when the velocity $U = 0$ while $dU/dt = U'$ is not zero, we obtain an infinite value for $\varepsilon$ as well as $\varepsilon Re$, indicating that the unsteady terms are very important at such times. Thus, although we have said that the unsteady term in the Navier–Stokes equation can usually be neglected, this is not true at times when the velocity is small (or zero) and the rate of change of velocity is not small. This occurs at the start of a single inhalation (as occurs with dry powder inhalers or metered dose inhalers) and at times between inhalation and exhalation during tidal breathing. To accurately capture the fluid dynamics at these times we need to include the unsteady terms in the equations.

Isabey et al. (1986) have obtained experimental data in the central airways in models of these airways and find that unsteadiness is important at the time of zero flow when exhalation stops and inhalation begins (or vice versa), as expected from the previous discussion. However, the time over which unsteadiness is important is only a small portion of the tidal breathing cycle so that this probably has negligible effect on the amount of aerosol depositing for pharmaceutical aerosols that are continuously supplied during inhalation (as with nebulizers). Of more concern is the case of dry powder inhalers and metered dose inhalers, which may have significant aerosol supplied at the start of inhalation when $U$ is low and $U'$ is large (indicating large $\varepsilon$). In this case, inclusion of unsteadiness in the fluid dynamics may sometimes be necessary to adequately model deposition with such aerosols, although this remains to be determined and would be strongly dependent on the precise moment at which the aerosol is delivered during the breath.

Even for tidal breathing, however, we must be careful in jumping to the conclusion that because unsteady effects are unimportant in the fluid equations, such effects are also unimportant in determining the deposition of pharmaceutical aerosols. In particular, we have considered the equation of motion of the fluid only and have considered neither the particle equation of motion nor the boundary conditions governing the problem. If we consider the boundary conditions governing the deposition of a particle in a particular lung generation, we realize that a particle being carried by the fluid is only present in a given generation for a time $\Delta t = L/U$, where $L$ is the length of a generation and $U$ is the average value of fluid velocity over the time the particle is in this generation. Then, in order to decide if unsteady fluid motion is important we could ask if the velocity of the fluid changes by a significant amount in this time $\Delta t$. If it does, then the particle is being exposed to significant unsteadiness while in this generation. If not, then the particle sees an effectively steady velocity field and we could then use the solution from the steady problem to predict particle deposition in this generation.

To examine this further, consider that in the time $\Delta t$, the fluid can undergo a velocity change

$$\Delta U = U' \, \Delta t \tag{6.7}$$

where $U'$ is a typical value of $dU/dt$ while the particle is in a particular generation. The time a particle is resident in a particular generation can be approximated as

$$\Delta t = L/U$$

where $L$ is the length of the generation and $U$ is the average fluid velocity in the generation when the particle is in that generation. The parameter

$$\Delta U/U = U' \, \Delta t/U = LU'/U^2 \tag{6.8}$$

is thus seen to determine whether unsteadiness in the fluid motion is important in predicting particle deposition or not. Comparing with Eq. (6.6) we see that $\Delta U/U$ is simply our earlier parameter $\varepsilon$ but with the airway diameter replaced with the airway length $L$. Using an estimate for $U'$ as $U' = U/(\tau/4)$ with $\tau = 5$ s, Fig. 6.3 shows the values of $\Delta U/U$ obtained in the lung model given in Chapter 5.

Small values of $\Delta U/U$ indicate that a particle sees little variation in the velocity field while traveling through that generation, and the deposition in that generation can be expected to be similar to that predicted from using a steady velocity equal to the value of the velocity at that time in the breathing cycle. If $\Delta U/U$ is not small, then one must consider using an unsteady velocity field to predict the deposition particles in that generation. From Fig. 6.3 we see that for typical tidal breathing patterns with nebulizers ($18\,\mathrm{l\,min^{-1}}$), unsteady effects may be important throughout the alveolar region, while for single breath patterns ($60\,\mathrm{l\,min^{-1}}$) they may be important only in the

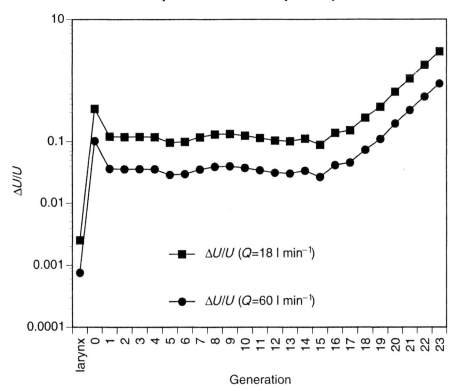

**Fig. 6.3** The parameter $\Delta U/U$ from Eq. (6.8), which indicates the importance of unsteadiness within each generation, plotted against lung generation for the idealized lung geometry of Chapter 5 and $U'$ estimated as $U' = U/(\tau/4)$ with $\tau = 5$ s.

last few generations. This has not been recognized by many previous deposition models, and may need to be corrected if deposition models are to more accurately reflect deposition *in vivo*. Eulerian deposition models that use a deposition rate obtained from the steady case, but use the instantaneous mean velocity in obtaining deposition rate from these equations, may deal with the unsteadiness adequately, but this needs to be determined.

## 6.3 Secondary flow patterns

From the preceding discussion it can be seen that in the upper half or so of the conducting airways, convection dominates the fluid motion over most of the breath cycle. The principal terms in the Navier–Stokes equations (Eq. 6.1) are then the pressure gradient term and the convective (nonlinear) term. With this knowledge, we can then use the following physical argument to suggest that the flow in these airways is not simply laminar, one-dimensional pipe flow, but instead contains swirling, secondary flow patterns induced by the curvature of the airways that occurs as each airway bifurcates.

This argument proceeds as follows. First, because we know the viscous term is small compared to the convection and pressure terms in the central airways (due to the large Reynolds number, where Reynolds number is a measure of the ratio of convective to viscous forces), let us ignore viscous forces in the dynamics. With this assumption, then the outward centrifugal force per unit mass acting on an infinitesimally small fluid element as it travels around the bend associated with a bifurcation at radius $r_a$ in this bend is $\rho v_a^2/r_a$, and must be balanced by an inward pressure gradient. If this fluid element is displaced to a larger radius $r_b > r_a$, then, if pressure is the only other force, one can show that we must have angular momentum conservation, so that the fluid element's velocity when it reaches radius $r_b$ will be $v_a r_a/r_b$ (since angular momentum per unit mass, $\rho v r$, is unchanged). However, the fluid element now experiences a centrifugal force $(\rho v^2/r)$ given by $\rho v_a^2\, r_a^2/r_b^3$, while the magnitude of the inward pressure gradient at radius $r_b$ is $\rho v_b^2/r_b$. As a result, if $(v_b r_b)^2 < (v_a r_a)^2$, the pressure gradient will not be sufficient to counteract the centrifugal force acting on the fluid element, and the displaced fluid will deviate further from its original position, i.e. the fluid is centrifugally unstable. This argument can be made rigorous (Drazin and Reid 1981); the essential quantity is the Rayleigh discriminant

$$\Phi = \frac{1}{r^3}\frac{d}{dr}\left[(rv)^2\right] \tag{6.9}$$

where $v$ is the streamwise velocity component of the fluid. For an inviscid flow with concentric circular streamlines, the flow will continue to remain as it is (i.e. it will be stable) if and only if $\Phi \geq 0$ throughout the entire flow. Thus, if the velocity decreases with radius faster than $1/r$, the flow is unstable.

   In the airways, without secondary flows the streamlines approximate concentric circles and we can use the above argument to suggest that these flows are unstable, since the velocity must drop to zero at the outside airway wall (see Fig. 6.4), possibly resulting in $\Phi < 0$.

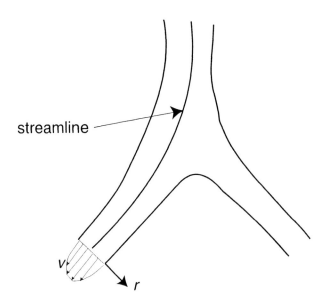

**Fig. 6.4** Fluid motion in a lung bifurcation gives streamlines that approximate concentric streamlines (in the absence of secondary flows) and which is unstable due to the decrease in velocity near the airway wall at the outside of the bend.

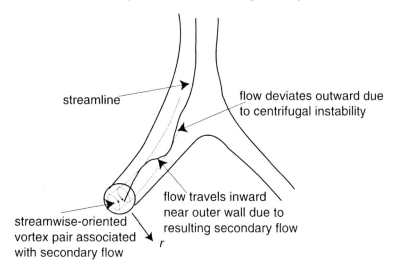

streamline

flow deviates outward due
to centrifugal instability

flow travels inward
near outer wall due to
resulting secondary flow

streamwise-oriented
vortex pair associated
with secondary flow

$r$

**Fig. 6.5** Schematic of secondary flow pattern that develops in a single lung bifurcation in the proximal half or so of the conducting airway generations during inhalation.

Of course, the above argument simply suggests that the fluid flowing around the bend associated with a lung bifurcation in the central airways is unstable, so that this flow is not simply a set of concentric streamlines. It does not tell us what sort of flow actually results due to this instability. For this we must resort to experiments or solutions of the nonlinear Navier–Stokes equations, which tell us that these secondary flows appear as the well-known streamwise-oriented vortices, indicated schematically in Fig. 6.5.

The presence of streamwise-oriented vortices like that shown in Fig. 6.5 is well documented by both experiments and numerical solution of the Navier–Stokes equations in idealized lung bifurcation geometries (see Pedley 1977, Pedley *et al.* 1979 or Chang 1989 for good reviews of earlier data, and Balásházy *et al.* 1996 or Zhao *et al.* 1997 for listings of more recent work). These secondary flow velocities can be quite strong in the first few lung generations, e.g. up to 50% of the streamwise velocity (Zhao and Lieber 1994a), so that the helical fluid streamlines may make an entire turn of a helix within one lung generation (since typical values of length/diameter are near $L/D = 3$, while the circumferential distance around a streamwise vortex can be approximated as $\pi D/4$, so that secondary flow velocities greater than 30% or so of the streamwise velocity can result in streamlines that complete one turn of a helix within a lung generation). The strength of these secondary flows decreases with Reynolds number (since the centrifugal term arises from the nonlinear term, whose strength decreases relative to the viscous term as the Reynolds number decreases). Indeed, our physical argument above neglected viscous forces in the dynamics. Since viscous forces become significant near the middle conducting airways, we should expect these secondary flows to be negligible beyond the middle generations of the conducting airways, although research is needed to confirm the parameter range over which these flows are important. (Although such flows do occur theoretically in curved tubes for all nonzero Reynolds numbers, their presence is so weak at low Reynolds numbers that they can essentially be ignored.)

It should be noted that the picture shown in Fig. 6.5 is representative only of the flow through a single bifurcation. In contrast, in the lung, the flow travels through many consecutive bifurcations. Unfortunately, little is known regarding the secondary flows in this case. Lee *et al.* (1996) have performed numerical simulations in a double bifurcation, and find the flow in the downstream bifurcation is more complex than occurs in a single bifurcation. In particular, the vortices formed in the first bifurcation are convected asymmetrically into the second bifurcation, and counter the formation of such vortices in one of the daughter branches, with the convected motion coming in from the upstream bifurcation actually dominating the secondary flow in the one daughter branch. Indeed, in the daughter bifurcation, secondary flow development would be expected to be much reduced due to the much lower velocity (and centrifugal force) compared to the parent bifucation, so that the secondary flow pattern that is carried into the daughter branch actually dominates the secondary flow there (and is thus not governed simply by the 'pristine' development of a streamwise-vortex pair like that shown in Fig. 6.5). Thus, the picture in Fig. 6.5 may not be very representative of the secondary flow patterns that actually occur in the multiple-branching geometry that is typical of the lung.

The strength of curvature-induced streamwise-oriented vortices is also determined by the ratio of the radius of curvature $R$ to tube diameter $D$ (which can be seen when we nondimensionalize the centrifugal term $\rho v^2 / R$ that is part of Eq. (6.1) when this equation is written in a curved coordinate system). Since there can be considerable variation in $R/(D/2)$ in different lung pathways and different individuals (Pedley 1977 suggests $R/(D/2)$ varies from 1 to 30, with typical values being between 5 and 10), we can expect considerable variations in the strength of the resulting secondary flows, adding to the already numerous caveats that are present in making general statements about fluid dynamics in the lung.

Upon expiration, the fluid travels back along paths that are just as curved as they were on inspiration, so we can expect secondary flows associated with streamwise vortices to develop for similar physical reasons as were discussed for inspiratory flow. However, during expiration, vortex pairs are generated by both airways that join to make the downstream airway. As a result, two pairs of vortices, for a total of four streamwise-oriented vortices, are found in the downstream airway, as has been seen by many authors (Pedley 1977, Chang 1989, Zhao and Lieber 1994b). However, secondary flow patterns during expiration in multiple bifurcation are likely quite different from those occurring in a single bifurcation (with the added complication of vortex stretching and its resultant intensification of vortices occurring because the flow accelerates from one generation to the next during expiration). Sarangapani and Wexler (1999) suggest that stronger secondary flows on expiration than inspiration play a principal role in the dispersion of inhaled aerosol boluses. Unfortunately, little is known about the nature of secondary flow in multiple-branching airways, and research is needed before a good understanding is possible.

## 6.4 Reduction of turbulence by particle motion

We have already discussed the fact that turbulence occurs in the upper airways. However, because particles move relative to the fluid, it is possible that they may affect turbulence themselves, either by reducing turbulence or increasing it (Crowe *et al.* 1998). Whether this occurs or not can be determined from simple consideration of the volume fraction of the aerosol and existing experimental data on this issue. It is known that for

**Table 6.1** Estimates of aerosol volume fractions for some inhaled pharmaceutical aerosols

| Device | Aerosol mass or volume | Inhaled volume | Aerosol volume fraction $\alpha$ |
|---|---|---|---|
| Nebulizer | 1 ml | 50 l | $2 \times 10^{-5}$ |
| Dry powder inhaler with lactose excipient | 20 mg | 1 l | $2 \times 10^{-5}$ |
| Dry powder inhaler without any excipient | 100 μg | 1 l | $10^{-7}$ |
| Metered dose inhaler sprayed directly into mouth | 50 μl | 1 l | $5 \times 10^{-5}$ |
| Metered dose inhaler with holding chamber | 50 μg | 1 l | $5 \times 10^{-8}$ |

volume fractions $\alpha$ below $10^{-6}$ the particles have negligible effect on the fluid turbulence (Crowe *et al.* 1996) and particle motion can be treated as having no effect on the fluid motion (i.e. a one-way coupled momentum treatment where the fluid affects the particle motion, but not vice versa, is adequate). Unfortunately, for pharmaceutical aerosols, volume fractions can be larger than $10^{-6}$ as can be seen in Table 6.1 where approximate estimates of $\alpha$ are given for a few typical pharmaceutical inhalation devices assuming a density of 1000 kg m$^{-3}$ for a particle. In Table 6.1 notice that values of $\alpha < 10^{-6}$ occur, for example, with some dry powder inhalers and with metered dose inhalers where the propellant has evaporated off in a holding chamber before inhalation begins. In these circumstances the effect of the particles on turbulence intensities can be expected to be negligible.

In cases where the particles may affect turbulence levels, it is worth knowing whether they can be expected to enhance or reduce turbulence intensities. The data of Gore and Crowe (1989) is useful in this regard. They find that if the ratio of particle diameter $d$ to the length scale $L$ of the most energetic turbulent eddies is such that $d/L < 0.1$, then the particles reduce turbulence intensities, but if $d/L > 0.1$, then the particles increase turbulence intensities. The most energetic eddies in internal flows occurring in inhalation devices can be expected to be of the order 0.5 times the diameter of the flow passages of these devices, giving $L = 5$ mm or so. Thus, the ratio $d/L$ can be expected to be less than 0.1 for inhaled pharmaceutical aerosols (which have $d < 100$ μm), and we expect the inhaled particles to reduce turbulence intensities. Since turbulence is likely important only in the upper airways, this effect can be expected to be of potential importance only in this region, where it can be expected to reduce the deposition of particles due to turbulent dispersion to the walls. Whether this plays a significant role in determining mouth–throat deposition of any inhaled pharmaceutical aerosols remains a topic for future research.

It should be noted that evaporating droplets (such as occurs with metered dose inhaler sprays and some nebulizers) can instead increase the turbulent kinetic energy in the fluid through the energy they exchange with the surrounding gas through two-way coupled heat and mass transfer (Mashayek 1998).

## 6.5 Temperature and humidity in the respiratory tract

Air inhaled into the respiratory tract is rapidly heated and humidified by heat and water vapor transfer from the airway walls. This heating and humidification is largely complete within the first few generations of the conducting airways (Daviskas *et al.* 1990), although this of course depends on the rate at which air is inhaled as well as on the temperature and humidity of the air being inhaled. However, we saw in Chapter 4 that the fate of hygroscopic droplets depends on the vapor concentration in the air that is carrying these droplets. Thus, in order to predict the size of hygroscopic droplets traveling through the airways it is necessary to know the temperature and water vapor concentration (or humidity) of the air in the airways. Because of the difficulties in actually measuring air temperature or humidity in the lung *in vivo*, mathematical models (Daviskas *et al.* 1990, Ferron *et al.* 1985, 1988) have been developed that appear to predict the limited amount of available *in vivo* data. A principal difficulty with developing such models is that the water vapor concentration and temperature at the airway wall surface is affected by the flow of airway surface fluid from the airway wall tissue, which is affected by the blood flow to this tissue. Thus, a complete model would have to include interaction of the blood flow rate with the rate at which heat and mass is transferred from the airway surface. This has not been done to the author's knowledge, and existing models instead specify the temperature and water vapor concentration at the airway wall surfaces *a priori*. Alternatively, *in vitro* replicas of airways (Eisner and Martonen 1989), have been used to study the temperature and humidity in the airways, although such studies are somewhat limited in the parameter space that can be explored because of the necessary complexity and time-consuming nature of such experiments.

## 6.6 Interaction of air and mucus fluid motion

So far we have been discussing only the fluid motion of the air in the respiratory tract. However, as mentioned in Chapter 5, the airways are lined with a liquid mucous layer. This layer is relatively thin (tens of microns thick) in normal subjects and would be expected to have little effect on airway fluid dynamics, since its motion is relatively independent of airway motion during inhalation of aerosols, and vice versa. However, in subjects with respiratory disease, the mucous layer can become much thicker in diseases with excess secretion or reduced clearance rates of airway surface fluid. In such cases, it is possible for the mucous layer to significantly affect the airway fluid dynamics, probably because of wave motion on the mucous layer surface (Clarke *et al.* 1970, King *et al.* 1982, Kim and Eldridge 1985, Kim *et al.* 1985, Chang 1989). This effect is probably due to interaction of turbulent structures in the air with motion of the airway surface fluid, and so is likely important only in the airways having turbulence present (which is usually only the first few proximal airway generations). Enhancement of turbulent energies leads to enhanced turbulent mixing, which results in increased frictional shear at the airway surface interface, thus giving increased pressure drops in the airways. This effect can be quite pronounced, with pressure drops in turbulent *in vitro* mucous-lined tubes being many times larger than in dry tubes, and also depending on the viscoelastic properties of the mucus. These effects are still not well understood, which is not surprising since the interaction of turbulent structures in air with interfacial motion of mucus–air surfaces is a very difficult two-phase flow problem that will not likely yield easily to future modeling efforts. Fortunately, turbulence is not present throughout much of the respiratory tract, and the effect of thickened mucus layers on the fluid dynamics in most of the airways reduces to simply having a reduced cross-sectional area for an essentially single-phase flow of air.

# References

Balásházy, I., Heistracher, T. and Hofmann, W. (1996) Air flow and particle deposition patterns in bronchial airway bifurcations: the effect of different CFD models and bifurcation geometries, *J. Aerosol Med.* **9**:287–301.

Chang, H. K. (1989) Flow dynamics in the respiratory tract, in *Respiratory Physiology: An Analytical Approach*, eds H. K. Chang and M. Paiva, Marcel Dekker.

Clarke, S. W., Jones, J. G. and Oliver, D. R. (1970) Resistance to two-phase gas-liquid flow in airways, *J. Appl. Physiol.* **29**:464–471.

Crowe, C. T., Troutt, T. R. and Chung, J. N. (1996) Numerical models for two-phase turbulent flows, *Ann. Rev. Fluid Mech.* **28**:11–43.

Crowe, C., Sommerfeld, M. and Tsuji, Y. (1998) *Multiphase Flow with Droplets and Particles*, CRC Press, Boca Raton.

Daviskas, E. Gonda, I. and Anderson, S. D. (1990) Mathematical modelling of heat and water transport in human respiratory tract, *J. Appl. Physiol.* **69**:362–372.

Drazin, P. G. and Reid, W. H. (1981) *Hydrodynamic Stability*, Cambridge University Press, Cambridge.

Eisner, A. D. and Martonen, T. B. (1989) Simulation of heat and mass transfer processes in a surrogate bronchial system developed for hygroscopic aerosol studies, *Aerosol Sci. Technol.* **11**:39–57.

Ferron, G. A., Haider, B. and Kreyling, W. G. (1985) A method for the approximation of the relative humidity in the upper human airways, *Bull. Math. Biol.* **47**:565–589.

Ferron, G. A., Haider, B. and Kreyling, W. G. (1988) Inhalation of salt aerosol particles – I. Estimation of the temperature and relative humidity of the air in the human upper airways, *J. Aerosol Sci.* **19**:343–363.

Gore, R. A. and Crowe, C. T. (1989) The effect of particle size on modulating turbulence intensity, *Int. J. Multiphase Flow* **15**:279–285.

Isabey, D., Chang, H. K., Delpuech, C., Harf, A. and Hatzfeld, C. (1986) Dependence of central airway resistance on frequency and tidal volume: a model study, *J. Appl. Physiol.* **61**:113–126.

Kim, C. S. and Eldridge, M. A. (1985) Aerosol deposition in the airway model with excessive mucus secretions, *J. Appl. Physiol.* **59**:1766–1772.

Kim, C. S., Abraham, W. A., Chapman, G. A. and Sackner, M. A. (1985) Influence of two-phase gas-liquid interaction of aerosol deposition in airways, *Am. Rev. Respir. Dis.* **131**:618–623.

King, M., Chang, H. B. and Weber, M. E. (1982) Resistance of mucus-lined tubes to steady and oscillatory airflow, *J. Appl. Physiol.* **52**:1172–1176.

Lee, J. W., Goo, J. H. and Chung, M. K. (1996) Characteristics of inertial deposition in a double bifurcation, *J. Aerosol Sci.* **27**:119–138.

Mashayek, F. (1998) Droplet-turbulence interactions in low-Mach-number homogeneous shear two-phase flows, *J. Fluid Mech.* **367**:163–203.

Panton, R. L. (1996) *Incompressible Flow*, Wiley.

Pedley, T. J. (1976) Viscous boundary layers in reversing flow, *J. Fluid Mech.* **74**:59–79.

Pedley, T. J. (1977) Pulmonary fluid dynamics, *Ann. Rev. Fluid Mech.* **9**:229–274.

Pedley, T. J., Schroter, R. C. and Sudlow, M. F. (1979) Gas flow and mixing in the airways, in *Bioengineering Aspects of the Lung*, ed. J. B. West, Marcel Dekker, New York.

Sarangapani, R. and Wexler, A. S. (1999) Modelling of aerosol bolus dispersion in human airways, *J. Aerosol Sci.* **30**:1345–1362.

Simone, A. F. and Ultmann, J. S. (1982) Longitudinal mixing by the human larynx, *Respir. Physiol.* **49**:187–203.

Tennekes, H. and Lumley, J. L. (1972) *A First Course in Turbulence*, MIT Press, Cambridge, MA.

Ultmann, J. S. (1985) Gas transport in the conducting airways, in *Gas Mixing and Distribution in the Lung*, eds L. A. Engel and M. Paiva, Marcel Dekker, New York, pp. 64–136.

White, F. M. (1999) *Fluid Mechanics*, 4th edn, McGraw-Hill.

Womersley, J. R. (1955) Method for the calcuation of velocity rate of flow and viscous drag in arteries when the pressure gradient is known, *J. Physiol. Lond.* **127**:553–563.

Zhao, Y. and Lieber, B. B. (1994a) Steady inspiratory flow in a model symmetric bifurcation, *J. Biomech. Eng.* **116**:488–496.

Zhao, Y. and Lieber, B. B. (1994b) Steady expiratory flow in a model symmetric bifurcation, *J. Biomech. Eng.* **116**:318–323.

Zhao, Y., Brunskill, C. T. and Lieber, B. B. (1997) Inspiratory and expiratory steady flow analysis in a model symmetrically bifurcating airway, *J. Biomech. Eng.* **119**:52–58.

# 7
# Particle Deposition in the Respiratory Tract

From discussions in earlier chapters we know that particle size plays an important role in determining where an inhaled pharmaceutical aerosol particle will deposit in the respiratory tract. However, several other factors, such as inhalation flow rate, affect particle deposition as well. By combining simplified lung geometries like those described in Chapter 5 with some basic fluid dynamics, it is possible to develop simple deposition models that quantitatively describe how these different factors affect particle deposition in the lung. Such models are useful in guiding the design of aerosol delivery devices, and their basis will be described in this chapter. However, because of the dramatic simplifications in lung geometry and fluid mechanics that are needed to make these models tractable, there are a number of factors that these models do not represent. In these cases, experimental data can help illuminate these effects, and such data will be presented in this regard in the present chapter.

Before getting down to details, it must be realized that because the actual geometry of the respiratory tract is so complicated, and because predicting particle trajectories throughout an entire lung is beyond prediction or measurement with current methods, our understanding of particle deposition in the respiratory tract is far from complete and remains a topic of current research. However, a reasonable understanding can be achieved by considering several simplified problems which we now turn to.

## 7.1 Sedimentation of particles in inclined circular tubes

The effect of gravity on particles inhaled into the respiratory tract can be understood to a certain extent by examining the deposition of particles in inclined circular tubes in which there is a laminar (i.e. one-dimensional, nonturbulent) air flow. Although this is a simplification of what actually occurs in the respiratory tract, it provides a starting point in understanding sedimentation of particles in the lung. The basic geometry is shown in Fig. 7.1.

What we want to determine is the fraction of particles of a given size that will deposit in a length $L$ of a circular tube if they enter the tube uniformly distributed across the entrance to the tube. To solve this problem, we must first know the fluid velocity field in the tube. A closed form solution for the velocity field in a circular tube, known as Poiseuille flow (named after the French physician Poiseuille who performed experiments on flow in tubes in the mid-1800s), is obtained by solving the Navier–Stokes equations

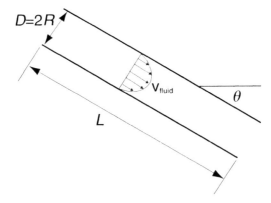

**Fig. 7.1** Fluid flow in a circular tube of diameter $D$, and length $L$ with the tube axis at an angle $\theta$ from the horizontal is a simple approximation for estimating particle sedimentation in the respiratory tract.

and is given by

$$v_{\text{fluid}} = 2\bar{U}(1 - r^2/R^2) \tag{7.1}$$

where $\bar{U}$ is the average velocity across the tube, $r$ is radial distance from the tube centerline, and $R = D/2$ is the radius of the tube. Poiseuille flow is valid only if the velocity of the fluid in the tube is steady, has only one component, and is independent of distance along the tube. These conditions will only be satisfied for laminar flow in straight tubes at distances, $x$, downstream from the inlet that satisfy the 'fully developed' condition that $x/D > 0.06Re$, where $Re$ is the Reynolds number $Re = \rho\bar{U}D/\mu$ (White 1999). Upstream of these locations, the velocity field is not well represented by Poiseuille flow. We saw in Chapter 6 that $Re \geq 1$ for most of the conducting airways, so that we can expect Poiseuille flow to be a poor approximation to the velocity field in most of these airways. Thus, only in the smallest conducting airways, and possibly more distal regions, do we expect the flow to be similar to Poiseuille flow. Note also that deep in the lung, where sedimentation is most important, airways are covered with alveoli, making the airways much different from circular tubes. Because such alveolated ducts lack the large surface area of containing walls that would normally result in slower velocities near the duct walls (Davis 1993), a uniform velocity field $v_{\text{fluid}} = \bar{U}$ (called 'plug' flow), directed along the tube axis, may be a better approximation to the actual velocity field than Poiseuille flow in such regions when estimating sedimentation.

Once we have specified a velocity field in the tube, then to determine what fraction of particles will deposit in the tube, we need to determine the trajectories of particles entering the tube at all points in the tube cross-section and see which particles deposit on the tube walls before they manage to exit the tube with the fluid. To determine particle trajectories, we can solve the equation of motion for each particle, which is given in Chapter 3 as

$$Stk\,\frac{\mathrm{d}}{\mathrm{d}t^*}\left(\frac{\mathbf{v}}{\bar{U}}\right) = \frac{v_{\text{settling}}}{\bar{U}}\,\hat{\mathbf{g}} - \frac{(\mathbf{v} - \mathbf{v}_{\text{fluid}})}{\bar{U}} \tag{7.2}$$

where we are using the average fluid velocity $\bar{U}$ as the fluid velocity scale, and $Stk$ is the Stokes number

$$Stk = \bar{U}\rho_{\text{particle}}d^2 C_c/18\mu D \tag{7.3}$$

and

$$v_{\text{settling}} = C_c\rho_{\text{particle}}gd^2/18\mu \tag{7.4}$$

is the settling velocity; $t^* = t/(D/\bar{U})$ is a dimensionless time, $v$ is particle velocity, while $\hat{g} = g/g$ is a unit vector in the direction of gravity.

The term in Eq. (7.2) with the Stokes number in front is responsible for deposition of particles by inertial impaction. However, from our discussions in Chapter 3, we know that sedimentation is an important deposition mechanism only in the more distal parts of the lung, where we know impaction is not an important deposition mechanism. So in terms of predicting amounts of particles sedimenting in the lung, as a first approximation it is common to neglect the impaction term in Eq. (7.2). In this case, Eq. (7.2) reduces to the following equation for particle velocity:

$$v = v_{\text{fluid}} + v_{\text{settling}}\hat{g} \tag{7.5}$$

Because we are assuming $v_{\text{fluid}}$ is parallel to the tube for both Poiseuille flow and plug flow, Eq. (7.5) predicts that all particles in a monodisperse aerosol will settle in the vertical direction at the same speed. Thus, neglecting particle–particle interactions (which, in Chapter 3, we decided was reasonable for many pharmaceutical aerosols), no particle can overtake another particle in a monodisperse aerosol. For this reason, we can draw a line in the cross-section of the tube entrance that divides those particles that will deposit in the tube from those particles that will travel through the tube to exit without depositing. Let us refer to this line as the 'sedimentation line'. Particles on the sedimentation line are said to follow 'limiting trajectories'. Once we determine the sedimentation line, we can determine the fraction of the particles entering the tube that deposit in the tube by calculating the mass flow rate over the two sections of the tube on either side of the sedimentation line.

## 7.1.1 Poiseuille flow

Using Eq. (7.5), the limiting trajectories and sedimentation line have been determined for horizontal tubes by Pich (1972), and for arbitrarily oriented tubes by Wang (1975) with the Poiseuille flow velocity. For Poiseuille flow, the fraction $P_s$ of particles depositing in the tube is given by (Wang 1975):

$$P_s = 1 - E - \Omega \tag{7.6}$$

where $E$ is the fraction of particles escaping the tube without depositing, and is given by

$$E = \begin{cases} \dfrac{2}{\pi}\left[\sqrt{\gamma(1-\gamma)}(1-2\gamma) + \arcsin\left(\sqrt{1-\gamma}\right)\right] & \text{for } -90° \le \theta \le 0° \text{ (i.e. uphill flow)} \\[2ex] \dfrac{2}{\pi}\arcsin\left(\sqrt{1-\eta^2}\right) \\[2ex] \quad -\dfrac{\sqrt{1-\eta^2}}{\pi\left(1+\dfrac{v_{\text{settling}}}{\bar{U}}\sin\theta\right)}\left[\dfrac{3v_{\text{settling}}L}{\bar{U}D}\cos\theta - \left(2+\dfrac{v_{\text{settling}}}{\bar{U}}\sin\theta\right)\eta\right] \\[2ex] \qquad\qquad \text{for } 0° < \theta \le 90° \text{ (i.e. downhill flow)} \end{cases} \tag{7.7}$$

and $\Omega$ is the fraction of particles retreating out of the tube due to gravitational settling (and thus not depositing), given by

$$\Omega = \begin{cases} 1 - \dfrac{1}{\pi}\left[3\sqrt{s(1-s)} + \arcsin\sqrt{1-s} + (1-9s^2)\arcsin\sqrt{\dfrac{1-s}{1+3s}}\right] & \text{for } -90° \le \theta \le 0° \\[2ex] 0 \quad \text{for } 0° < \theta \le 90° \end{cases} \tag{7.8}$$

Recall that $\theta$ is the angle of the tube from the horizontal as shown in Fig. 7.1. The parameters appearing in Eqs (7.7) and (7.8) are

$$\gamma = \frac{\left(\dfrac{3v_{\text{settling}}L}{4\bar{U}D}\cos\theta\right)^{2/3}}{1 - \dfrac{v_{\text{settling}}}{2\bar{U}}\sin\theta} \tag{7.9}$$

$$\eta = \left[\frac{\dfrac{v_{\text{settling}}}{\bar{U}}\left(6\dfrac{L}{D}\cos\theta + \sqrt{4\left(\dfrac{v_{\text{settling}}}{\bar{U}}\right)\sin^3\theta + 36\left(\dfrac{L}{D}\right)^2\cos^2\theta}\right)}{16}\right]^{1/3}$$

$$\quad -\frac{\left(\dfrac{v_{\text{settling}}}{\bar{U}}\right)^{2/3}\sin\theta}{2^{2/3}\left(6\dfrac{L}{D}\cos\theta + \sqrt{4\left(\dfrac{v_{\text{settling}}}{\bar{U}}\right)\sin^3\theta + 36\left(\dfrac{L}{D}\right)^2\cos^2\theta}\right)^{1/3}} \tag{7.10}$$

$$s = \frac{\dfrac{v_{\text{settling}}}{6\bar{U}}\sin\theta}{1 - \dfrac{v_{\text{settling}}}{2\bar{U}}\sin\theta} \tag{7.11}$$

The parameter $\eta$ given in Eq. (7.10) is the solution to a third-order polynomial equation given in Wang (1975), for which Wang gives an approximate solution, but for which an exact solution can be obtained as given in Eq. (7.10).

Equations (7.6)–(7.11) are rather cumbersome. Thus, simplifications to these equations are useful. Under the condition that $v_{\text{settling}}\sin\theta \ll \bar{U}$, Heyder and Gebhart (1977) show that Eqs (7.6)–(7.11) reduce to

$$P_s = \frac{2}{\pi}\left[2\kappa\sqrt{1-\kappa^{2/3}} - \kappa^{1/3}\sqrt{1-\kappa^{2/3}} + \arcsin(\kappa^{1/3})\right] \tag{7.12}$$

where

$$\kappa = \frac{3}{4}\frac{v_{settling}}{\bar{U}}\frac{L}{D}\cos\theta \tag{7.13}$$

and use has been made of the result

$$1 - \frac{2}{\pi}\arcsin\sqrt{1-\kappa^{2/3}} = \frac{2}{\pi}\arcsin\kappa^{1/3} \tag{7.14}$$

Equation (7.12) is symmetric about $\theta = 0$, so that deposition with this equation is independent of whether the flow is uphill or downhill.

The condition $v_{settling}\sin\theta \ll \bar{U}$, which is required for Eq. (7.12) to be valid, can be written as a restriction on particle size using the definition of settling velocity in Eq. (7.4) and average flow velocities in the simplified lung geometry presented in Chapter 5, yielding the result that Eq. (7.12) is a good approximation to Eqs (7.6)–(7.11) for particles of diameter $d$ satisfying

$$d \ll d_s \text{ where } d_s = \sqrt{\frac{72Q\mu}{2^n \sin\theta\, \rho_p g\pi D_n^2}} \tag{7.15}$$

where $Q$ is the flow rate at the trachea, $n$ is the generation number ($n = 0$ in the trachea), and $D_n$ is the diameter of the $n$th generation airway. A plot of $d_s$ against generation number is shown in Fig. 7.2 with $\theta = 38.24°$ (which is a commonly used tube orientation in lung model sedimentation, as we will see shortly).

It can be seen in Fig. 7.2 that only in the last few alveolar generations of the lung do we expect there to be any difficulty in satisfying Eq. (7.15) for typical inhaled pharmaceutical aerosols (which have particle diameters normally between 1 and 10 microns or so). However, we have already suggested that Poiseuille flow is probably not a particularly good approximation to the flow field deep in the lung anyway, so that Eq. (7.12) is reasonable for the small conducting airways where we expect the flow to be similar to Poiseuille flow.

Note that Eq. (7.12) gives complex numbers for the deposition fraction when $\kappa > 1$, so that is usual to set $P_s = 1$ if $\kappa > 1$, since when $\kappa > 3/4$ the time needed for a particle to move one tube diameter perpendicular to the flow streamlines (because of sedimentation) is less than the time it takes for the average flow velocity to travel the length of the tube.

### 7.1.2 Laminar plug flow

For plug flow, simple geometric consideration of the area between overlapping ellipses is all that is needed to separate depositing from nondepositing particles and determine the sedimentation line (since the fluid velocity is the same everywhere in the tube and the particles occupy an elliptical region that settles at constant velocity inside the vertical cross-section of the tube, which is also an ellipse). These considerations are given in Heyder (1975). The fraction of particles depositing, $P_s$, in a circular tube in this case is given by

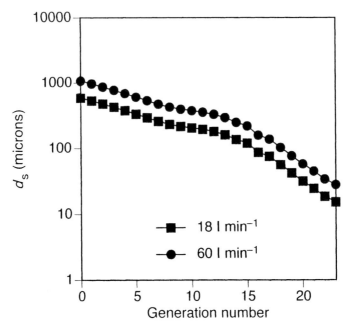

**Fig. 7.2** The parameter $d_s$ in Eq. (7.15) is shown for inhalation flow rates of 18 l min$^{-1}$ and 60 l min$^{-1}$ at the various generations in the idealized lung geometry of Chapter 5 with particle density $\rho_p = 1000$ kg m$^{-3}$. For particles with diameter $d \ll d_s$ the simplified version of Poiseuille flow sedimentation (Eq. (7.12)) is a good approximation to the more complex Eqs (7.6)–(7.11).

$$P_s = 1 - \frac{2}{\pi} \left[ \arccos\left(\frac{4}{3}\kappa\right) - \frac{4}{3}\kappa\sqrt{1 - \left(\frac{4}{3}\kappa\right)^2} \right] \qquad (7.16)$$

where $\kappa$ is given in Eq. (7.13). Note that $P_s$ is a real number only for $4\kappa/3 \leq 1$, so that if $\kappa \geq 3/4$ it is usual to set $P_s = 1$, since if $\kappa > 3/4$ then, as mentioned in Section 7.1.1, the time for a particle to move one tube diameter perpendicular to the flow streamlines is less than the time it takes for the average flow velocity to travel the length of the tube.

### 7.1.3 Well-mixed plug flow

The velocity field in the central airways is probably not well approximated by simple flow fields such as Poiseuille flow or plug flow. This is because of secondary flows associated with inertial effects in the curved regions of bifurcations of the larger airways as discussed in Chapter 6. The development of exact models would require simulation of the Navier–Stokes equations to predict sedimentation in these regions rigorously. However, an approximation for sedimentation in these regions can be made by assuming that the effect of the secondary flows is to produce a well-mixed aerosol in the tube cross-section. (This is in contrast to the sedimentation results above for Poiseuille and plug flow where the entire aerosol settles with a well-defined upper boundary and no mixing occurs between the aerosol-free and aerosol-containing regions.) With a well-mixed aerosol, the problem then reduces to estimating sedimentation in a plug flow where the aerosol is assumed to have a uniform number density in the tube cross-section. The rate of deposition in such a flow can be obtained from steady mass conservation:

$$\int_S M\mathbf{v} \cdot d\mathbf{S} = 0 \tag{7.17}$$

where $M$ is the mass of aerosol per unit volume, and $S$ is the surface bounding the volume containing the aerosol under consideration. This equation is simply a statement of the fact that the rates at which aerosol mass enters or leaves the tube along its cylindrical sides or ends must sum to zero because of mass conservation.

Since we are assuming a well-mixed aerosol at each tube cross-section, $M$ varies only with distance $x$ along the tube. The mass flux of aerosol through the tube entrance is

$$\int_{entrance} M\mathbf{v} \cdot d\mathbf{S} = M(x)(\bar{U} + v_{settling} \sin \theta)\pi \frac{D^2}{4} \tag{7.18}$$

and the mass flux of aerosol exiting through the tube exit for an infinitesimally short length of tube, $dx$, is

$$\int_{exit} M\mathbf{v} \cdot d\mathbf{S} = M(x + dx)(\bar{U} + v_{settling} \sin \theta)\pi \frac{D^2}{4} \tag{7.19}$$

Realizing that no aerosol deposits on the upper side of the tube due to sedimentation, the mass flux of aerosol depositing on the sides of the tube, $DE$, is

$$DE = \int_{bottom\ half\ of\ tube} M\mathbf{v} \cdot d\mathbf{S} \tag{7.20}$$

With the tube axis oriented at an angle $\theta$ downhill from the horizontal, geometrical considerations yield

$$\mathbf{v} \cdot d\mathbf{S} = (v_{settling} \cos \theta)(R\,d\phi \sin \phi)dx \tag{7.21}$$

where $\phi$ denotes angular distance around a circular cross-section of the tube. Equation (7.20) can thus be written

$$DE = M\left(x + \frac{dx}{2}\right)v_{settling} \cos \theta\,dx \int_0^\pi \sin \phi\, R\,d\phi \tag{7.22}$$

Integrating yields

$$DE = M\left(x + \frac{dx}{2}\right)v_{settling} \cos \theta\, D\,dx \tag{7.23}$$

Putting Eqs (7.18), (7.19) and (7.23) into Eq. (7.17) yields

$$[M(x + dx) - M(x)](\bar{U} + v_{settling} \sin \theta)\frac{\pi D^2}{4} = -M\left(x + \frac{dx}{2}\right)v_{settling} \cos \theta D\,dx \tag{7.24}$$

Expanding $M(x + dx)$ and $M(x + dx/2)$ in Taylor series about $x$, dividing through by $dx$, and taking the limit as $dx \to 0$, we finally obtain

$$\frac{dM}{dx} = -M\frac{4v_{settling} \cos \theta}{(\bar{U} + v_{settling} \sin \theta)\pi D} \tag{7.25}$$

Integrating Eq. (7.25), we obtain the mass of aerosol per unit volume as a function of distance $x$ along the tube:

$$M(x) = M_0 \exp\left[-\frac{4v_{settling}\cos\theta}{(\bar{U} + v_{settling}\sin\theta)\pi D}x\right] \tag{7.26}$$

where $M_0$ is the aerosol mass per unit volume at the tube entrance. The fraction of mass depositing in a tube of length $L$ is given by

$$P_s = \frac{M_0 - M_L}{M_0} \tag{7.27}$$

where $M_L = M(x = L)$. Using Eq. (7.26) to evaluate $M_L$, Eq. (7.27) gives us our final result for the fraction of aerosol depositing in an inclined tube assuming the aerosol concentration remains uniform over the tube cross-section:

$$P_s = 1 - \exp\left[-\frac{4}{\pi}\frac{v_{settling}\cos\theta}{(\bar{U} + v_{settling}\sin\theta)}\frac{L}{D}\right] \tag{7.28}$$

In the central and upper airways (where we expect secondary flows to give well-mixed aerosols) we have already seen that Eq. (7.15) implies $v_{settling}\sin\theta \ll \bar{U}$, so that Eq. (7.28) can be well approximated in these regions by

$$P_s = 1 - \exp\left(-\frac{16}{3\pi}\kappa\right) \tag{7.29}$$

which is a result given by other authors for horizontal tubes (Morton 1935, Fuchs 1964). Here, $\kappa$ is given in Eq. (7.13).

The fraction of particles $P_s$ depositing with the different types of flow we have considered are shown in Fig. 7.3 at various $\kappa$.

It can be seen in Fig. 7.3 that the assumed velocity field in the tube affects deposition fractions. Each of the three approximations shown in Fig. 7.3 might be a reasonable

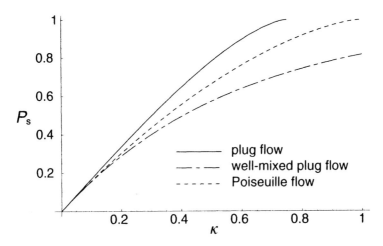

**Fig. 7.3** The fraction $P_s$ of aerosol sedimenting in a tube is shown for the different velocity and aerosol fields including plug flow (Eq. (7.16)), well-mixed plug flow (Eq. (7.29)), and Poiseuille flow (Eq. (7.12)) as a function of the parameter $\kappa$ in Eq. (7.13).

approximation for different parts of the respiratory tract, with Eq. (7.29) (well-mixed plug flow) probably the most reasonable of the three in the central conducting airways, Eq. (7.12) (Poiseuille flow) applying in the small conducting airways, and Eq. (7.16) (plug flow) applying in the alveolated airways, although none of them will exactly duplicate sedimentational deposition in real lung geometry since the flow there is neither strictly Poiseuille nor plug flow.

### 7.1.4 Randomly oriented circular tubes

The above equations allow estimation of the fraction of aerosol depositing in a tube at a known angle $\theta$. However, the different airways in the lung are oriented in many different directions. One approach to determining where particles would deposit in the lung due to sedimentation is to track many different individual particles through many different individual paths through the lung (using a Monte Carlo approach to give the orientation of each tube, in which a random number is used to select an orientation from a distribution of tube orientations), using the above equations to approximate the amount depositing in each generation due to sedimentation. This is the approach taken by Koblinger and Hofmann (1990), who used Eq. (7.29) for the sedimentation probability in nonalveolated airways.

An alternative approach to dealing with sedimentation in the many different orientations $\theta$ of airways in the lung is to treat the airways as a collection of randomly oriented tubes. The average fraction of aerosol depositing in one of a randomly oriented set of tubes is then given by

$$\bar{P}_s = \frac{\displaystyle\int_{-\pi/2}^{\pi/2} P_s f(\theta) d\theta}{\displaystyle\int_{-\pi/2}^{\pi/2} f(\theta) d\theta} \tag{7.30}$$

Here, $P_s$ is the fraction of aerosol depositing in a tube at known angle $\theta$ and is given by one of the various approximations considered above, while $f(\theta)d\theta$ is the probability of finding a tube at an angle $\theta$. An expression for $f(\theta)$ is obtained by realizing that an infinite number of randomly oriented tubes of length $L$ having one end centered at the origin will fill a sphere of radius $L$. The fraction of these tubes that are oriented between angles $\theta$ and $\theta + d\theta$ is then proportional to the surface area of a ring on the sphere between these two angles as shown in Fig. 7.4.

The area of this ring is simply the arc length $L\,d\theta$ multiplied by the circumference of the ring $2\pi L \cos\theta$, while the total surface area of the sphere is $4\pi L^2$, so the fraction of tubes oriented at angles between $\theta$ and $\theta + d\theta$ is then

$$f(\theta)d\theta = \frac{2\pi L^2 \cos\theta\, d\theta}{4\pi L^2} \tag{7.31}$$

which simplifies to

$$f(\theta)d\theta = \frac{\cos\theta\, d\theta}{2} \tag{7.32}$$

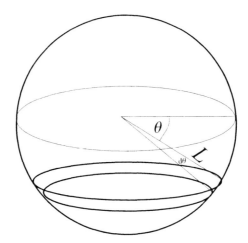

**Fig. 7.4** The area of a ring on the surface of a sphere is proportional to the fraction of randomly oriented tubes that are directed at angle $\theta$ from the horizontal.

Putting Eq. (7.32) into Eq. (7.30), we obtain

$$\bar{P}_s = \frac{1}{2} \int_{-\pi/2}^{\pi/2} P_s \cos\theta \, d\theta \qquad (7.33)$$

We can evaluate $\bar{P}_s$ using Eq. (7.33) with the various different equations we have developed for $P_s$. However, the only case in which an exact integration of Eq. (7.33) is possible is for well-mixed plug flow (Eq. (7.29)), for which we obtain:

$$\bar{P}_s = 2 + \pi I_1\left(\frac{4t'}{\pi}\right) - 2\,{}_1F_2\left[1, \left(\begin{array}{c} 0.5 \\ 1.5 \end{array}\right), \frac{4t'^2}{\pi^2}\right] \qquad (7.34)$$

where

$$t' = \frac{v_{\text{settling}}}{\bar{U}}\frac{L}{D} \qquad (7.35)$$

is the time it takes a particle to travel one tube length divided by the time the fluid takes to settle one tube diameter. Here, $I_1$ is the first-order modified Bessel function of the first kind (Mathews and Walker 1970), while ${}_1F_2$ is a generalized hypergeometric function (Gradshteyn and Ryzhik 1980). Hypergeometric functions are defined as the solutions of an ordinary differential equation called the hypergeometric equation. A series expansion solution to the hypergeometric equation gives the following series expansion for the hypergeometric function in Eq. (7.34):

$$\,{}_1F_2\left[1, \left(\begin{array}{c} b_1 \\ b_2 \end{array}\right), x\right] = \sum_{k=0}^{\infty} \frac{x^k}{(b_1)_k(b_2)_k} \qquad (7.36)$$

where the notation $(\alpha)_k$ indicates the following product:

$$(\alpha)_k \equiv \begin{cases} \alpha(\alpha+1)(\alpha+2)\cdots(\alpha+k-1) & \text{for } k \geq 1 \\ 1 & \text{for } k = 0 \end{cases} \qquad (7.37)$$

For both plug flow (Eq. (7.16)) and Poiseuille flow (Eq. (7.12)), numerical integration of Eq. (7.33) is necessary in order to accommodate the definition that $P_s = 1$ when

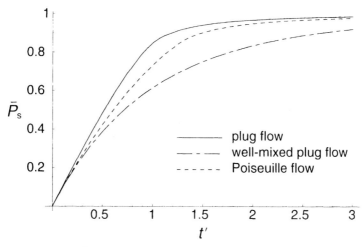

**Fig. 7.5** The average fraction $\bar{P}_s$ of aerosol sedimenting in a tube with randomly chosen angle from the horizontal is shown as a function of the parameter $t'$ in Eq. (7.35) by evaluating Eq. (7.33) for the different velocity and aerosol fields, including plug flow (Eq. (7.16)), well-mixed plug flow (Eq. (7.29)), and Poiseuille flow (Eq. (7.12)).

$\kappa \geq 3/4$ (for Eq. (7.16)) or $\kappa \geq 1$ (for Eq. (7.12)). Values for the average fraction of aerosol $\bar{P}_s$ depositing in a randomly oriented circular tube with the various different types of flows using Eq. (7.33) are shown in Fig. 7.5 as a function of the parameter $t'$ defined in Eq. (7.35).

It can be seen in Fig. 7.5 that the different flow fields give reasonably similar values to $\bar{P}_s$. This fact combined with the knowledge that all of these flow fields are only approximations to the actual flow field in the airways, and the fact that most inhaled pharmaceutical aerosols are polydisperse (resulting in a wide range of $t'$) reduces the importance of the differences between the different $\bar{P}_s$ shown in Fig. 7.5.

Because it is somewhat cumbersome to use Eq. (7.34), or to integrate Eq. (7.33) numerically for the different sedimentation functions $P_s$ given earlier, approximations to these equations are sometimes used. In particular, for $\kappa \ll 1$ it can be shown (Pich 1972) that $P_s$ for Poiseuille flow reduces to

$$P_s = \frac{4}{\pi}\kappa \qquad \text{(Poiseuille flow with } \kappa \ll 1 \text{)} \tag{7.38}$$

Substituting Eq. (7.38) into Eq. (7.33) one obtains

$$\bar{P}_s = \kappa \qquad \text{(Poiseuille flow with } \kappa \ll 1 \text{)} \tag{7.39}$$

However, Eq. (7.39) can be obtained by observing that coincidentally, for $\kappa \ll 1$

$$\bar{P}_s = P_s|_{\theta=\arccos(\pi/4)} \tag{7.40}$$

For this reason, a commonly used empirical approximation for $\bar{P}_s$ for Poiseuille flow is to use Eq. (7.40) for all $\kappa$, not just $\kappa \ll 1$ (Heyder and Gebhart 1977). In this case Eq. (7.12) is evaluated using a constant value of $\theta = 38.24° = \arccos(\pi/4)$ for all airways in the lung.

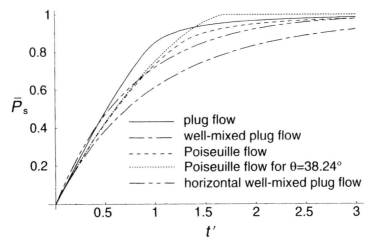

**Fig. 7.6** The two simplified equations, Eq. (7.40) (Poiseuille flow for $\theta = 38.24°$) and Eq. (7.41) (horizontal well-mixed plug flow), for the average fraction $\bar{P}_s$ of aerosol sedimenting in a circular tube with randomly chosen angle from the horizontal are shown with the more rigorously derived values shown already in Fig. 7.5.

For well-mixed plug flow, a sometimes used empirical approximation for $\bar{P}_s$ is

$$\bar{P}_s = P_s|_{\theta=0} \qquad \text{(horizontal well-mixed plug flow)} \qquad (7.41)$$

which simply assumes that all the airways are horizontally oriented. The two simplified sedimentation equations in Eqs (7.40) and (7.41) are shown in Fig. 7.6 together with the more rigorous sedimentation results obtained by integrating Eq. (7.33) numerically for Poiseuille flow (Eq. (7.12)), plug flow (Eq. (7.16)) and well-mixed plug flow (Eq. (7.29)).

It can be seen that Eq. (7.40) is a reasonable approximation to the more general Poiseuille flow result for most $t'$, while Eq. (7.41) is a reasonable approximation to either of the two more general plug flow results, especially for $t' < 0.5$.

## Example 7.1

Estimate the probability that a 3 µm particle of density 1000 kg m$^{-3}$ entering the 20$^{th}$ generation of the idealized lung geometry given in Chapter 5 will deposit in that generation by sedimentation if the inhalation flow rate is 50 l min$^{-1}$.

## Solution

An average value of this probability can be estimated by assuming this generation is randomly oriented with respect to the horizontal and using one of the equations developed above for randomly oriented tubes and shown in Fig. 7.6. For this purpose we need to evaluate the parameter $t'$ given in Eq. (7.35):

$$t' = \frac{v_{\text{settling}}}{\bar{U}} \frac{L}{D} \qquad (7.35)$$

To do this, we must evaluate the settling velocity, which we know from Chapter 3 is given by

$$v_{settling} = \rho_{particle}gd^2/18\mu$$

where $\mu$ is the viscosity of air ($\mu = 1.8 \times 10^{-5}$ kg m$^{-1}$ s$^{-1}$). Thus, we obtain

$$v_{settling} = 1000 \text{ kg m}^{-3} \times 9.81 \text{ m s}^{-2} \times (3 \times 10^{-6} \text{ m})^2/(18 \times 1.8 \times 10^{-5} \text{ kg m}^{-1} \text{ s}^{-1})$$
$$= 0.273 \text{ mm s}^{-1}$$

Also, we evaluate $\bar{U}$ from the volume flow rate $Q$, since $Q = \bar{U} \times$ (cross-sectional area of $20^{th}$ generation), so that

$$\bar{U} = Q/(2^{20} \times \pi D^2/4)$$

But from Chapter 5, we know that generation 20 of our idealized lung model has diameter $D = 0.033$ cm, so

$$\bar{U} = 50 \text{ l min}^{-1} \times 1000 \text{ cm}^3 \text{ l}^{-1} \text{ (min/60 s)}/(2^{20} \times \pi \times (0.033 \text{ cm})^2/4)$$
$$= 0.929 \text{ cm s}^{-1}$$

Using the airway length $L = 0.068$ cm from Chapter 5 we thus obtain

$$t' = (0.0273 \text{ cm s}^{-1}/0.929 \text{ cm s}^{-1}) \times (0.068 \text{ cm}/0.033 \text{ cm})$$
$$t' = 0.061$$

If we assume a randomly oriented tube, then from Fig. 7.6 we obtain estimates for the average sedimentation probability as

$$\bar{P}_s = 0.059 \text{ for randomly oriented Poiseuille flow}$$
$$\bar{P}_s = 0.059 \text{ for well-mixed plug flow}$$
$$\bar{P}_s = 0.061 \text{ for plug flow}$$

We also obtain $\bar{P}_s = 0.075$ for horizontal plug flow and for Poiseuille flow oriented at 38.24°. We see that the three randomly oriented results differ little, while the two simplified equations ((7.40) and (7.41)) both overestimate deposition by approximately 25%. However, we must decide if we really think representing the tube as a circular duct is reasonable, an issue that we now turn to.

## 7.2 Sedimentation in alveolated ducts

Because the conducting airways bear a resemblance to cylindrical tubes over much of their lengths, we have some confidence that basic aspects of sedimentation in the conducting airways can be approximated using the equations we have developed above. However, in the alveolar region of the lung, the respiratory bronchioles and alveolar ducts are covered with alveoli that might be expected to cause sedimentation to be different from that occurring in cylindrical tubes. For this reason, various authors have proposed equations to predict sedimentation rates that are specific to the alveolated parts of the lung. Since the alveoli have a roughly spherical shape, these equations are sometimes based on considerations of sedimentation of particles in stationary fluid in spherically shaped containers (Taulbee and Yu 1975, Egan and Nixon 1985). However, to estimate sedimentation with such an approach it is necessary to make assumptions on the amount of aerosol that is present in the alveoli compared to the amount in the core of the alveolated duct, with a common assumption being that the concentrations in the duct and the alveoli are the same. However, such assumptions are not likely to be valid for typical particle sizes seen with inhaled pharmaceutical aerosols, since Brownian diffusion is small for such particles and the principal mechanism causing the particles to enter the alveoli is sedimentation, so that uniform concentrations of aerosol across the duct and alveoli would not be expected. Indeed, simulations with the Navier–Stokes equations and particle equations of motion in three-dimensional alveolar duct-like geometries (Darquenne and Pavia 1996) show that sedimentation is considerably overestimated in a horizontal alveolar duct with such assumptions.

Although the most rigorous approach for predicting sedimentation in alveolated ducts would be based on simulations of the Navier–Stokes and particle equations of motion, the complex, detailed, time-dependent geometry of the alveoli has prevented much work from being done with such an approach. Future work may change this, but for now it is desirable to have a simple method of predicting sedimentation in alveolar regions. For this purpose, various authors have instead simply used the sedimentation formulas given above for circular ducts but with the diameter set equal to the diameter of the alveolar duct. Such an approach is justified for randomly oriented ducts by data obtained by Tsuda *et al.* (1994), who show that the fraction of aerosol depositing in randomly oriented two-dimensional alveolated ducts differs little from that in randomly oriented circular tubes. (This result is somewhat coincidental, since deposition in a circular tube at a given gravity orientation can be quite different from that in two-dimensional alveolated ducts, but when averaged over all gravity angles these differences approximately cancel out.)

Another issue, which we have already raised in Chapter 6, is that unsteadiness in the flow may need to be included when estimating sedimentation in some of the alveolar region. This is because changes in the flow velocity during a particle's transit of a generation in this region can be large compared to the flow velocity itself. The equations for sedimentation given above are for steady flow, and it remains to be determined what effect unsteadiness has on their accuracy in estimating sedimentation in the lung for pharmaceutical inhalation applications.

## Example 7.2

Darquenne and Pavia (1996) numerically solve the particle equations of motion in a three-dimensional representation of a horizontal alveolar duct of inner (lumen) diameter 0.3 mm and length 0.6 mm in which the fluid motion was given by solving the fluid equations with an assumption of axisymmetry and a Poiseuille flow profile at the duct inlet. The fluid flow rate through the duct was $Q = 2.4 \times 10^{-4}$ cm$^3$ s$^{-1}$ and the particles had a density of 1000 kg m$^{-3}$. They find that 38.79% of 5 µm diameter spherical particles entering the duct are deposited in the duct. When sedimentation was instead predicted with a simple model using an assumption of uniform aerosol concentration and modeling the alveoli as portions of spheres (mentioned above), they found that 93.77% of the particles were predicted to deposit, indicating the simple model is inadequate. What percentage of particles are predicted to deposit in the duct if Eq. (7.12) is used instead (i.e. sedimentation in Poiseuille flow) with the lumen diameter used as the circular tube diameter in these equations?

## Solution

This problem requires us to evaluate the sedimentation fraction using the result we derived earlier for sedimentation in Poiseuille flow through a circular tube for the special case when the tube is oriented horizontally ($\theta = 0$). Thus, Eq. (7.15) is readily satisfied, so that Eq. (7.12) is appropriate. Equation (7.12) is

$$P_s = \frac{2}{\pi}\left[2\kappa\sqrt{(1 - \kappa^{2/3})} - \kappa^{1/3}\sqrt{1 - \kappa^{2/3}} + \arcsin(\kappa^{1/3})\right] \tag{7.12}$$

where

$$\kappa = \frac{3}{4}\frac{v_{\text{settling}}}{\bar{U}}\frac{L}{D}\cos\theta \tag{7.13}$$

To use these equations we must first evaluate $\kappa$ in Eq. (7.13). For this purpose, we must evaluate the settling velocity, which we know from Chapter 3 is given by

$$v_{\text{settling}} = \rho_{\text{particle}}gd^2/18\mu$$
$$= 1000 \text{ kg m}^{-3} \times 9.81 \text{ m s}^{-2} \times (5 \times 10^{-6} \text{ m})^2/(18 \times 1.8 \times 10^{-5} \text{ kg m}^{-1} \text{ s}^{-1})$$
$$= 0.757 \text{ mm s}^{-1}$$

Also, we evaluate $\bar{U}$ from the volume flow rate $Q$, since $Q = \bar{U} \times$ (cross-sectional area of lumen), so that

$$\bar{U} = Q/(\pi D^2/4)$$
$$= 2.4 \times 10^{-4} \text{ cm}^3 \text{ s}^{-1}/(\pi \times (0.03 \text{ cm})^2/4)$$
$$= 0.34 \text{ cm s}^{-1}$$

We can now evaluate $\kappa$ from Eq. (7.13) with $\theta = 0$ (since the duct is horizontal) as

$$\kappa = (3/4)(0.0757/0.34)(0.6/0.3)$$
$$= 0.334$$

Putting $\kappa = 0.334$ into Eq. (7.12) gives

$$P_s = 0.476$$

Thus, 47.6% of the particles entering the duct will deposit by sedimentation. This compares relatively favorably with the value of 38.79% obtained by Darquenne and Pavia (1996) in their detailed numerical simulation, and gives us further confidence in using the circular tube sedimentation equations as a simple way of approximating sedimentation in the alveolated airways.

## 7.3 Deposition by impaction in the lung

The other main mechanism that causes inhaled pharmaceutical aerosols to deposit in the respiratory tract is inertial impaction. We might at first think that we could proceed as we did with sedimentation and develop relatively simple approximate equations for estimating impaction from theoretical solutions of the governing equations. However, inertial impaction in the lung occurs because particles are unable to follow the curved streamlines that the air follows in passing through bifurcations. Thus, to develop a model that might predict impaction with reasonable accuracy we must at least have a fluid velocity field that duplicates the curved nature of the streamlines that occur in the lung. Thus, we cannot use such simple straight tube flows as plug flow or Poiseuille flow as we did in estimating sedimentation. Instead, more complex flows must be considered.

However, before considering impaction in such flows, it is worthwhile deciding what parameters will be important in determining impaction. From our discussions in Chapter 3 on similarity of particle motion, we know that the motion of a spherical particle in a given geometry with low particle Reynolds number is affected only by the following parameters: Stokes number $Stk$, flow Reynolds number $Re_{flow}$ and nondimensional settling velocity. These parameters are given by

$$Stk = U_0 \rho_{particle} d^2 C_c / 18 \mu D$$

$$Re_{flow} = \rho_{fluid} U_0 D / \mu$$

nondimensional settling velocity:  $v_{settling} / U_0$

Here, $U_0$ and $D$ are a characteristic flow velocity and linear dimension, respectively, in the given geometry; $d$ is particle diameter, and $C_c$ is the Cunningham slip correction factor. In addition to these three parameters, geometric parameters (e.g. branching angle, parent/daughter diameter ratio) and the actual geometrical shape of an airway can affect impaction. If all these parameters are important in determining deposition, it would be difficult to develop simple formulas for estimating impaction. However, from Chapter 3 we know that if we compare the Stokes number to the nondimensional settling velocity, we obtain an estimate of the importance of impaction vs. sedimentation. This comparison is readily done by examining the parameter $Fr$, where

$$Fr = Stk / (v_{settling} / U_0)$$

or

$$Fr = U_0^2 / (Dg) \tag{7.42}$$

$Fr$ is generally referred to as the Froude number. Its value is shown in Fig. 7.7 throughout the idealized lung geometry of Chapter 5 for two different inhalation flow rates.

From Fig. 7.7 it is apparent that sedimentation is negligible compared to inertial impaction throughout the conducting airways at an inhalation flow rate of 60 l min$^{-1}$, while at the tidal breathing flow rate of 18 l min$^{-1}$ this is true only in the large airways. In these regions impaction is the dominant mechanism of deposition and we expect impaction to be independent of the sedimentation parameter $v_{settling} / U_0$, so that the only parameters affecting impaction then are the dynamical parameters $Stk$ and $Re_{flow}$, in addition to nondimensional geometrical parameters that characterize the bifurcation geometry. However, numerous experimental and numerical studies (Schlesinger et al. 1977, Chan and Lippman 1980, Gurman et al. 1984, Kim and Iglesias 1989a,b, Balásházy et al. 1991, Balásházy and Hofmann 1993a,b, Kim et al. 1994, Balásházy and Hofmann 1995, Kim and Fisher 1999) have measured deposition of particles in airway replicas and casts. These authors find that for typical airway geometries in normal lungs and for the Reynolds numbers encountered in the regions where impaction is important (typically $Re_{flow} > 1$), inertial impaction is only weakly dependent on both $Re_{flow}$ and the various geometrical parameters. Thus, we come to the empirical conclusion that for inhaled pharmaceutical aerosols, for which sedimentation and

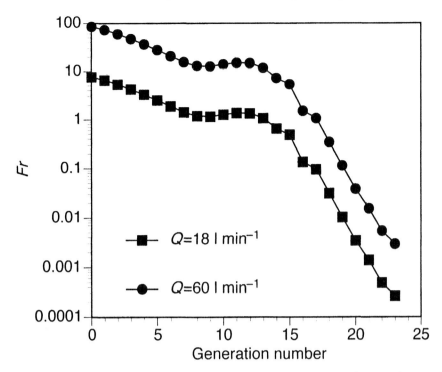

**Fig. 7.7** The Froude number $Fr$ from Eq. (7.42), defined here as the ratio of Stokes number to nondimensional settling velocity, is shown as a function of airway generation in the idealized lung geometry of Chapter 5 for two inhalation flow rates. Large values of $Fr$ correspond to regions where sedimentation can be expected to be small compared to impaction, while small values of $Fr$ correspond to regions where impaction can be expected to be small compared to sedimentation.

impaction are the dominant deposition mechanisms, deposition by inertial impaction in the airways can be approximated as being only a function of the Stokes number.

The result that inertial impaction can be approximated as being dependent on only the Stokes number for normal airway geometries is a major simplification. However, obtaining the functional dependence of impaction on Stokes number requires us to duplicate the curvature of streamlines in the airways that leads to this impaction. A variety of flows have been considered by various authors for this purpose, varying from the simple (e.g. flow in a bent circular tube), to the complex (e.g. simulations of the Navier–Stokes equations in bifurcations resembling human airways), and also include experiments on the deposition of monodisperse aerosols in models and casts of human airways. These studies give various empirical correlations that approximate the variation of impaction efficiency with Stokes number, as given in Table 7.1 and shown in Fig. 7.8.

From Fig. 7.8 it is clear that the different equations in Table 7.1 give quite a range of impaction probabilities. However, most of these equations are based on experiments or theory using only a single airway generation, without consideration of the effect that upstream generations, including the larynx, have on the fluid dynamics in the generation. Since parent generations create secondary flow patterns that can affect deposition in daughter generations, this is of some concern. Only the equations of Chan and Lippmann (1980) and ICRP (1994) are based on data that include the effect of a larynx and multiple generations (both are based on experiments in casts of airways), so that

**Table 7.1**   Various formulas for inertial impaction found in the literature are shown. Note that *Stk* is the Stokes number in the airway that impaction is occurring (the 'daughter airway'). Several of the formulae were originally given instead in terms of the Stokes number in the parent airway, $Stk_p$. For these cases, $Stk_p$ has been replaced by *Stk* using $Stk_p = 2\,Stk\,DR^3$ by assuming symmetrical branching, where *DR* is the diameter ratio (i.e. $DR = D_d/D_p$ where $D_d$ = diameter of daughter airway and $D_p$ = diameter of parent airway)

| Formula | | Source |
|---|---|---|
| $P_i = 0$ if $Stk < 0.02$, otherwise<br>   $= -0.0394 + 3.7417(2\,Stk\,DR^3)^{1.16}$ for $DR = 0.8\text{--}1.0$ | (7.43) | Kim *et al.* (1994) |
| $P_i = -0.1299 + 1.5714(2\,Stk\,DR^3)^{0.62}$ for $DR = 0.64$ | (7.44) | Kim *et al.* (1994) |
| $P_i = a\,Stk$ | (7.45) | Cai and Yu (1988) |
| where $a = f(\beta, DR)$ and $a = 1.53473$ for Poiseuille flow<br>and branching angle of $\beta = 35°$, $DR = 0.7853$ | | |
| $P_i = b\,Stk/(1 + b\,Stk)$ | (7.46) | Landahl (1950) |
| where $b = 4DR^3 \sin\beta$ and $b = 1.1111$ for $\beta = 35°$, $DR = 0.7853$ | | |
| $P_i = 1 - \dfrac{2}{\pi}\arccos(\beta\,Stk) + \dfrac{1}{\pi}\sin[2\arccos(\beta\,Stk)]$ | (7.47) | Yeh and Schum (1980) |
| Note: $\beta = 0.568977$ for 32.6° average branching angle | | |
| $P_i = 1.606\,Stk + 0.0023$ | (7.48) | Chan and Lippmann (1980) |
| $P_i = 1.3(Stk - 0.001)$ | (7.49) | Taulbee and Yu (1975) |
| $P_i = 6.4\,Stk^{1.43}$ generations 1–3<br>   $= 1.78\,Stk^{1.25}$ generations 4–5 | (7.50)<br>(7.51) | ICRP (1994) |
| $P_i = 0$ if $Stk < 0.1$, otherwise<br>   $= 4(Stk - 0.1)/(Stk + 1)$ | | Ferron *et al.* (1988)<br>(7.52) |

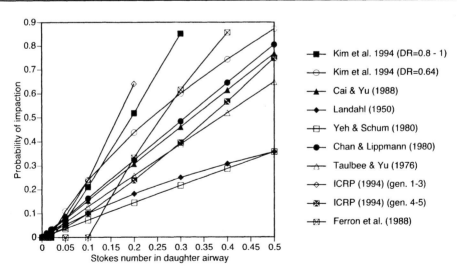

**Fig. 7.8**  The formulas in Table 7.1 giving the probability of impaction $P_i$ as a function of Stokes number *Stk* in the daughter airway are shown.

these equations probably represent typical values that would be expected in an actual lung. Of course, since these are empirical equations, any equation in Fig. 7.8 that comes close to these equations can be expected to give reasonable impaction probabilities as well. From an entirely empirical viewpoint, the data of Chan and Lippmann (1980) is seen to give impaction probabilities in the middle of the range of the various different equations in Table 7.1, so that this equation typifies representative values of impaction probabilities.

## Example 7.3

Use Eq. (7.48) (Chan and Lippmann 1980) in Table 7.1 to estimate the probability that a 3 μm diameter particle of density 1000 kg m$^{-3}$ entering the $10^{th}$ generation of the idealized lung geometry given in Chapter 5 will deposit in that generation by impaction if the inhalation flow rate is 30, 60 and 90 l min$^{-1}$.

## Solution

Equation (7.48) is

$$P_i = 1.606Stk + 0.0023$$

To use this equation we must first calculate the Stokes number

$$Stk = U_0 \rho_{particle} d^2 C_c / 18 \mu D$$

Here, we must evaluate $U_0$ from the volume flow rate $Q$, since $Q = U_0 \times$ (cross-sectional area of $10^{th}$ generation), so that

$$U_0 = Q/(2^{10} \times \pi D^2/4)$$

But from Chapter 5, we know that generation 10 of our idealized lung model has diameter $D = 0.162$ cm, so

$$U_0 = Q \, \text{l min}^{-1} \times 1000 \text{ cm}^3 \text{ l}^{-1} \times (1 \text{ min}/60 \text{ s})/(2^{10} \pi \times 0.162^2 \text{ cm}^2/4)$$
$$= 0.007896 \, Q \text{ m s}^{-1}, \text{ where } Q \text{ is in l min}^{-1}$$

Putting this into our definition of $Stk$ and approximating $C_c = 1$, then

$$Stk = 0.007896 \, Q \text{ m s}^{-1}(1000 \text{ kg m}^{-3})(3 \times 10^{-6} \text{ m})^2$$
$$\div (18 \times 1.8 \times 10^{-5} \text{ kg m}^{-1} \text{ s}^{-1} \times 0.00162 \text{ m})$$
$$= 0.000135 \, Q \text{ where } Q \text{ is in l min}^{-1}$$

Thus, we have

$$Stk = 0.0041 \text{ for } Q = 30 \text{ l min}^{-1}$$
$$Stk = 0.008 \text{ for } Q = 60 \text{ l min}^{-1}$$
$$Stk = 0.012 \text{ for } Q = 90 \text{ l min}^{-1}$$

Putting these into Eq. (7.48), we have the probability of impaction in this generation as

$$P_i = 0.0088 \text{ for } Q = 30 \text{ l min}^{-1}$$
$$P_i = 0.0153 \text{ for } Q = 60 \text{ l min}^{-1}$$
$$P_i = 0.0218 \text{ for } Q = 90 \text{ l min}^{-1}$$

We see that the probability of impaction is linearly dependent on flow rate, so that an aerosol consisting of particles of this size would deposit significant amounts in the conducting airways at the higher flow rates, but would penetrate well into the alveoli at low flow rates. This is partly why there are no strict criteria as to what an appropriate particle size is for inhaled pharmaceutical aerosols – deposition in the airways is dependent not just on particle size, but also on flow rate, which can vary considerably in patients.

## 7.4 Deposition in cylindrical tubes due to Brownian diffusion

We have already seen in Chapter 3 that molecular diffusion (i.e. Brownian motion) can play a role in the deposition of small diameter inhaled pharmaceutical aerosols in the respiratory tract. To proceed to rigorously estimate diffusion in the respiratory tract would require us to solve the Navier–Stokes equations for the fluid flow and then solve either the equations of motion for a particle moving in this flow field with a Brownian motion superposed on the trajectory, or else solve a convection-diffusion equation for the aerosol concentration with this velocity field. With the latter approach, we need to solve a standard convection-diffusion equation for the aerosol concentration (Fuchs 1964):

$$\frac{\partial n}{\partial t} + \nabla \cdot (n\mathbf{v}) = D_d \nabla^2 n \tag{7.53}$$

subject to the boundary condition that $n = 0$ at walls, and appropriate initial conditions. Here $D_d$ is the diffusion coefficient, which we gave in Chapter 3 for spherical particles as

$$D_d = \frac{kTC_c}{3\pi\mu d} \tag{7.54}$$

and $\mathbf{v}$ is the bulk velocity field of the particulate phase (usually assumed equal to the fluid velocity for the simplified case here of a uniform aerosol concentration and deposition due to diffusion alone).

Because of the complexity of solving the Navier–Stokes equations and Eq. (7.53) in such a complicated geometry as the respiratory tract, approximations are desirable. Two such approximations are obtained by solving Eq. (7.53) with either an assumption of a Poiseuille flow velocity field (Eq. 7.1) or a uniform, plug flow velocity field. For Poiseuille flow, several authors have solved Eq. (7.53) with various simplifying asymptotic approximations to the resulting infinite series (Townsend 1900, Nusselt 1910, Gormley and Kennedy 1949, Ingham 1975) to obtain expressions for the average deposition probability $P_d$ in a cylindrical tube with Poiseuille flow. For example, Ingham (1975) gives

$$P_d = 1 - 0.819\,e^{-14.63\Delta} - 0.0967\,e^{-89.22\Delta} - 0.0325\,e^{-228\Delta} - 0.0509\,e^{-125.9\Delta^{2/3}} \tag{7.55}$$

while Gormley and Kennedy (1949) give the commonly used result accurate for $\Delta < 0.1$:

$$P_d = 6.41\Delta^{2/3} - 4.8\Delta - 1.123\Delta^{4/3} \tag{7.56}$$

where in both Eqs (7.55) and (7.56)

$$\Delta = \frac{kTC_c}{3\pi\mu d}\frac{L}{\bar{U}}\frac{1}{4R^2} \tag{7.57}$$

and $k = 1.38 \times 10^{-23}$ J K$^{-1}$ is Boltzmann's constant, $T$ is the gas temperature, $\mu$ is the gas viscosity, $C_c$ is the Cunningham slip correction factor, $d$ is particle diameter, $R$ is the airway radius, $L$ is airway length and $\bar{U}$ is average flow velocity in the tube.

If a plug flow velocity field is assumed, Eq. (7.53) reduces to the same equation as that for a stationary aerosol residing in a cylindrical tube for a time $t = L/\bar{U}$. This equation is simply the time-dependent diffusion equation, which can be solved analytically by straightforward separation of variables to obtain (Buchwald 1921, Fuchs 1964)

$$P_d = 1 - 4\sum_{m=1}^{\infty}\frac{1}{\lambda_m^2}e^{-4\lambda_m^2\Delta} \tag{7.58}$$

where $\lambda_m$ is the $m$th zero of the zero-order Bessel function $J_0$. To aid in evaluating Eq. (7.58), the following algebraic approximation for $\lambda_m$ for large $m$ is useful (Abramowitz and Stegun 1981):

$$\lambda_m = \beta + \frac{1}{8\beta} - \frac{124}{(8\beta)^3} + \frac{120928}{15(8\beta)^5} - \frac{401743168}{105(8\beta)^7} + O\left(\frac{1}{\beta^9}\right), \text{ where } \beta = \left(m - \frac{1}{4}\right)\pi \tag{7.59}$$

Approximating $\lambda_m$ using only the first five terms on the right-hand side of Eq. (7.59) gives values that typically differ from the exact zeros by less than $10^{-9}$ for $m \geq 100$.

Unfortunately, the sum in Eq. (7.58) converges slowly for small values of $\Delta$, so that a large number of terms in the sum are required for reasonable accuracy when $\Delta$ is small. For example, compared to the result obtained by truncating the infinite sum in Eq. (7.58) after 100,000 terms, the error incurred by truncating after 200 terms is only 1.8% for $\Delta = 10^{-6}$, but for $\Delta = 10^{-8}$ the error increases dramatically so that even if we keep 10,000 terms in this sum the error is still 7.7%, and for $\Delta = 10^{-9}$ the error for 10,000 terms increases to 34%. This can be a problem for inhaled pharmaceutical aerosols, since such aerosols often have quite small values of $\Delta$. Thus, an empirical approximation to Eq. (7.58) is useful in order to avoid the need for such lengthy summations when $\Delta$ is small. The following expression differs by less than 2% from Eq. (7.58) for the range $10^{-9} < \Delta < 0.3$:

$$P_d = \begin{cases} 0.164385\Delta^{1.15217}\exp[3.94325\,e^{-\Delta} + 0.219155(\ln\Delta)^2 + 0.0346876(\ln\Delta)^3 \\ \quad + 0.00282789(\ln\Delta)^4 + 0.000114505(\ln\Delta)^5 + 1.81798 \times 10^{-6}(\ln\Delta)^6] \\ \qquad\qquad\qquad\qquad\qquad\qquad\qquad\qquad\qquad\qquad\text{if } \Delta \leq 0.16853 \\ 1 \quad \text{if } \Delta > 0.16853 \end{cases} \tag{7.60}$$

A comparison of $P_d$ for Poiseuille flow using Eqs (7.55) and (7.56), as well as the result for plug flow (Eq. (7.60)) is shown in Fig. 7.9. It can be seen that Eqs (7.55) and (7.56) differ negligibly, but plug flow (Eq. (7.60)) gives considerably higher deposition probabilities than either of these equations at low values of $\Delta$.

Because the flow in the airways and alveolar regions of the lung is neither plug flow nor Poiseuille flow, none of Eqs (7.55), (7.56) or (7.60) will exactly predict diffusional

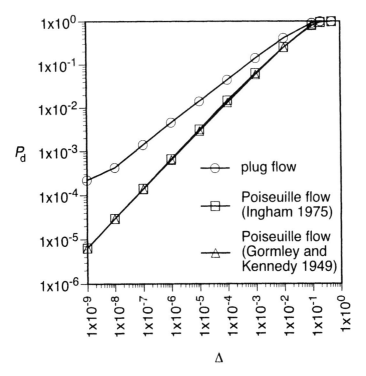

**Fig. 7.9** Probability of deposition due to diffusion, $P_d$, is shown for plug flow (Eq. (7.60)) and Poiseuille flow (Eqs (7.55) and (7.56)) in a cylindrical tube as a function of the parameter $\Delta$ in Eq. (7.57).

deposition in the lung airways. For this reason, various authors have examined deposition in geometries more closely resembling airways and developed alternative models for diffusional deposition, some of which account for the nonplug/nonPoiseuille flow nature in the conducting airways (Martin and Jacobi 1972, Cohen and Asgharian 1990, Martonen 1993, ICRP 1994, Yu and Cohen 1994, Martonen et al. 1995, 1997, and also see Brockmann 1993 for a review of other literature), as well as the effect of alveoli (Taulbee and Yu 1975, Tsuda et al. 1994, Darquenne and Pavia 1996). However, much of this work on diffusional deposition is aimed at deposition in the conducting airways or at particles that are much smaller than those occurring in inhaled pharmaceutical aerosols, so that at this point it is worth examining how important diffusion is expected to be as a deposition mechanism in the different parts of the lung for inhaled pharmaceutical aerosols. We have already examined this in an order of magnitude manner in Chapter 3, but let us re-examine this issue more specifically. In particular, we can use Eqs (7.55) or (7.60) to compare diffusional deposition probabilities to sedimentation and impaction probabilities from the equations we saw earlier in this chapter. Shown in Figs 7.10 and 7.11 is the ratio of the probability of diffusional deposition to the probability of deposition by either impaction or sedimentation. In Fig. 7.11, $\kappa \leq 0.3$ for the parameter range shown, so that Eq. (7.40) is a reasonable approximation for sedimentation probability in a randomly oriented tube with Poiseuille flow, while Eq. (7.41) is a reasonable approximation for sedimentation probability in a randomly oriented tube with plug flow in Fig. 7.10.

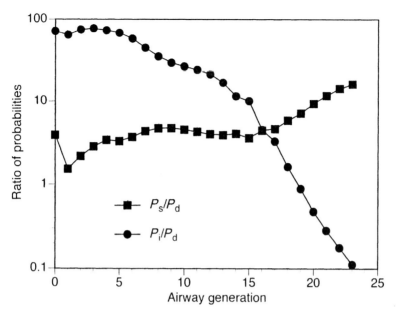

**Fig. 7.10** The ratios of the probability of deposition of a 3.5 µm diameter particle due to sedimentation and diffusion ($P_s/P_d$) and impaction and diffusion ($P_i/P_d$) are shown in each generation of the idealized lung geometry of Chapter 5 for an inhalation flow rate of 60 l min$^{-1}$. Plug flow is assumed in the different lung airways in calculating sedimentation and diffusion. Thus, $P_s$ is from Eq. (7.41), and $P_d$ is from Eq. (7.60). $P_i$ is from Eq. (7.48).

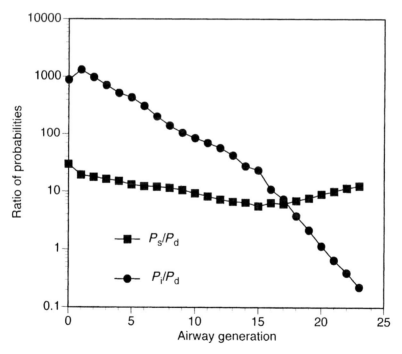

**Fig. 7.11** Same as Fig. 7.10, but now for a 2.0 µm diameter particle and Poiseuille flow is used for the calculation of sedimentation and diffusion probabilities. Thus, $P_s$ is from Eq. (7.40), and $P_d$ is from Eq. (7.55), while $P_i$ is still from Eq. (7.48).

From Figs 7.10 and 7.11, we see that during inhalation at 60 l min$^{-1}$, diffusion starts to become nonnegligible only in the alveolar region (generations 15 and higher). This fact makes the empirical correlations developed for diffusional deposition in the conducting airways (ICRP 1994, Yu and Cohen 1994) less useful for inhaled pharmaceutical aerosols than for occupational exposure aerosols, since these equations apply only to the conducting airways (where Figs 7.10 and 7.11 show that diffusion is unimportant for inhaled pharmaceutical aerosols).

Figure 7.10 also shows that for plug flow, diffusion begins to become nonnegligible in the alveolar region once particles have diameters smaller than approximately 3.5 μm, but Fig. 7.11 shows that for Poiseuille flow we need particle diameters below approximately 2.0 μm for this to occur. At first sight this may not be apparent from Figs 7.10 and 7.11, but once it is realized that the ratios $P_i/P_d$ and $P_s/P_d$ both increase with particle size, it is seen that the particle sizes shown in Figs 7.10 and 7.11 represent the approximate critical particle sizes where both these ratios together drop below 10 over a significant part of the lung. The difference in critical particle size below which diffusion becomes important (i.e. 3.5 μm for plug flow and 2.0 μm for Poiseuille flow) is due to the much larger diffusional deposition probabilities that occur with plug flow (see Fig. 7.9) at small $\Lambda$. Recall that in Chapter 3 we determined that diffusion would become important relative to sedimentation for particle sizes on the order of 3.5 μm for an inhalation flow rate of 60 l min$^{-1}$, which agrees well with our result here for plug flow.

Note that at lower flow rates, diffusional probabilities decrease in importance relative to sedimentation as discussed in Chapter 3, so that smaller particle sizes are needed for diffusion to become important at lower flow rates (e.g. in Chapter 3 at 18 l min$^{-1}$ we estimated that diffusion was negligible for particles larger than approximately 3 μm in diameter).

It should be noted that Figs 7.10 and 7.11 give the relative importance of diffusion while air is flowing through the lung airways. If there is a breath-hold at the end of inhalation, then because the amount of aerosol sedimenting in a given time is approximately proportional to time and because the time occupied by inhalation is often considerably less than a typical breath-hold, the amount of aerosol sedimenting during inhalation is much smaller than that sedimenting during the breath-hold. In this case, diffusion becomes less important as a deposition mechanism, as mentioned in Chapter 3. Indeed, we can calculate probabilities for sedimentation or diffusion during a breath-hold by replacing the particle residence time $L/\bar{U}$ with $t_{breath-hold}$ in the equations developed for sedimentation or diffusion probabilities. Doing so, we find that diffusion probabilities with either Eq. (7.55) or Eq. (7.60) are less than 1/3 the sedimentation probabilities in Fig. 7.5 for particles larger than 1 μm with a 10 s breath-hold.

In conclusion, we see that diffusion is not generally a dominant mechanism of deposition for inhaled pharmaceutical aerosols when a breath hold is used, but can become nonnegligible during the act of inhalation itself in the more distal parts of the lung (where impaction probabilities are small). For this reason, diffusion is usually included when modeling the fate of inhaled pharmaceutical aerosols, even though its importance may not be large for many such aerosols, particularly if breath holding is performed. It should be noted that in low gravity environments where sedimentation becomes negligible, Figs 7.10 and 7.11 indicate that diffusion is likely to be the dominant mechanism of deposition in the distal parts of the lung.

## Example 7.4

Darquenne and Pavia (1996) solve the axisymmetric Navier–Stokes equations to obtain the velocity field in a three-dimensional idealized alveolar duct geometry, and then use this velocity field in their numerical solution of the three-dimensional equations of motion for particle trajectories in this geometry. For 0.01 μm diameter particles they calculate a deposition probability of 70.22%, which can be assumed to be nearly entirely due to diffusion. They use a Poiseuille flow inlet condition to the alveolar duct with a flow rate of $2.4 \times 10^{-4}$ cm$^3$ s$^{-1}$. The duct has a length of 0.6 mm. The inner diameter of the duct (which resembles a 'chicken wire' cylinder made from all the entrances to alveoli) is 0.3 mm and the outer diameter (formed by the outsides of the alveoli) is 0.45 mm. Calculate the probability of diffusional deposition predicted by Eq. (7.56) for Poiseuille flow by assuming the alveolar duct is a cylinder of diameter

(a) 0.3 mm,
(b) 0.45 mm,

and compare these values to the value obtained more rigorously by Darquenne and Pavia.

## Solution

Diffusional deposition can be calculated for Poiseuille flow using either Eq. (7.55) or (7.56). However, to use these equations we must calculate the value of

$$\Delta = \frac{kTC_c}{3\pi\mu d} \frac{L}{\bar{U}} \frac{1}{4R^2} \tag{7.57}$$

where $k = 1.38 \times 10^{-23}$ J K$^{-1}$, $T = 310$ K, $\mu = 1.8 \times 10^{-5}$ kg m$^{-1}$ s$^{-1}$ is the viscosity of air, $C_c = 23.06$ is the Cunningham slip correction factor (Willeke and Baron 1993) for particle diameter $d = 0.01 \times 10^{-6}$ m, $R$ is the airway radius, $L = 600 \times 10^{-6}$ m is airway length and $\bar{U}$ is average flow velocity in the tube. However, since $\bar{U} = Q/(\pi R^2)$, we can rewrite Eq. (7.57) as

$$\Delta = \frac{kTC_c}{12\mu d} \frac{L}{Q} \tag{7.61}$$

From this equation we see that the cylinder diameter does not appear in our calculations, and the answers to parts (a) and (b) of this question are identical and independent of tube diameter.

Putting the numbers into Eq. (7.61), we obtain $\Delta = 0.114$, and substituting this value into Eq. (7.55) we obtain $P_d = 0.85$ for both (a) and (b) of this equation. This is a somewhat higher probability than the value of 0.7022 calculated by Darquenne and Pavia (1996), but considering the simplicity of Eq. (7.55) compared to the calculations of Darquenne and Pavia, the error is surprisingly small.

# 7.5 Simultaneous sedimentation, impaction and diffusion

So far we have been examining deposition due to the three principal mechanisms as if the probability of deposition for each mechanism could be calculated independently of the others. However, the particle equation of motion (Newton's second law) is in general not a linear equation in particle velocity or position, so that it is not rigorous to simply superpose the particle motion or velocity due to each mechanism, since such superposition is valid only when the governing equation is linear. The rigorous solution to this dilemma is to determine particle trajectories under the simultaneous action of gravity, inertia and diffusion. However, this would require three-dimensional numerical simulations that are not practical if our goal is to develop simple estimation procedures for predicting deposition in the respiratory tract. Instead, it is common to use an empirical approach.

One such approach is to combine deposition probabilities due to the individual deposition mechanisms in the form

$$P = (P_i^p + P_s^p + P_d^p)^{1/p} \tag{7.62}$$

with different authors suggesting different values of $p$; e.g. Asgharian and Anjilvel (1994) suggest $p = 3$ for straight tubes, but $p = 1.4$ for bifurcating airways, while ICRP (1994) suggests $p = 2$, and other authors use $p = 1$ (the validity of which is examined by Balásházy et al. 1990a in the absence of diffusion). When only diffusion and gravitation are present, Heyder et al. (1985) suggest using

$$P = P_d + P_s - \frac{P_s P_d}{P_s + P_d} \tag{7.63}$$

When diffusion is ignored, Balásházy et al. (1990b) examine the validity of using

$$P = P_i + P_s - P_i P_s \tag{7.64}$$

Other expressions have also been suggested, and with so many different empirical approaches it would at first appear to be difficult to choose the most appropriate one. However, the differences between the different approximations are largest for 'transitional' particle sizes near 1 μm in diameter (typically between 0.5 and 2 μm) that have low mobility (i.e. these are particles that are not small enough for Brownian diffusion to be strong, but also not large enough for sedimentation and impaction to be strong either). For transitional particle sizes, the form of empirical approach that is chosen to represent the combined effects can be important. However, outside the transitional range, the differences resulting from the different empirical approaches to superposing them is less important and probably give errors of the same or higher order as the other assumptions that are made in developing the equations for the individual deposition probabilities, as can be seen in the following example.

## Example 7.5

Calculate the deposition probabilities $P_d$, $P_i$ and $P_s$ due to diffusion, impaction and sedimentation for spherical, stable particles (assume a density of 1000 kg m$^{-3}$) ranging in size from 0.5 to 5 μm in diameter, and estimate the total combined deposition probability $P$ due to the simultaneous presence of these three deposition mechanisms using the so-called '$L_p$-norm' from Eq. (7.62)

$$P = (P_s^p + P_i^p + P_d^p)^{1/p}$$

for $p = 1, 2$ and 3. Do the calculations for generations 0, 15, and 23 of the idealized lung geometry from Chapter 5 (Table 5.1) for an inhalation flow rate of 60 l min$^{-1}$. For simplicity, use the equations developed for plug flow when calculating $P_d$ (i.e. Eq. (7.60)) and well-mixed horizontal plug flow for $P_s$ (i.e. Eq. (7.41)) for all lung generations, and use Eq. (7.48) for impaction probabilities.

## Solution

This problem reduces to substituting the numbers into the appropriate equations using the tube dimensions from Chapter 5. The equation for sedimentation probability that we are told to use is Eq. (7.41), which gives (from Eq. (7.29))

$$P_s = 1 - \exp\left(-\frac{16}{3\pi}\kappa\right)$$

where we set $\theta = 0$ in the definition

$$\kappa = \frac{3}{4}\frac{v_{settling}}{\bar{U}}\frac{L}{D}\cos\theta$$

from Eq. (7.13) in order that we consider horizontal well-mixed plug flow. Here, the settling velocity must be calculated from its definition in Chapter 3:

$$v_{settling} = C_c \rho_{particle} g d^2/18\mu$$

where $C_c$ is the Cunningham slip factor, which can be approximated for the present particle size range by $C_c = 1 + 2.52\lambda/d$ where $\lambda \approx 0.07$ μm is the molecular mean free path of air. Here, recall that $\mu = 1.8 \times 10^{-5}$ kg m$^{-1}$ s$^{-1}$ is the viscosity of air in the lung. The flow velocity $\bar{U}$ is obtained from the inhalation flow rate $Q$ using

$$\bar{U} = Q/(2^n \pi D^2)$$

where $D$ is the diameter of the lung generation $n$ being considered from Chapter 5 (where we gave $D$ as 1.81 cm, 0.049 cm, and 0.024 cm for generations 0, 15 and 23, respectively). From Table 5.1, the lengths $L$ of generations 0, 15 and 23 are 12.456 cm, 0.134 cm and 0.073 cm, respectively.

The impaction probability $P_i$ is to be calculated from Eq. (7.48):

$$P_i = 1.606 Stk + 0.0023$$

where $Stk$ is the Stokes number of the particle, defined in Chapter 3 as

$$Stk = \bar{U}\rho_{particle} d^2 C_c/18\mu D$$

Finally, the diffusional deposition probability is to be calculated from Eq. (7.60):

$$P_d = \begin{cases} 0.164385\Delta^{1.15217} \exp[3.94325\,e^{-\Delta} + 0.219155(\ln\Delta)^2 + 0.0346876(\ln\Delta)^3 \\ \quad + 0.00282789(\ln\Delta)^4 + 0.000114505(\ln\Delta)^5 + 1.81798 \times 10^{-6}(\ln\Delta)^6] \\ \hfill \text{if } \Delta \leq 0.16853 \\ 1 \quad \text{if } \Delta > 0.16853 \end{cases}$$

which requires calculating the value of $\Delta$ from Eq. (7.57):

$$\Delta = \frac{kTC_c}{3\pi\mu d}\frac{L}{\bar{U}}\frac{1}{4R^2}$$

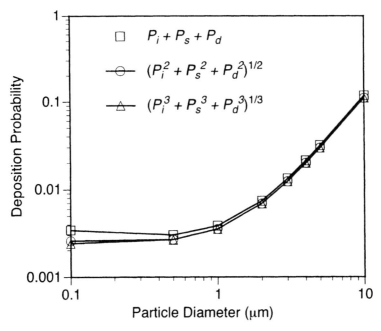

**Fig. 7.12** Different values of $p$ ($p$ = 1, 2, 3) in Eq. (7.62) are used to combine the individual deposition probabilities due to impaction, $P_i$, from Eq. (7.48), sedimentation, $P_s$, from Eq. (7.41) and diffusion from Eq. (7.60). Data is shown for generation 0 (trachea) of the idealized lung geometry from Chapter 5 for an inhalation flow rate of 60 l min$^{-1}$ for various particle sizes.

where the tube radius is $R = D/2$, $T \approx 310$ K is the air temperature in the given lung generation, and $k = 1.38 \times 10^{-23}$ J K$^{-1}$ is Boltzmann's constant. Having calculated $P_s$, $P_i$ and $P_d$ from these equations, we then combine these values in Eq. (7.62) with $p = 1, 2$ and 3 for the different particle sizes and lung generations. The results of these calculations are shown in Figs 7.12–7.14.

From Figs 7.12–7.14 it is seen that differences between the empirical methods of combining the individual deposition probabilities are present, but these differences become significant only for particles near 1 μm in diameter (assuming densities of 1000 kg m$^{-3}$). This is a concern for applications in which this is the primary particle size of interest. However, for many pharmaceutical inhalation applications, somewhat larger particles are expected and the differences between the different approaches for combining the individual deposition mechanisms are reduced. As mentioned earlier, none of the various approaches is rigorously correct. However, Figs 7.12–7.14 along with our previous examples suggest that the errors incurred by combining the individual deposition probabilities with the various approaches is of similar or higher order to that produced by the assumptions we have made in obtaining these individual deposition probabilities (such as assuming uniform or Poiseuille flow for $P_d$ and $P_s$, or using empirical results for $P_i$ based on experiments in casts of the larger airways).

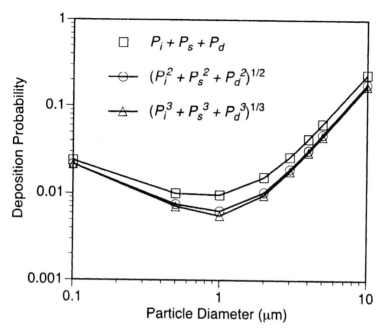

**Fig. 7.13** Same as Fig. 7.12, but for generation 15 (respiratory bronchioles) of the idealized lung geometry of Chapter 5.

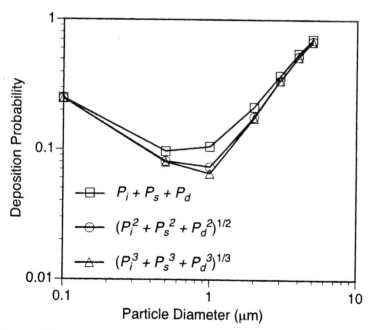

**Fig. 7.14** Same as Fig. 7.12, but for generation 23 (the most distal generation) of the idealized lung geometry of Chapter 5.

## 7.6 Deposition in the mouth and throat

Equations for sedimentation, impaction and diffusion like those given in the previous sections form the basis for current models that predict the amounts of an inhaled aerosol that will deposit in the lung. However, a significant portion of an inhaled aerosol may never reach the lung because of the filtering effect of the mouth and throat (the nose is an even better filter, and for this reason, inhaled pharmaceutical aerosols whose target is the lung are not generally used with nasal inhalation because of the high losses in the nose). Thus, if we are to develop a model for estimating amounts of inhaled aerosol that deposit in the lung, we must know how much of the aerosol, and what particle sizes, are able to travel past the mouth and throat.

Unfortunately, this information is hard to come by for two major reasons. First, the geometry of the mouth and throat does not resemble any simple idealized shape (see Chapter 5) and varies considerably between individuals, so that simple analyses (like the straight tube deposition equations we have looked at in modeling deposition in the lung) do not provide reasonable accuracy. Second, the fluid mechanics in this geometry is normally turbulent (see Chapter 6) and dependent on the fluid dynamics created by the inhalation device placed at the mouth (DeHaan and Finlay 1998, 2001). For these reasons, and because of the poor performance of turbulence models in complex geometries (Finlay *et al.* 1996b, Stapleton *et al.* 2000), general equations that allow prediction of deposition in the mouth and throat have not yet been developed for oral inhalation with pharmaceutical aerosol devices, but are a topic of ongoing and future research.

At present, models for mouth–throat deposition exist only for unencumbered inhalation of aerosols and are meant for modeling the fate of environmental and occupational aerosols (ICRP 1994). These models are empirical and provide equations that fit experimental data on *in vivo* mouth–throat deposition in human subjects inhaling via straight tubes inserted into the mouth. Such models can be expected to accurately predict mouth–throat deposition only for pharmaceutical inhalation devices that supply aerosol to the mouth via a geometry that resembles that of a relatively long, straight tube. Most pharmaceutical inhalation devices do not resemble straight tubes in their fluid mechanics (the most notable exception being some nebulizer designs, which do supply nearly a straight tube for inhalation), so that existing mouth–throat deposition models are inappropriate for quantitative prediction of mouth–throat deposition with many pharmaceutical inhalation devices (DeHaan and Finlay 1998, 2001).

Despite this, it is useful to examine the predictions of such models, since they can provide us with a qualitative understanding of certain features of mouth–throat deposition that can be expected to carry over to pharmaceutical inhalation devices. As demonstrated by Figs 7.7, 7.10 and 7.11, impaction is by far the dominant deposition mechanism in the mouth and throat for inhaled pharmaceutical aerosols, so that only the impactional deposition part of such models need be considered for our purposes. With this in mind, an important observation that we can draw from existing empirical models of impactional deposition is their dependence not just on the aerosol properties, but also on a geometrical flow rate effect (as was mentioned in Example 3.5 in Chapter 3). This effect arises because the mouth–throat geometry of a given individual changes with increasing inhalation flow rate (e.g. the glottal opening widens). Probably the most commonly used model incorporating this effect is the empirical model of Rudolf *et al.* (1990), which is used in the ICRP (1994) model. This model gives the impactional deposition in the mouth–throat region as

$$P_i = 1 - \frac{1}{1.1 \times 10^{-4}(d_{ae}^2 Q^{0.6} V_t^{-0.2})^{1.4} + 1} \tag{7.65}$$

where $d_{ae}$ is the particle's aerodynamic diameter in μm (defined in Chapter 3), $Q$ is the inhalation flow rate in $cm^3 s^{-1}$, and $V_t$ is tidal volume in $cm^3$.

If impaction in the mouth–throat was solely determined by the particle's impaction properties, we would expect particle diameter to appear via a term proportional to $d_{ae}^2 Q$ (since this is the form of the Stokes number, which determines impaction). Instead, flow rate appears to the power 0.6 in the term $d_{ae}^2 Q^{0.6}$. This fact, and the appearance of tidal volume in Eq. (7.5) are empirical effects of flow rate and tidal volume changes to the mouth–throat geometry.

Despite the fact that Eq. (7.65) was not designed to be predictive of mouth–throat deposition with inhalation devices attached at the mouth, two qualitative features of Eq. (7.65) remain valid for inhaled pharmaceutical aerosol, and these are the increase in mouth–throat deposition with particle size and with inhalation flow rate. Indeed, the well-known rule of thumb that suggests particles larger than approximately 5 μm in aerodynamic diameter are not suitable for efficient inhalation delivery is derived from the rapid rise in mouth–throat deposition that occurs with increasing particle size and seen in Eq. (7.65). Note that this rule of thumb does not give allowance to the effect of inhalation flow rate, nor does it account for the differing fluid dynamics that occurs with different inhalers at the mouth, so that this rule of thumb should be used with due caution.

## 7.7 Deposition models

By combining equations and analyses like those presented in the preceding sections of this chapter, it is possible to develop relatively simple models for predicting the amount of an inhaled aerosol that will deposit in the different lung regions. A plethora of such models has been presented in the archival literature (Heyder and Rudolf 1984 list 27 models prior to 1984), dating back more than 65 years to Findeisen's work (Findeisen 1935).

Recent deposition models can be categorized into three types, based on the approach they take to the fluid and particle motion in the lung: empirical models, Lagrangian dynamical models and Eulerian dynamical models. In the two types of dynamical models, equations governing the dynamics of the aerosol are solved to predict the amount of aerosol depositing in the different parts of the respiratory tract. In Lagrangian models the aerosol is examined in a reference frame that moves with the aerosol, while in Eulerian models the aerosol is examined in a stationary frame. Table 7.2 classifies a few of the many deposition models presented in the archival literature into the three basic types of models. More will be said about each of the three types of models shown in Table 7.2.

It should be noted that all respiratory tract deposition models to date suffer from their inability to model mouth–throat deposition with most pharmaceutical inhalation devices, as mentioned in the previous section. Indeed, nearly all such models were originally developed to model the deposition of inhaled occupational and environmental aerosols during tidal breathing, and all three approaches are capable of matching subsets of laboratory *in vivo* data with such aerosols. If mouth–throat deposition is given

**Table 7.2**   A sampling of respiratory tract deposition models in the archival literature, classified by type

| Empirical models | Lagrangian dynamical models | Eulerian dynamical models |
|---|---|---|
| ICRP (1994), Yu *et al.* (1992), Rudolf *et al.* (1990), Rudolf *et al.* (1986), Davies (1982) | Finlay and Stapleton (1995), Darquenne and Paiva (1994), Koblinger and Hofmann (1990), Ferron *et al.* (1988), Persons *et al.* (1987), Martonen (1983), Yeh and Schum (1980), Gerrity *et al.* (1979), ICRP (1966), Beeckmans (1965), Landahl (1950), Findeisen (1935) | Edwards (1995), Scott and Taulbee (1985), Egan and Nixon (1985), Taulbee *et al.* (1978), Taulbee and Yu (1975) |

empirically, such models are probably also capable of modeling intrathoracic (lung) deposition with most inhaled pharmaceutical aerosols in normal subjects, although this is a topic of ongoing research.

Empirical models are the simplest models. These models give a set of algebraic equations that fit a set of experimental *in vivo* data. Empirical models do not give explicit consideration to the particle and fluid dynamics, and so do not rely on dynamical analyses like those used to develop the equations given earlier in this chapter. Empirical models are usually relatively easy to implement and have low computational demands. However, they do not lend themselves well to extrapolation outside the parameter space of the experimental data, nor to inclusion of dynamical effects like hygroscopicity. This limits their applicability in some cases.

### 7.7.1 Lagrangian dynamical models

The next level of conceptual complexity beyond empirical models is achieved by Lagrangian dynamical models. In current versions of these models that treat the entire respiratory tract, particles are followed in one dimension (i.e. depth into the lung) through an idealized lung geometry in which the fluid flow in each lung generation is specified (usually either plug flow or Poiseuille flow). The particles are simply convected through the idealized lung geometry at the average local flow velocity in each lung generation. The probability of deposition as a particle travels through each generation of this geometry is estimated using equations like those given earlier in this chapter. Symmetrical lung geometries are most commonly used, although asymmetric geometries can be considered, for example with the addition of Monte Carlo techniques (Koblinger and Hofmann 1990). Two-way and one-way coupled hygroscopic effects (Persons *et al.* 1987, Ferron *et al.* 1988, Finlay and Stapleton 1995), can also be readily incorporated into these models, since the equations governing these effects are written very naturally in a Lagrangian form.

The principal limitations of current Lagrangian models stem from their use of only one spatial dimension and the assumed fluid dynamics (although these same features also lead to the attractive simplicity and low computational requirements of these models). These limitations are responsible for the difficulty such models have in simulating the axial dispersion of an aerosol bolus (i.e. a short burst of inhaled aerosol spreads out axially as it travels through the lung, which is not readily captured with current Lagrangian dynamical models, although Sarangapani and Wexler (2000) propose an approximate approach if a parabolic velocity profile is assumed). Although generalizing these models to more than one spatial dimension would allow circumvention of this difficulty, it would also make them far more computationally demanding. A second limitation of Lagrangian models is the difficulty they have in treating flow rates and aerosol concentrations that vary during a breath, a feature that can be important with some inhaled pharmaceutical aerosols, such as dry powder inhalers.

## 7.7.2 Eulerian dynamical models

Bolus dispersion and time-dependence can be more easily implemented using the third framework mentioned above, the Eulerian approach. The general concept of a Eulerian model is to solve a convection-diffusion equation for the aerosol in an idealized version of the lung geometry, using ideas first developed for modeling gas transport in the lung (Taulbee and Yu 1975, Taulbee et al. 1978, Egan and Nixon 1985). The aerosol is convected through the lung by the air motion, but also diffuses relative to the air due to Brownian motion. As with the Lagrangian models, all present versions of Eulerian deposition models that treat the entire respiratory tract are one-dimensional, with depth into the lung being the spatial dimension that is used. Such one-dimensional models can be derived by considering the equation for mass conservation of the aerosol particles, which in integral form can be written

$$\frac{\partial}{\partial t}\int_V n\,dV + \int_S n\mathbf{v}\cdot d\mathbf{S} = \int_S D_d\nabla n\cdot d\mathbf{S} \qquad (7.66)$$

where $V$ contains the volume of aerosol under consideration and $S$ bounds this volume. Here $\mathbf{v}$ is the velocity of the aerosol when treated as a continuum and $D_d$ is the molecular diffusion coefficient given in Eq. (7.54).

This equation can be reduced to one dimension as follows. First, introduce a coordinate system $(x, y, z)$ with $x$ representing depth into the lung and $y, z$ representing the cross-section of the air-filled portions of all parts of the lung at a depth $x$. Next, we consider a short section of lung from depth $x$ to depth $x + dx$. We integrate the first term on the left-hand side over the two spatial dimensions $y$ and $z$, and break up the other two terms into integrations over two types of surfaces: airway surfaces $S_{yz}$ and airway lumen $S_x$, as shown in Fig. 7.15.

In this way we can rewrite Eq. (7.66) as

$$\frac{\partial}{\partial t}\int_x^{x+dx} A_T\bar{n}\,dx + \int_{S_x} n\mathbf{v}\cdot d\mathbf{S} + \int_{S_{yz}} n\mathbf{v}\cdot d\mathbf{S} = \int_{S_x} D_d\nabla n\cdot d\mathbf{S} + \int_{S_{yz}} D_d\nabla n\cdot d\mathbf{S} \qquad (7.67)$$

where $A_T$ is the total cross-sectional area of the airways at depth $x$, and $\bar{n}$ is the average aerosol number concentration over this cross-section.

To simplify this equation further, we can write

$$\int_{S_x} D_d\nabla n\cdot d\mathbf{S} = \left(D_d A_A \frac{\partial\bar{n}}{\partial x}\right)\Bigg|_{x+dx} - \left(D_d A_A \frac{\partial\bar{n}}{\partial x}\right)\Bigg|_x \qquad (7.68)$$

where $A_A$ is the total area in the cross-section of the airway passages that make up $S_x$ and we have assumed for simplicity, as is normally done, that the average value of aerosol

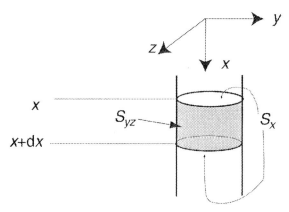

**Fig. 7.15** A portion of a lung airway at depth $x$ in the lung is shown. $S_x$ is the airway lumen surface at depth $x$ and $S_{yz}$ is the airway surface itself.

concentration over $A_A$ is the same as over $A_T$. In addition, we can write:

$$\int_{S_x} n\mathbf{v}\cdot d\mathbf{S} = \left[(\bar{n}A_A u)|_{x+dx} - (\bar{n}A_A u)|_x\right] + F\,dx \tag{7.69}$$

where $u$ is the average velocity in the $x$-direction over the area $A_A$, and the term $F\,dx$ is a flux correction (White 1999) that arises because of the nonlinear integrand in Eq. (7.69), so this equation is exact with $F = 0$ only in the case of plug flow with a uniform aerosol concentration across the airway cross-section. In general, $F \neq 0$. In order to evaluate $F$ exactly we would need to know the value of the aerosol number concentration and velocity across all cross-sections $S_x$. Because obtaining this information defeats the simplicity of a one-dimensional model, it is usual to introduce the following approximation, in which the flux correction $F$ is represented as an effective diffusive flux with diffusion coefficient $D_F$:

$$F = -\frac{\partial}{\partial x}\left(A_A D_F \frac{\partial \bar{n}}{\partial x}\right) \tag{7.70}$$

where $D_F$ must be specified in some empirical manner.

The difference between $A_A$, appearing in Eqs (7.68) and (7.69), and $A_T$ appearing in Eq. (7.67) arises because the terms in Eqs (7.68) and (7.69) represent convective and diffusive transport of aerosol along the $x$-direction. Because direct connection between successive lung generations occurs through the airway passages, not via alveoli, such transport occurs only across the area $A_A$ (the cross-sectional area of the airway passages), not across the entire lung cross-section area $A_T$ (which includes alveoli). In contrast, $A_T$ arises because we have taken an average over the entire lung cross-section at depth $x$.

At this point we also need to realize the meaning of the last term on each side of Eq. (7.67). Together these two terms represent the rate at which the aerosol number concentration changes due to deposition of aerosol at the airway surface walls. Defining $L$ as this deposition rate per unit length, then we can define $L$ as

$$L\,dx = \int_{S_{yz}} D_d \nabla n \cdot d\mathbf{S} - \int_{S_{yz}} n\mathbf{v}\cdot d\mathbf{S} \tag{7.71}$$

Substituting Eqs (7.68)–(7.71) into Eq. (7.67), dividing by d$x$ and taking the limit as d$x \to 0$ we obtain the following equation for the average aerosol concentration, $\bar{n}$, at depth $x$:

$$\frac{\partial}{\partial t}(A_T \bar{n}) + \frac{\partial}{\partial x}(A_A \bar{n}u) = \frac{\partial}{\partial x}\left(A_A D_{\text{eff}} \frac{\partial \bar{n}}{\partial x}\right) - L \tag{7.72}$$

where $D_{\text{eff}}$ is an effective diffusion coefficient given by

$$D_{\text{eff}} = D_d + D_F \tag{7.73}$$

and $D_d$ is the molecular diffusion coefficient from Eq. (7.54), while $D_F$ is from Eq. (7.70).

Although Eq. (7.72) is exact as written, it cannot be solved without knowing $A_T$, $A_A$, $u$, $L$ and $D_F$, which require approximation, as follows.

The areas $A_T$ and $A_A$ can be approximated using an idealized lung geometry like those described in Chapter 5. (Note that the areas $A_T$ and $A_A$ can be made time-dependent to allow for lung expansion during inhalation; Taulbee et al. 1978, Egan and Nixon 1985).

For the chosen idealized lung geometry and a given inhalation flow rate, the velocity $u$ in Eq. (7.72) can be approximated as the mean air velocity of the airway cross-section by using simple mass conservation of the air.

To obtain the deposition rate $L$, it is useful to realize that the number of aerosol particles depositing per unit time throughout a given lung generation can be approximated by $\bar{n}QP$, where $Q$ is the air flow rate through the lung generation, and $P$ is the total deposition probability in a given lung generation due to impaction, sedimentation and diffusion (e.g. from Eq. (7.62)). The deposition rate $L$ per unit time and length is then given simply by

$$L = \frac{\bar{n}QP}{l_m} \tag{7.74}$$

where $l_m$ is the length of an airway generation at depth $x$ in the idealized lung geometry being used. Equation (7.74) allows $L$ to be estimated using the approximate expressions for the deposition probabilities $P_i$, $P_d$, $P_s$ that we developed earlier in this chapter.

One of the most empirical aspects of one-dimensional Eulerian models lies in the specification of the effective diffusion coefficient $D_F$ defined in Eqs (7.70) and (7.73). Most authors follow Scherer et al. (1975), who performed experiments on the dispersion of gases in a glass tube model of the first five generations of a Weibel A lung geometry. For dispersion of gases they found $D_F$ had the form

$$D_F = \alpha u d_m \tag{7.75}$$

where $\alpha$ is a factor of order 1 (they found $\alpha = 1.08$ during inhalation, while $\alpha = 0.37$ for exhalation). Also, $u$ is the average air velocity in the $m$th lung generation and $d_m$ is the diameter of this lung generation. For aerosols, various authors have suggested using Eq. (7.75) with various values of $\alpha$ in the range 0.1–1.0, in addition to empirical forms different than Eq. (7.75) (Taulbee and Yu 1975, Taulbee et al. 1978, Darquenne and Pavia 1994, Edwards 1995). Edwards (1995) suggests using

$$D_F = \alpha u l_m - \beta \frac{L}{\bar{n}} l_m^2 \tag{7.76}$$

where $\beta$ is a second empirically specified coefficient, with a value of $\beta = 6$ giving the best agreement in Edwards' (1995) comparison to bolus dispersion experiments. Note that by substituting Eq. (7.74) into Eq. (7.76), this equation can instead be written as

$$D_F = (\alpha u - \beta Q P) l_m \tag{7.77}$$

Given that different velocity profiles occur in different parts of the lung, from Eqs (7.69) and (7.70) we can see that it is not unreasonable to expect different forms of $D_F$ in different parts of the lung (Lee *et al.* 2000b), and indeed Li *et al.* (1998) suggest an alternative equation for $D_F$ in the mouth–throat.

At present, there is little detailed direct experimental evidence for any of these forms for $D_F$ in aerosol deposition models (evidence is usually given in indirect comparisons of dispersion predictions of respiratory tract models where many other assumptions are made, so that direct scrutiny of the validity of Eqs (7.75) or (7.76) is not possible). Further research is needed in this area.

One limitation of standard Eulerian models is the difficulty they have in including two-way coupled hygroscopic effects, since a Lagrangian viewpoint is more natural in predicting these effects. Removal of this limitation is possible using the approach given by Lange and Finlay (2000).

Generalizing Eulerian models to more than one spatial dimension is difficult because inertial impaction is more naturally dealt with by tracking individual particles (using a Lagrangian approach). However, as mentioned earlier, generalizing a purely Lagrangian approach to multiple spatial dimensions is also difficult because of dispersion of the aerosol bolus. For these reasons, future respiratory tract models with more than one spatial dimension may resort to mixed models, in which the fluid phase is treated in a Eulerian manner and the particles are tracked in a Lagrangian manner.

## 7.8 Understanding the effect of parameter variations on deposition

One of the most useful features of respiratory tract deposition models is their ability to show the effect of how different parameters affect deposition in the lung. Indeed, we can obtain a qualitative understanding of these effects simply by examining the equations we developed earlier in this chapter on deposition in simplified geometries. From these equations, it becomes clear that deposition due to the three principal deposition mechanisms (i.e. impaction, sedimentation and diffusion) increases monotonically in a given lung generation with three parameters, as follows.

From Fig. 7.8 we see that the different inertial impaction equations all increase monotonically with Stokes number

$$Stk = \bar{U} \rho_{particle} d^2 C_c / 18 \mu D \tag{7.78}$$

From Figs 7.5 and 7.6 we see that the different sedimentation equations all increase monotonically with the parameter

$$t' = \frac{v_{settling}}{\bar{U}} \frac{L}{D} \tag{7.79}$$

where

$$v_{settling} = C_c \rho_{particle} g d^2 / 18 \mu \tag{7.80}$$

Finally, from Fig. 7.9 we see that the different diffusion equations all increase monotonically with the parameter

$$\Delta = \frac{kTC_c}{3\pi\mu d}\frac{L}{\bar{U}}\frac{1}{4R^2} \tag{7.81}$$

Thus, even though different deposition models may use different equations governing each of the three principal deposition mechanisms, all these deposition models predict that deposition in a given lung generation will increase if $Stk$, $t'$ or $\Delta$ is increased. From this fact, we can draw an understanding of how different variables affect deposition.

Before doing this, however, let us make the flow rate $Q$ appear explicitly in Eqs (7.78)–(7.81), by realizing that flow rate is related to flow velocity $\bar{U}$ and airway diameter $D$ by

$$Q = \bar{U}\pi D^2/4 \tag{7.82}$$

We can thus rewrite Eqs (7.78)–(7.81) as

$$Stk = \frac{C_c}{72\pi\mu}\frac{\rho d^2 Q}{D^3} \tag{7.83}$$

$$t' = \frac{C_c g\pi}{72\mu}\frac{\rho d^2 LD}{Q} \tag{7.84}$$

$$\Delta = \frac{kTC_c}{12\mu}\frac{L}{Qd} \tag{7.85}$$

From these equations we see that particle size (appearing as $d^2$) and airway diameter (appearing as $D^3$) are the only variables that appear to a power other than unity. Thus, of all the variables, these two have the potential to have the largest effect on deposition, since changes in $d$ or $D$ are amplified by being raised to a nonunitary integer power. Indeed, Phalen et al. (1990) show that in their Lagrangian dynamical deposition model, airway dimension is one of the parameters that has the strongest influence on tracheo-bronchial deposition. Since particle size can be controlled when designing a delivery device, while airway dimension cannot, it is for this reason that particle size is the single most important parameter in pharmaceutical aerosol delivery.

Equations (7.83) and (7.84) show that deposition in a given lung generation will increase with particle diameter (since the decrease in diffusional deposition with particle diameter seen in Eq. (7.85) usually plays only a secondary role in deposition of inhaled pharmaceutical aerosols). However, this conclusion is strongly altered by the fact that if more aerosol deposits in one generation, then less aerosol will reach downstream generations. Indeed, because the mouth–throat filters out particles before they reach the tracheo-bronchial region, while the tracheo-bronchial region filters out particles before they reach the alveolar region, actual deposition in the tracheo-bronchial region and alveolar region decreases with particle size for larger particles due to this effect, as will be seen in the next section.

From Eqs (7.83)–(7.85), we see that inhalation flow rate increases impactional deposition, but decreases sedimentational and diffusional deposition. Thus, tracheo-bronchial deposition can increase with inhalation flow rate if the flow rate is high enough that impaction is the dominant mechanism there (although again we must be careful because of the filtering effect of upstream regions, since we saw that mouth–throat deposition increases with flow rate as well, so that less aerosol reaches the tracheo-bronchial region at higher flow rates and it is possible to actually have a decrease in tracheo-bronchial deposition with increased flow rate due to this effect).

The effect of increasing flow rate on alveolar deposition is clearer, since if impaction increases in the mouth–throat and tracheo-bronchial region due to flow rate effects, then less is available to deposit in the alveolar regions, and in addition, Eqs (7.84) and (7.85) show that even less of this available aerosol will deposit since the primary deposition mechanisms in the alveolar region (i.e. sedimentation and diffusion) diminish with increased flow rate.

The effect of changes in airway diameter on deposition is dramatic in Eq. (7.83), since any such changes are raised to the third power there (due to the combination of increased velocity that occurs at fixed flow rate in a smaller diameter tube and changes in the ratio of stopping distance to tube diameter). As a result, impaction in the mouth–throat and tracheo-bronchial airways decreases rapidly if airway dimensions are made larger (e.g. as a person grows through childhood). This allows more aerosol to reach the alveolar region, which acts together with the increase in sedimentation that occurs in Eq. (7.84) as $D$ increases, resulting in dramatic increases in alveolar deposition if airway diameters are made larger.

At first sight, the increase in sedimentation with airway diameter in Eq. (7.84) seems counterintuitive, since an increase in tube diameter will increase the distance a particle must settle before depositing (so that one might think that sedimentation should decrease with increased tube diameter). However, this effect is outweighed by the fact that the particle has more time to deposit in the tube because flow velocity varies inversely with the square of tube diameter.

Equations (7.83)–(7.85) show that increases in airway length cause increases in sedimentational and diffusional deposition, resulting in increased alveolar deposition if the remaining variables are in the usual range where impaction is the dominant deposition mechanism in the conducting airways. Since increases in airway length occur in concert with increases in airway diameter during progression through childhood, in this case, such increases in alveolar deposition with increased airway length would add to the increased alveolar deposition that we have already seen occurs with increases in airway diameter.

## 7.9  Respiratory tract deposition

A principal reason for the existence of respiratory tract deposition models is to provide an understanding of how different factors affect this deposition, as we have just seen. However, such an understanding can also be partly gained by examining experimental data in which deposition of aerosol particles has been measured in human subjects. The largest and most systematic such experimental data sets have been obtained during tidal breathing of subjects breathing monodisperse aerosols delivered via a tube inserted into the mouth, much of which is summarized by Stahlhofen *et al.* (1989) and shown in Figs 7.16–7.19 (from Stahlhofen *et al.* 1989).

Although mouth–throat deposition with pharmaceutical inhalation devices will in general be different than that given in Fig. 7.16 (due to the previously mentioned effect of the presence of the device at the mouth), this figure clearly shows the previously mentioned increase in oropharyngeal (mouth–throat) deposition with particle size and

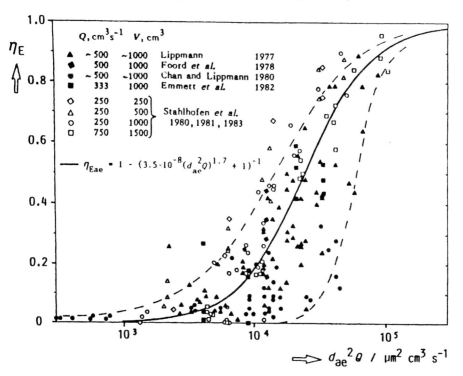

**Fig. 7.16** From Stahlhofen *et al.* (1989). Mouth–throat ('extrathoracic') deposition efficiency (i.e. fraction of inhaled aerosol depositing in mouth–throat) in human subjects measured during mouth breathing shown as a function of $d_{ae}^2 Q$, where $d_{ae}$ is aerodynamic diameter and $Q$ is inhalation flow rate. The solid curve is an empirical fit to the average of all of the datapoints while the dashed lines indicate the approximate range of the data from Lippmann (1977) and Chan and Lippmann (1980). Reprinted with permission.

inhalation flow rate. Indeed, for particles $> 10$ μm, deposition is more than 90% at an inhalation rate of 60 l min$^{-1}$, and more than 50% at an inhalation rate of 18 l min$^{-1}$.

Figures 7.17 and 7.18 suggest that there is a broad maximum in lung (i.e. intrathoracic) deposition (which is near 6 μm for tracheo-bronchial deposition but near 3 μm for alveolar deposition at 30 l min$^{-1}$ inhalation flow rate). These maxima arise because of the combination of effects mentioned in the previous section, where increases in deposition with particle size are offset by increased filtering in upstream regions. It is because of these maxima, and the dramatic increase in mouth–throat deposition at higher particle sizes, that rules-of-thumb regarding optimal particle sizes for inhalation have arisen, with a commonly used such rule-of-thumb being that inhaled pharmaceutical aerosols must be in the 1–5 μm range to reach the lung. However, because of the flow rate dependence of particle deposition and the lack of abrupt cut-offs in actual deposition curves like those in Figs 7.17 and 7.18, such rules-of-thumb should be used with caution (Finlay *et al.* 1997a).

Figure 7.19 shows total respiratory tract deposition, which is the sum of the individual regional deposition fractions shown in Figs 7.16–7.18. The well-known minimum in total deposition seen in this figure for submicron particle sizes occurs because sedimentation and impaction decrease with decreasing particle size, but diffusion increases with

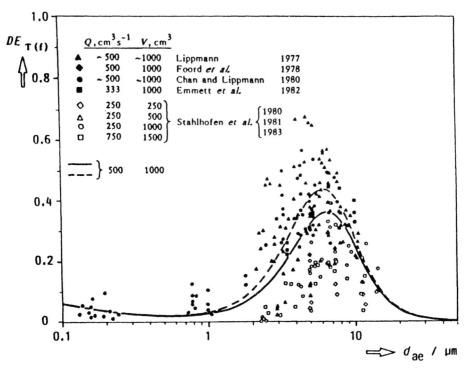

**Fig. 7.17** From Stahlhofen *et al.* (1989). Fast-cleared deposition efficiency (i.e. fraction of inhaled aerosol depositing in the lung that is cleared within a day or so) against aerodynamic particle diameter, measured in human subjects during tidal mouth breathing. Tracheo-bronchial deposition consists of this fast-cleared deposition, plus a small (unknown) portion of the slow-cleared deposition in Fig. 7.18 The solid curve is an empirical fit to the average of all of the datapoints, while the dashed line is an average of the data from Lippman (1977) and Chan and Lippmann (1980), both with tidal volume of 1 l, inhalation flow rate of 30 l min$^{-1}$. Reprinted with permission.

decreasing particle size, resulting in a crossover between these different deposition mechanisms where the minimum in total deposition occurs.

### 7.9.1 Slow-clearance from the tracheo-bronchial region

One of the principal difficulties in understanding experimental data on deposition within the lung (i.e. intrathoracic deposition) in human subjects is the difficulty of obtaining accurate measures of deposition in the different anatomical regions of the lung. This difficulty arises because the resolution of radionuclide imaging methods used in previous such experiments does not allow accurate discrimination of individual airways, so that quantitative mapping of the radioactivity onto anatomical airway generations, for example, has not been possible. Although progress in this direction has occurred (Fleming *et al.* 1995, 2000, Lee *et al.* 2000a), at present, experimental data cannot accurately determine deposition at the generational level throughout the lung. Instead, the most commonly used approach has been to measure the so-called 'fast-cleared' and 'slow-cleared' fractions of lung deposition like those shown in Figs 7.17 and 7.18. The

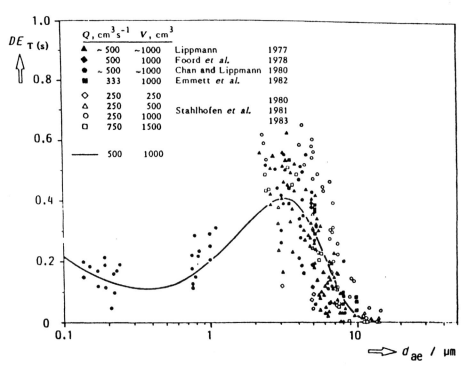

**Fig. 7.18** From Stahlhofen *et al.* (1989). Slow-cleared deposition efficiency (i.e. fraction of inhaled aerosol depositing in the lung that isn't cleared within a day or so) against aerodynamic particle diameter, measured in human subjects during tidal mouth breathing. The majority of this deposition is alveolar deposition, but a small (unknown) portion is due to tracheo-bronchial deposition. The solid curve is an empirical fit to the average of the data assuming a tidal volume of 1 l, inhalation flow rate of 30 l min$^{-1}$. Reprinted with permission.

rationale for these measurements is based on the presence of cilia in the tracheo-bronchial airways, and the absence of cilia in the alveolar regions. As a result, mucociliary clearance causes clearance of most particles depositing within the tracheo-bronchial region within one day or so, while particles depositing in the alveolated regions are cleared much more slowly. Thus, alveolar deposition has commonly been taken to be equal to the 'slow-cleared' fraction, while tracheo-bronchial deposition is obtained by subtracting this slow-cleared fraction from the initial total lung deposition (giving the 'fast-cleared' fraction, which has often been equated with tracheo-bronchial deposition).

Unfortunately, there is evidence that some particles depositing in the tracheo-bronchial region are actually cleared slowly (see ICRP 1994 for a good review of this evidence). Indeed, in the clearance model proposed by the ICRP, a fraction of particles depositing in the tracheo-bronchial region is cleared slowly, as shown in Fig. 7.20. It should be noted that considerable uncertainty remains as to the amount of slow-clearance that occurs, as is seen by the large confidence interval associated with the values proposed in the ICRP (1994) model.

Figure 7.20 suggests that a reasonable portion of the 'slow-cleared' deposition in Fig. 7.18 in the range of interest for inhaled pharmaceutical aerosols (e.g. 1–5 µm) is actually due to tracheo-bronchial deposition, particularly for the smaller particles. However, because of their small mass (so that impaction and sedimentation are small), but not too

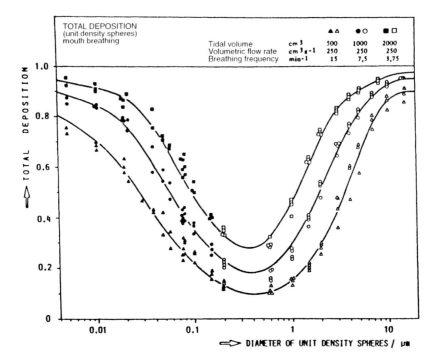

**Fig. 7.19** From Stahlhofen *et al.* (1989). Total respiratory tract deposition as a function of the diameter of unit density spheres for several tidal volumes and breathing frequencies, measured in human subjects during tidal mouth breathing. Reprinted with permission.

small diameter (so that diffusion is still small), these smaller particles have relatively low deposition probabilities in the tracheo-bronchial region compared to the alveolar region. As a result, the actual amount of slow-cleared aerosol from the tracheo-bronchial region is relatively small compared to alveolar deposition, and the correction that is needed to equate slow-cleared fractions with alveolar deposition is relatively minor (ICRP 1994). Instead, the principal difficulty arises for particles with diameter $<5$ μm when equating the fast-cleared fraction with tracheo-bronchial deposition, since for these particles, tracheo-bronchial deposition can be significantly underestimated by the fast-cleared fraction (e.g. by an average factor of 0.5 for particles with diameter $<2.5$ μm according to the ICRP clearance model).

The exact mechanism of slow-clearance from the tracheo-bronchial airways is still not fully understood, but several factors have been suggested, including burrowing of particles in the epithelium associated with minimizing their interfacial free energy (cf. Gehr *et al.* 1990), ingestion of particles by macrophages (Stirling and Patrick 1980), and 'also the fact that not all of the tracheobronchial airway surface is lined with ciliated epithelium . . . or, that not all of the ciliated epithelium is covered with mucus all the time' (Stalhofen *et al.* 1989). A possible way to circumvent several of these mechanisms in experiments with humans may be to use water droplets (instead of solid particles) containing radiolabeled ultrafine particles, as proposed by Finlay *et al.* (1998a), although further research would be useful to verify to what extent slow-clearance from the tracheo-bronchial region can be prevented in this manner.

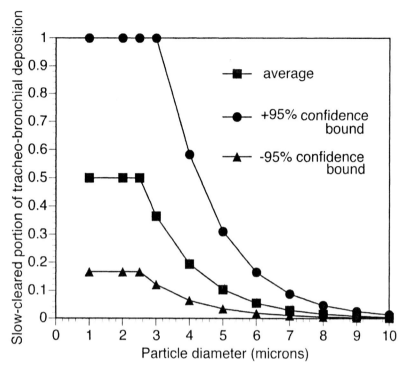

**Fig. 7.20** The average fraction $f$ of tracheo-bronchial deposition that is 'slow-cleared' in the ICRP (1994) clearance model, given by $f = 0.5\,e^{-0.63(d-2.5)}$ for $d > 2.5$ μm and $f = 0.5$ for $d = 2.5$ μm, where $d$ is in microns, is shown. Particle diameter here is the diameter of a sphere having the same volume as the particle (i.e. volume equivalent diameter). The upper and lower curve bound the suggested 95% confidence interval in the ICRP model (which are 3 times and 1/3 times the average value).

### 7.9.2 Intersubject variability

Despite the uncertainty in our ability to interpret the experimental data on regional deposition of inhaled particles (because of our uncertainty in clearance rates), one fact is made very clear by these data: for a given aerosol, there is tremendous variation from individual to individual in the amount of aerosol that will deposit in the different regions of the respiratory tract. Figures 7.16–7.18 amply demonstrate this fact. Although empirical formulas for this variation have been developed for tidal breathing through tubes (Rudolf *et al.* 1990, Stapleton 1997), methods for predicting this variation based on the underlying dynamics and physics have not been developed. This is understandable given that the factors responsible for this intersubject variability remain poorly characterized. Based on deposition experiments in humans inhaling aerosols from tubes inserted in the mouth, the two major factors responsible for this variability are variations between individuals in breathing pattern (Bennett 1988) and lung geometry (Heyder *et al.* 1988). Each factor is certainly responsible for some of this variation, although which is the major contributor remains a topic of discussion. Indeed, it is probable that different parts of the lung are affected more by one of these two factors than the other, with variations in deposition in the larger airways possibly being more due to intersubject variations in lung geometry, while variability in deposition in the peripheral lung regions may be more due to variations in breathing pattern (Bennett 1990).

Regardless of the reason for the large intersubject variability, dynamical respiratory deposition models usually predict only average values of deposition. Use of deposition models should be tempered with this realization.

### 7.9.3 Comparison of models with experimental data

Most deposition models based on the concepts described earlier in this chapter give predictions that are in reasonable agreement with data similar to those shown in Figs 7.17, 7.18 and 7.19, although the uncertainty in clearance rates in the tracheo-bronchial region make comparisons of regional deposition more difficult. This difficulty can be removed by including the slow-clearance fraction shown in Fig. 7.20 in the model predictions, allowing direct comparison of slow-cleared and fast-cleared fractions in deposition models with experimentally measured values like those shown in Figs 7.17 and 7.18. Such a comparison is presented in ICRP (1994) and shown in Figs 7.21 and 7.22, where the deposition model predictions are from the Eulerian dynamical model of Egan *et al.* (1989).

Figures 7.21 and 7.22 are typical of the many comparisons of models to experiment that exist in the literature in that reasonable agreement is seen between model and

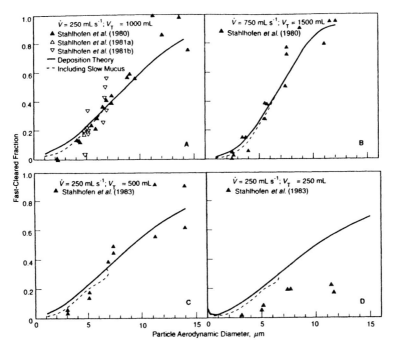

**Fig. 7.21** Reprinted from ICRP (1994) with permission. Comparisons of fast-cleared fraction of lung deposition in human subjects measured experimentally (symbols) and predicted tracheo-bronchial deposition (solid line) and fast-cleared (dashed line, incorporating the slow-cleared fraction from Fig. 7.20) using the deposition model of Egan *et al.* (1989) at various flow rates ($\dot{V}$) and tidal volumes ($V_T$). In contrast to Figs 7.16–7.19, note that the fractions here are given as fractions of aerosol entering the trachea, not as fractions of inhaled aerosol.

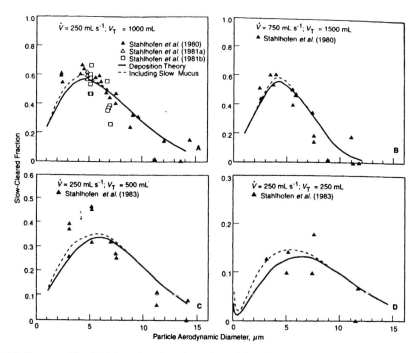

**Fig. 7.22** Same as Fig. 7.21 but showing slow-cleared fraction. Reprinted from ICRP (1994) with permission.

experimental data. Although these comparisons provide some confidence in these deposition models, existing such models lack the ability to predict deposition with many inhaled pharmaceutical aerosols because of the previously mentioned lack of models for predicting mouth–throat deposition when inhalation occurs with different inhalation devices in the mouth. Indeed, Clark and Egan (1994) find a model based on that of Egan *et al.* (1989) is unable to adequately predict lung deposition with various dry powder inhalers, underpredicting mouth–throat deposition and overpredicting lung deposition by a factor of 1.5–2. This is partly explained by DeHaan and Finlay (1998, 2001) as being the result of enhanced deposition occurring with these inhalers due to fluid dynamics that is not accounted for by existing empirical mouth–throat models based on inhalation of laminar flow from straight tubes (such as Eq. (7.65), which Clark and Egan 1994 use).

Because nebulizers often give mouth–throat fluid dynamics that more closely resembles that occurring during inhalation from a tube, more success has been had with dynamical deposition models in predicting deposition with nebulizers than dry powder inhalers or metered dose inhalers (Finlay *et al.* 1996a, 1997b, 1998a,b). However, as we have seen earlier in this chapter, there are a number of simplifying assumptions made in developing such deposition models, so that further research would be useful to fully understand the importance of these assumptions under a variety of circumstances. Indeed, Finlay *et al.* (2000) show that the choice of idealized lung geometry can have a large effect on regional lung deposition predictions with nebulized aerosols, with the Weibel A lung geometry tending to overestimate tracheo-bronchial deposition and underestimate alveolar deposition compared to more recent lung geometries, largely because of the smaller diameters of the tracheo-bronchial airways in the Weibel A geometry (leading to higher flow velocities and overestimation of impaction, as was seen in our examination of Eq. (7.83)). This may partly explain the data of Hashish *et al.* (1998), who find similar over- and under-prediction in their comparison of model data using a Weibel A lung model to experimental data with nebulized aerosols.

## 7.10 Targeting deposition at different regions of the respiratory tract

By combining observations from experiments like that presented above with deposition model predictions, it is possible to achieve a reasonably good understanding of how different parameters affect respiratory tract deposition. With this understanding it is possible to have some control over where in the respiratory tract an aerosol will deposit. Such targeting of the respiratory tract can be of importance with therapeutic agents where efficacy is thought to depend in part on where the drug deposits in the lung, such as with drugs intended for systemic uptake into the blood through the alveolar epithelium, or with antimicrobial agents delivered to regional sites of infection. It should be noted though that such targeting will be fairly broad, for a number of reasons.

First, it is not possible to adequately control the regions to which air carries aerosol in the lung (i.e. regional ventilation is not controllable in any precise way, due to a combination of the stochastic nature of the lung, dispersion, chaotic mixing and diffusion). For example, we cannot target only the tracheo-bronchial region since the pathways to some terminal bronchioles are much shorter than others and will start to fill alveolar regions before the air has even reached the terminal bronchioles in other pathways. (Although targeting of the first few lung generations can be achieved by delivering aerosol only during the very early part of a breath (Scheuch and Stahlhofen 1988), such targeting becomes less precise in the smaller airways for the above mentioned reasons.) We also cannot have air reach only the alveolar region since it must first travel through the tracheo-bronchial region and the mouth–throat.

Second, the factors that affect deposition change gradually from region to region in the lung. Thus, for example, if we choose particles that deposit mainly by sedimentation, such particles will deposit mainly in the alveolar regions. However, as we saw in Fig. 7.7, sedimentation is also operational in the small bronchioles, so that we cannot entirely avoid deposition in the tracheo-bronchial region in this manner.

Finally, variations between subjects (i.e. intersubject variability) in the parameters that control deposition will also broaden the regions actually reached by attempts at targeting specific lung regions.

Despite these difficulties, broad targeting of aerosol is possible. In particular, it is clear that if we increase inhalation flow rate, mouth–throat deposition will increase according to Fig. 7.16, while the total dose to the lung will decrease. In addition, if flow rate is already relatively high (so that Fig. 7.7 indicates impaction is the dominant deposition mechanism in the tracheo-bronchial region), then further increases in flow rate will increase impaction in the tracheo-bronchial region so that whatever aerosol does manage to make it past the mouth–throat will tend to deposit in the tracheo-bronchial region[1]. Conversely, reducing inhalation flow rate will shift deposition away from the mouth–throat and into the lung. If inhalation flow rate is very low, then impaction in the larger airways can be mostly avoided, resulting in small airway and/or alveolar deposition, depending on particle size (Camner *et al.* 1997). It should be realized that it is probably unrealistic to expect patients to consistently control inhalation flow rate very precisely on their own, so that the use of specific inhalation flow rates for targeting is probably best achieved with a delivery device that either delivers aerosol only when the patient supplies a certain flow rate range (Schuster *et al.* 1997) or that causes patients to supply such flow rates (perhaps through feedback).

Probably the most commonly used approach to targeting aerosol deposition is through particle size. We have seen clearly throughout this chapter how important particle size is in determining deposition, with larger particles impacting and sedimenting more readily than smaller ones, while smaller ones deposit more readily by diffusion. We saw that increasing particle size causes increased mouth–throat deposition, resulting in less aerosol reaching the lung. Within the lung, Figs 7.17 and 7.18 show that during tidal breathing at $30\,l\,min^{-1}$ and $1\,l$ tidal volumes, alveolar deposition is maximized at smaller particle sizes (near $3\,\mu m$) than is tracheo-bronchial deposition (near $6\,\mu m$). However, the maximum in deposition vs. particle diameter plot is quite broad so that when one of these regional depositions is maximized, the other is still near 50% of its maximum (Figs 7.17 and 7.18). Thus, although some targeting can be achieved using particle size (and indeed, this is the principal approach used in existing pharmaceutical inhalation devices, particularly when targeting the alveolar region by using particle sizes near $1-3\,\mu m$), it should be realized that such targeting is typically quite broad.

Flow rate and particle size targeting are usually the most common means for achieving targeting of inhaled pharmaceutical aerosols. However, inhalation volume can affect deposition, since for smaller volumes the conducting airways occupy more of the inhaled volume, resulting in a relative shift toward more tracheo-bronchial than alveolar deposition at low inhalation volumes, and an opposite shift at higher lung volumes (particularly with breath-holding so that all aerosol reaching the alveolar regions deposits there).

In addition, particle density affects deposition, since impaction and sedimentation are affected by $pd^2$ in Eqs (7.83) and (7.84), so that aerodynamic particle size, $d_{ae}$, rather than particle size itself governs these mechanisms, as discussed in Chapter 3. Thus, the use of porous particles in dry powder inhalers (Edwards *et al.* 1997) allows targeting via particle density reductions in a way that parallels particle size targeting.

Airway dimensions also strongly affect deposition (as already mentioned in discussing Eqs (7.83)–(7.85)), with smaller airway diameters leading to enhanced tracheo-bronchial deposition due to increased flow rates and impaction (Kim *et al.* 1983, Martonen *et al.* 1995a), although control over airway dimensions is probably not a practical approach to targeting. However, alterations in airway dimensions can occur in a given patient due to changes in disease state or age, which can change the intended target of an inhaled aerosol, as discussed in the following sections.

One of the most helpful uses of aerosol deposition models is in aiding methods for targeting aerosols in the respiratory tract in pharmaceutical aerosol research and

---

[1] At lower flow rates, impaction and sedimentation respond oppositely to flow rate in the tracheo-bronchial region as seen in Eqs (7.83) and (7.84) since sedimentation decreases with increases in flow rate due to reduced residence times, while impaction increases with flow rate. As a result, flow rate targeting is not as effective at moderate flow rates where neither impaction nor sedimentation is dominant.

development. Indeed, serious attempts at targeting are often guided by deposition models since these models allow quantitative exploration of the effects of different variables on respiratory tract deposition targeting in a manner that is much more detailed than is possible with the general discussion presented above. When deposition models are combined with models for estimating the thickness of the mucus in the individual tracheo-bronchial airways, it is possible to estimate the concentration of drug in the mucus in the various lung generations (Finlay *et al.* 2000a), which can be useful in the design of delivery systems for drugs intended for local action in the airways.

## 7.11 Deposition in diseased lungs

Most of our knowledge of respiratory tract aerosol deposition has been obtained from data on subjects and deposition models where the lungs are 'normal', i.e. not altered by any disease state. However, many existing inhaled pharmaceutical aerosols are intended to be used in the treatment of various lung diseases in which the lung passages are altered by the presence of the disease being treated. Unfortunately, our current understanding of the dynamics of aerosol deposition in diseased lungs remains in its infancy, largely due to the inadequacy of our knowledge of the detailed geometry of diseased lungs (as mentioned in Chapter 5), and our inability to measure deposition at a detailed enough level in human subjects. However, a number of studies have been done that indicate certain affects that are worth mentioning here.

First, obstructions in the airways can dramatically increase deposition, particularly in the larger airways where the Reynolds number is high (so that separated, possibly turbulent, flow patterns can occur with such obstructions). For example, in experiments in obstructed straight glass tubes at $Re = 140$–$2800$ (which are Reynolds numbers typically occurring proximal to the segmental bronchi), Kim *et al.* (1984) find that three different types of obstructions all give similar increases in deposition (increasing to levels that are orders of magnitude higher than the unobstructed case), with deposition correlating with pressure drop across the obstruction. Also at high Reynolds numbers, excessive mucus secretions may dramatically enhance deposition, due to two-phase turbulent structure interactions mentioned in Chapter 6, as found in glass tubes by Kim and Eldridge (1985) and in sheep (Kim *et al.* 1985). Kim *et al.* (1989) suggest that excess mucus secretions throughout the conducting airways may be responsible for the 33% increase in total respiratory tract deposition they observed in sheep given pilocarpine intravenously (to induce such secretions).

Besides the direct effect of causing enhanced local deposition, airway obstructions can also alter ventilation patterns in the lung, so that the region of the lung distal to an obstruction receives less aerosol than other regions, potentially leading to reduced amounts depositing in such distal regions (Kim *et al.* 1983, 1989). In addition, because the obstruction causes reduced flow rates in its distal regions, impaction in distal generations may be reduced. However, sedimentation may increase due to increased residence times, which can partly counterbalance this latter effect.

In addition to static airway obstructions, it is possible for transient airway obstructions to occur, even in lungs without static obstructions, since it is well known that partial collapse of large airway walls can occur during coughing in normals and during normal expiration in some lung diseases due to abnormal pressure differences between the inside and outside of the airways (combined with altered material properties of the airway walls in the case of diseased lungs). Such a constriction of an airway is called a 'flow-limiting segment', and can result in significant enhancement of particle deposition in these segments (Smaldone and Messina 1985), presumably due to similar effects to those occurring in static, obstructed tubes mentioned above.

Flow-limiting segments, mucus secretions and static airway obstructions have been proposed as the principal causes of the patchy, central deposition patterns seen in scintigraphic studies of diseased subjects during tidal breathing of aerosols. It should be noted that with most inhaled pharmaceutical aerosols (excluding nebulizers), there is little deposition on exhalation because of breath-holding, so that such 'flow-limiting' effects may not be important for typical single-breath inhalation devices, so that static obstructions and mucus secretions may be the dominant mechanisms causing deposition with these devices in diseased subjects to be different from that seen in normals.

Unfortunately, our understanding of diseased lungs is not yet sufficient to allow predictive dynamical modeling of the above effects in determining regional respiratory tract deposition. In addition, the development of empirical models based on regional anatomical deposition is hampered by the lack of knowledge of the effect of disease on slow-clearance fractions from the tracheo-bronchial region (mucociliary clearance is often altered in a disease-specific manner, sometimes making the extrapolation of data on mucociliary clearance in normals inappropriate).

Despite our present lack of knowledge on the dynamics of regional deposition in disease, quantitative data on total deposition quite clearly shows that disease can have a strong effect on deposition. For example, Kim *et al.* (1988) found that total deposition of 1 μm particles inhaled over five breaths in subjects with chronic obstructive pulmonary disease was 31% ± 9% of the value seen in normals.

The development of a mechanistic understanding of deposition in diseased lungs awaits the gathering of data on lung geometry and the dynamics of such lungs. This task is made difficult by the differing effects of different diseases on the lung, and will likely occupy future researchers for some time to come.

## 7.12 Effect of age on deposition

We have already noted that airway dimensions and length are important in determining deposition in the lung, as is clearly seen in Eqs (7.83)–(7.85). With this in mind, and the knowledge that these dimensions change considerably during growth from birth to adulthood, then it is logical to suggest that deposition in children is different from that in adults. Although experimental data on deposition in children is much scarcer than in adults because of radiological exposure concerns in pediatric scintigraphic studies, the available data do indeed suggest that total deposition is higher in children than in adults, nearly doubling at age 5 vs. 25 years even if inhalation volume, flow rate and particle size are the same (Yu *et al.* 1992).

Because of the scarcity of experimental data, our present understanding of regional deposition in children is based on deposition models in which the idealized lung geometry has been altered to reflect measured morphometric data on casts of pediatric lungs (Hofmann 1982, Phalen *et al.* 1985, Xu and Yu 1986, Hofmann *et al.* 1989, ICRP 1994, Finlay *et al.* 1999). The effect of age on regional deposition from one such model is shown in Figs 7.23 and 7.24 for tidal breathing.

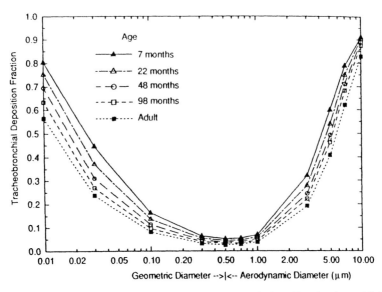

**Fig. 7.23** Reprinted from Hofmann *et al.* (1989) with permission. Tracheobronchial deposition (given as a fraction of the aerosol entering the trachea) is shown as a function of particle size and age as calculated by a Lagrangian dynamical model.

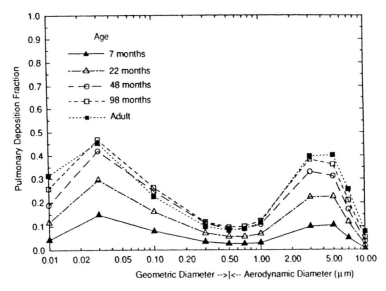

**Fig. 7.24** Reprinted from Hofmann *et al.* (1989) with permission. Alveolar deposition (given as a fraction of the aerosol entering the trachea) is shown as a function of particle size and age as calculated by a Lagrangian dynamical model.

Figures 7.23 and 7.24 show the large effect that age-related airway dimension changes have on regional deposition during tidal breathing. For particles of interest in pharmaceutical aerosols (i.e. $> 1$ μm), tracheo-bronchial deposition decreases with age largely due to the inverse cubic relation with airway diameter in Eq. (7.83), while alveolar deposition increases with age largely due to the quadratic relation with airway dimensions in the numerator of Eq. (7.84). These changes are made even larger when age-related changes in mouth–throat deposition are included. (Figs 7.23 and 7.24 show deposition as a fraction of the aerosol entering the trachea rather than the mouth.) In fact, deposition in the mouth and throat can be considerably larger in children than adults because of increased impaction in the smaller pediatric mouth–throat (this is reflected, for example, in the empirical scaling factor suggested in ICRP 1994 to modify Eq. (7.65) for use in children). Thus, although total deposition may be higher in children, amounts depositing in the lung can actually be lower even if the same amount of aerosol is inhaled as in adults. This partially explains the findings in clinical studies with pharmaceutical inhalation devices in children (Chrystyn 1999, Dolovich 1999) where much lower lung doses (as a percentage of the drug initially in the device) are usually seen in young children than in adults, although part of this reduction is instead due to lower amounts of aerosol inhaled by young children (depending on device type) in *in vivo* studies.

In addition to differences caused by smaller airway dimensions, inhalation flow rates are generally lower in children than adults (see Chapter 5). From Eqs (7.83)–(7.85) we see that flow rate changes affect deposition in an opposite manner to airway dimension changes. However, for impactional deposition (Eq. (7.83)), airway dimensions appear to the third power, while flow rate appears only linearly, while for sedimentational deposition (Eq. (7.84)), airway dimensions appear quadratically (via the product $LD$), while again flow rate appears only to the first power. Thus, although decreased flow rates tend to counter the effect of decreased airway dimensions, the effects of airway dimension changes with age apparently outweigh flow rate effects, as is seen in Figs 7.23 and 7.24.

## 7.13 Conclusion

By analyzing aerosol deposition in simplified geometries that resemble parts of the respiratory tract, we have seen that relatively simple deposition models can be developed, from which a good understanding of deposition in the respiratory tract can be produced. Such models produce results that are in generally good agreement with experimental results, and can be useful in the development of inhaled pharmaceutical aerosol devices, particularly when targeting of such aerosols is important. However, there are a number of simplifying assumptions that go into these models, most of which we have pointed out in this chapter, that limit such models in their applicability. The most severe of these limitations from the point of view of inhaled pharmaceutical aerosols is the inability of these models to predict deposition in diseased lungs. In addition, such models have not been validated in young children in their regional deposition predictions. These are topics for future research.

## References

Abramowitz, M. and Stegun, I. A. (1981) *Handbook of Mathematical Functions*, Dover, NY.
Asgharian, B. and Anjilvel, S. (1994) A Monte Carlo calculation of the deposition efficiency of inhaled particles in lower airways, *J. Aerosol Sci.* **4**:711–721.

Balásházy, I. and Hofmann, W. (1993a) Particle deposition in airway bifurcations – I. Inspiratory Flow, *J. Aerosol Sci.* **24**:745–772.

Balásházy, I. and Hofmann, W. (1993b) Particle deposition in airway bifurcations – II. Expiratory Flow, *J. Aerosol Sci.* **24**:773–786.

Balásházy, I. and Hofmann, W. (1995) Deposition of aerosols in asymmetric airway bifurcations, *J. Aerosol Sci.* **26**:273–292.

Balásházy, I., Martonen, T. B. and Hofmann, W. (1990a) Inertial impaction and gravitational deposition of aerosols in curved tubes and airway bifurcations, *Aerosol Sci. Technol.* **13**:303–321.

Balásházy, I., Martonen, T. B. and Hofmann, W. (1990b) Simultaneous sedimentation and impaction of aerosols in two-dimensional channel bends, *Aerosol Sci. Technol.* **13**:20–34.

Balásházy, I., Hofmann, W. and Martonen, T. B. (1991) Inspiratory particle deposition in airway bifurcation models, *J. Aerosol Sci.* **22**:15–30.

Beeckmans, J. M. (1965) The deposition of aerosols in the respiratory tract, *Can. J. Physiol. Pharm.* **43**:157–172.

Bennett, W. (1988) Human variation in spontaneous breathing deposition fraction: a review, *J. Aerosol Med.* **1**:67–80.

Bennett, W. (1990) Response to Joachim Heyder's letter on 'Intersubject variability of intrapulmonary deposition', *J. Aerosol Med.* **3**:218–220.

Brockmann, J. (1993) Sampling and transport of aerosols, in K. Willeke and P. A. Baron, eds, *Aerosol Measurement, Principles, Techniques and Applications*, Van Nostrand Reinhold, NY.

Buchwald, E. (1921) *Ann. Physik* **66**:1.

Cai, F. S. and Yu, C. P. (1988) Inertial and interceptional deposition of spherical particles and fibers in a bifurcating airway, *J. Aerosol Sci.* **19**:679–688.

Camner, P., Anderson, M., Philipson, K., Bailey, A., Hashish, A., Jarvis, N., Bailey, M. and Svartengren, M. (1997) Human bronchiolar deposition and retention of 6-, 8- and 10-μm particles, *Exp. Lung Res.* **23**:517–535.

Chan, T. L. and Lippmann, M. (1980) Experimental measurement and empirical modelling of the regional deposition of inhaled particles in humans, *Am. Ind. Hyg. Assoc. J.* **41**:399–409.

Chrystyn, H. (1999) Anatomy and physiology in delivery: can we define our target?, *Allergy* **54**:82–87.

Clark, A. R. and Egan, M. (1994) Modelling the deposition of inhaled powdered drug aerosols, *J. Aerosol Sci.* **25**:175–186.

Cohen, B. S. and Asgharian, B. (1990) Deposition of ultrafine particles in the upper airways: An empirical analysis, *J. Aerosol Sci.* **21**:789–797.

Darquenne, C. and Pavia, M. (1994) One-dimensional simulation of aerosol transport and deposition in the human lung, *J. Appl. Physiol.* **77**:2889–2898.

Darquenne, C. and Pavia, M. (1996) Two- and three-dimensional simulations of aerosol transport and deposition in alveolar zone of human lung, *J. Appl. Physiol.* **80**:1401–1414.

Davies, C. N. (1982) Deposition of particles in the human lung as a function of particle size and breathing pattern: an empirical model, *Ann. Occup. Hyg.* **26**:119–135.

Davis, A. M. J. (1993) Periodic blocking in parallel shear or channel flow at low Reynolds number, *Phys. Fluids* **5**:800–809.

DeHaan, W. H. and Finlay, W. H. (1998) Effect of mouthpiece on aerosol deposition in the mouth and throat, in *Respiratory Drug Delivery VI*, Hilton Head Island, South Carolina, May 3–7, 1998. Interpharm Press, pp. 307–309.

DeHaan, W. H. and Finlay, W. H. (2001) In vitro monodisperse aerosol deposition in a mouth and throat with six different inhalation devices, *J. Aerosol Med.* (in press).

Dolovich, M. (1999) Aerosol delivery to children, *Ped. Pulm. Supplement* **18**:79–82.

Edwards, D. (1995) The macrotransport of aerosol particles in the lung: aerosol deposition phenomena, *J. Aerosol Sci.* **26**:293–317.

Edwards, D. A., Hanes, J., Caponetti, G., Hrkach, J., Lotan, N., Ben-Jebria, A. and Langer, R. (1997) Large porous biodegradable particles for pulmonary drug delivery, *Science* **276**:1868–1871.

Egan, M. J. and Nixon, W. (1985) A model of aerosol deposition in the lung for use in inhalation dose assessments, *Rad. Prot. Dos.* **11**:5–17.

Egan, M. J., Nixon, W., Robinson, N. I., James, A. C. and Phalen, R. T. (1989) Inhaled aerosol transport and deposition calculations for the ICRP Task Group, *J. Aerosol Sci.* **20**:1305–1308.

Emmett, P. C., Aitken, R. J. and Hannan, W. J. (1978) Measurements of the total and regional deposition of inhaled particles in the human respiratory tract, *J. Aerosol Sci.* **13**:549–560.

Ferron, G. A., Kreyling, W. G. and Haider, B. (1988) Inhalation of salt aerosol particles -- II. Growth and deposition in the human respiratory tract, *J. Aerosol Sci.* **19**:611–631.

Findeisen, W. (1935) Über das Absetzen kleiner, in der Luft suspendierter Teilchen in der menschlichen Lunge, *Pflüger Arch. F.d. ges. Physiol.* **236**:367–379.

Finlay, W. H. and Stapleton, K. W. (1995) The effect on regional lung deposition of coupled heat and mass transfer between hygroscopic droplets and their surrounding phase, *J. Aerosol Sci.* **26**:655–670.

Finlay, W. H., Stapleton, K. W., Chan, H. K., Zuberbuhler, P. and Gonda, I. (1996a) Regional deposition of inhaled hygroscopic aerosols: in vivo SPECT compared with mathematical deposition modelling, *J. Appl. Physiol.* **81**:374–383.

Finlay, W. H., Stapleton, K. W. and Yokota, J. (1996b) On the use of computational fluid dynamics for simulating flow and particle deposition in the human respiratory tract, *J. Aerosol Med.* **9**:329–342.

Finlay, W. H., Stapleton, K. W. and Zuberbuhler, P. (1997a) Fine particle fraction as a measure of mass depositing in the lung during inhalation of nearly isotonic nebulized aerosols, *J. Aerosol Sci.* **28**:1301–1309.

Finlay, W. H., Stapleton, K. W. and Zuberbuhler, P. (1997b) Predicting lung dosages of a nebulized suspension: Pulmicort [R] (Budesonide), *Particulate Sci. Technol.* **15**:243–251.

Finlay, W. H., Hoskinson, M. and Stapleton, K. W. (1998a) Can models be trusted to subdivide lung deposition into alveolar and tracheobronchial fractions?, in *Respiratory Drug Delivery VI*, Hilton Head Island, South Carolina, May 3–7. Interpharm Press, pp. 235–242.

Finlay, W. H., Stapleton, K. W., Zuberbuhler, P. (1998b) Variations in predicted regional lung deposition of salbutamol sulphate between 19 nebulizer models, *J. Aerosol Med.* **11**:65–80.

Finlay, W. H., Lange, C. F., Li, W.-I. and Hoskinson, M. (2000) Validating deposition models in disease: what is needed?, *J. Aerosol Med.* **13**:381–385.

Finlay, W. H., Lange, C. F., King, M. and Speert, D. (2000a) Lung delivery of aerosolized dextran, *Am. J. Resp. Crit. Care Med.* **160**:1–7.

Fleming, J. S., Nassim, M. A., Hashish, A. H., Bailey, A. G., Conway, J., Holgate, S. T., Halson, P., Moore, E. and Martonen, T. B. (1995) Description of pulmonary deposition of radiolabelled aerosol by airway generation using a conceptual three-dimensional model of lung morphology, *J. Aerosol Med.* **3**:341–356.

Fleming, J. S., Conway, J. H., Holgate, S. T., Bailey, A. G. and Martonen, T. B. (2000) Comparison of methods for deriving aerosol deposition by airway generation from three-dimensional radionuclide imaging, *J. Aerosol Sci.* **31**:1251–1259.

Foord, N., Black, A. and Walsh, M. (1978) Regional deposition of 2.5–7.5 µm diameter inhaled particles in healthy male non-smokers, *J. Aerosol Sci.* **9**:323–357.

Fuchs, N. A. (1964) *The Mechanics of Aerosols*, p. 264, Dover, New York.

Gehr, P., Schürch, S., Berthiaume, Y., Im Hof, V. and Geiser, M. (1990) Particle retention in airways by surfactant, *J. Aerosol Med.* **3**:27–43.

Gerrity, T. R., Lee, P. S., Hass, F. J., Marinelli, A., Werner, P. and Lourenço, R. V. (1979) Calculated deposition of inhaled particles in the airway generations of normal subjects, *J. Appl. Physiol.* **47**:867–873.

Gormley, P. G. and Kennedy, K. (1949) Diffusion from a stream flowing through a cylindrical tube, *Proc. Roy. Irish Soc.* **52A**:163.

Gradshteyn, S. and Ryzhik, I. M. (1980) *Table of Integrals, Series and Products*, Academic Press, New York.

Gurman, J. L., Lippman, M. and Schlesinger, R. B. (1984) Particle deposition in replicate casts of the human upper tracheobronchial tree under constant and cyclic inspiratory flow. I. Experimental, *Aerosol Sci. Technol.* **1**:245–252.

Hashish, A. H., Fleming, J. S., Conway, J., Halson, P., Moore, E., Williams, T. J., Bailey, A. G., Nassim, J. and Holgate, S. T. (1998) Lung deposition of particles by airway generation in healthy subjects: three-dimensional radionuclide imaging and numerical model prediction, *J. Aerosol Sci.* **29**:205–215.

Heyder, J. (1975) Gravitational deposition of aerosol particles within a system of randomly oriented tubes, *J. Aerosol Sci.* **6**:133–137.

Heyder, J. and Gebhart, J. (1977) Gravitational deposition of particles from laminar aerosol flow through inclined circular tubes, *J. Aerosol Sci.* **8**:289–295.

Heyder, J. and Rudolf, G. (1984) Mathematical models of particle deposition in the human respiratory tract, in *Lung Modelling for Radioactive Materials*, pp. 17–38, eds. H. Smith and G. Gerber, EUR 9834 EN. Commission of the European Communities, Luxembourg.

Heyder, J., Gebhart, J. and Scheuch, G. (1985) Interaction of diffusional and gravitational transport in aerosols, *Aerosol Sci. Tech.* **4**:315–326.

Heyder, J., Gebhart, J. and Scheuch, G. (1988) Influence of human lung morphology on particle deposition, *J. Aerosol Med.* **1**:81–88.

Hofmann, W. (1982) Mathematical model for the postnatal growth of the human lung, *Respir. Physiol.* **49**:115–367.

Hofmann, W., Martonen, T. B. and Graham, R. C. (1989) Predicted deposition of nonhygroscopic aerosols in the human lung as a function of subject age, *J. Aerosol Med.* **2**:49–68.

ICRP Task Group on Lung Dynamics (1966) Deposition and retention models for internal dosimetry of the human respiratory tract, *Health Phys.* **12**:173–207.

ICRP (International Commission on Radiolocial Protection) (1994) Human respiratory tract model for radiological protection, *Annals of the ICRP*, ICRP Publication 66, Elsevier, New York.

Ingham, D. B. (1975) Diffusion of aerosols from a stream flowing through a cylindrical tube, *J. Aerosol Sci.* **6**:125–132.

Kim, C. S. and Eldridge, M. A. (1985) Aerosol deposition in the airway model with excessive mucus secretions, *J. Appl. Physiol.* **59**:1766–1772.

Kim, C. S. and Fisher, D. M. (1999) Deposition characteristics of aerosol particles in sequentially bifurcating airway models, *Aerosol Sci. Technol.* **31**:198–220.

Kim, C. S. and Iglesias, A. J. (1989a) Deposition of inhaled particles in bifurcating airway models: I. Inspiratory deposition, *J. Aerosol Med.* **2**:1–14.

Kim, C. S. and Iglesias, A. J. (1989b) Deposition of inhaled particles in bifurcating airway models: II. Expiratory deposition, *J. Aerosol Med.* **2**:15–27.

Kim, C. S., Brown, L. K., Lewars, G. G. and Sackner, M. A. (1983) Deposition of aerosol particles and flow resistance in mathematical and experimental airway models, *J. Appl. Physiol.* **55**:154–163.

Kim, C. S., Lewars, G. G., Eldridge, M. A. and Sackner, M. A. (1984) Deposition of aerosol particles in a straight tube with an abrupt obstruction, *J. Aerosol Sci.* **15**:167–176.

Kim, C. S., Abraham, W. A., Chapman, G. A. and Sackner, M. A. (1985) Influence of two-phase gas–liquid interaction of aerosol deposition in airways, *Am. Rev. Respir. Dis.* **131**:618–623.

Kim, C. S., Lewars, G. A. and Sackner, M. A. (1988) Measurement of total lung aerosol deposition as an index of lung abnormality, *J. Appl. Physiol.* **64**:1527–1536.

Kim, C. S., Eldridge, M. A., Garcia, L. and Wanner, A. (1989) Aerosol deposition in the lung with asymmetric airways obstruction: in vivo observation, *J. Appl. Physiol.* **67**:2579–2585.

Kim, C. S., Fisher, D. M., Lutz, D. J. and Gerrity, T. R. (1994) Particle deposition in bifurcating airway models with varying airway geometry, *J. Aerosol Sci.* **25**:567–581.

Koblinger, L. and Hofmann, W. (1990) Monte Carlo modeling of aerosol deposition in human lungs. Part I. Simulation of particle transport in a stochastic lung structure, *J. Aerosol Sci.* **21**:661–674.

Landahl, H. D. (1950) On the removal of air-borne droplets by the human respiratory tract: I. The lung, *Bull. Math. Biophys.* **12**:43–56.

Lange, C. F. and Finlay, W. H. (2000) Introducing new dimensions in the modelling of pharmaceutical aerosols, in R. N. Dalby, P. R. Byro and S. J. Farr, eds, *Respiratory Drug Delivery VII*, May 14–18, 2000, Palm Harbor, Florida. Serentec Press, Raleigh, NC, pp. 569–572.

Lee, Z. L., Berridge, M. S., Finlay, W. H. and Heald, D. L. (2000a) Mapping PET-measured Triamcinolone Acetonide (TAA) Aerosol distribution into deposition by airway generation, *Int. J. Pharm.* **199**:7–16.

Lee, J. W., Lee, D. Y. and Kim, W. S. (2000b) Dispersion of an aerosol bolus in a double bifurcation, *J. Aerosol Sci.* **31**:491–505.

Li, W.-I., Perzl, M., Ferron, G. A., Batycky, R., Heyder, J. and Edwards, D. A. (1998) The macrotransport properties of aerosol particles in the human oral-pharyngeal region, *J. Aerosol Sci.* **29**:995–1010.

Lippmann, M. (1977) Regional deposition of particles in the human respiratory tract, in D. H. K. Lee *et al.*, eds, *Handbook of Physiology – Reaction to Environmental Agents*, American Physiological Society, Bethesda, MD, pp. 213–232.

Martin, D. and Jacobi, W. (1972) Diffusion deposition of small-sized particles in the bronchial tree, *Health Phys.* **23**:23–29.

Martonen, T. B. (1983) On the fate of inhaled particles in the human: a comparison of experimental data with theoretical computations based on a symmetric and asymmetric lung, *Bull. Math. Biol.* **45**:409–424.

Martonen, T. B. (1993) Mathematical models for the selective deposition of inhaled pharmaceuticals, *J. Pharm. Sci.* **82**:1191–1199.

Martonen, T. B., Katz, I. and Cress, W. (1995a) Aerosol deposition as a function of airway disease: cystic fibrosis, *Pharm. Res.* **12**:96–102.

Martonen, T., Zhang, Z., Yang, Y. and Bottei, G. (1995) Airway surface irregularities promote particle diffusion in the human lung, *Rad. Prot. Dos.* **59**:5–14.

Martonen, T., Zhang, Z. and Yang, Y. (1997) Particle diffusion from developing flows in rough-walled tubes, *Aerosol Sci. Technol.* **26**:1–11.

Mathews, J. and Walker, R. L. (1970) *Mathematical Methods of Physics*, Benjamin/Cummings, Menlo Park, CA.

Morton, W. B. (1935) The settling of a suspension flowing along a tube, *Proc. Roy. Irish Acad.* **43**:1.

Nusselt, W. (1910) *Z. Ver. Deutsch. Ing.* **54**:1154.

Persons, D. D., Hess, G. D., Muller, W. J. and Scherer, P. W. (1987) Airway deposition of hygroscopic heterodispersed aerosols: results of a computer calculation, *J. Appl. Physiol.* **63**:1195–1204.

Phalen, R. F., Oldham, M. J., Beaucage, C. B., Crocker, T. T. and Mortensen, J. D. (1985) Postnatal enlargement of human tracheo-bronchial airways and implications for particle deposition, *Anat. Rec.* **212**:368–380.

Phalen, R. F., Schum, G. M. and Oldham, M. J. (1990) The sensitivity of an inhaled aerosol tracheo-bronchial deposition model to input parameters, *J. Aerosol Med.* **3**:271–282.

Pich, J. (1972) Theory of gravitational deposition of particles from laminar flows in channels, *J. Aerosol Sci.* **3**:351–361.

Rudolf, G., Gebhart, J., Heyder, J., Schiller, Ch. F. and Stahlhofen, W. (1986) An empirical formula describing aerosol deposition in man for any particle size, *J. Aerosol Sci.* **17**:350–355.

Rudolf, G., Köbrich, R. and Stahlhofen, W. (1990) Modelling and algebraic formulation of regional and aerosol deposition in man, *J. Aerosol Sci.* **21**, Suppl. 1: S403–S406.

Sarangapani, R. and Wexler, A. S. (2000) The role of dispersion in particle deposition in human airways, *Toxicol. Sci.* **54**:229–236.

Scherer, P. W., Shendalman, L. H., Greene, N. M. and Bouhuys, A. (1975) Measurement of axial diffusivities in a model of the bronchial airways, *J. Appl. Physiol.* **38**:719–723.

Scheuch, G. and Stahlhofen, W. (1988) Particle deposition of inhaled aerosol boluses in the upper human airways, *J. Aerosol Med.* **1**:29–36.

Schlesinger, R. B., Bohning, D. E., Chan, T. L. and Lippmann, M. (1977) Particle deposition in a hollow cast of the human tracheo-bronchial tree, *J. Aerosol Sci.* **8**:429–445.

Schuster, J. A., Rubsamen, R. M., Lloyd, P. M. and Lloyd, L. J. (1997) The AERx aerosol delivery system, *Pharm. Res.* **3**:354–357.

Scott, W. R. and Taulbee, D. B. (1985) Aerosol deposition along the vertical axis of the lung, *J. Aerosol Sci.* **16**:323–333.

Smaldone, G. C. and Messina, M. S. (1985) Enhancement of particle deposition by flow-limiting segments in humans, *J. Appl. Physiol.* **59**:509–514.

Stahlhofen, W., Gebhart, J. and Heyder, J. (1980) Experimental determination of the regional deposition of aerosol particles in the human respiratory tract, *Am. Ind. Hyg. Assoc. J.* **41**:385–398a.

Stahlhofen, W., Gebhart, J. and Heyder, J. (1981) Biological variability of regional deposition of aerosol particles in the human respiratory tract, *Am. Ind. Hyg. Assoc. J.* **42**:348–352.

Stahlhofen, W., Gebhart, J., Heyder, J. and Scheuch, G. (1983) New regional deposition data of the human respiratory tract, *J. Aerosol Sci.* **14**:186–188.

Stahlhofen, W., Rudolf, G. and James, A. C. (1989) Intercomparison of experimental regional aerosol deposition data, *J. Aerosol Med.* **2**:285–308.

Stapleton, K. W. (1997) Deposition of medical aerosols in the human respiratory tract, PhD thesis, University of Alberta.

Stapleton, K. W., Guentsch, E., Hoskinson, M. K. and Finlay, W. H. (2000) On the suitability of k-ε turbulence modelling for aerosol deposition in the mouth and throat: a comparison with experiment, *J. Aerosol Sci.* **31**:739–749.

Stirling, C. and Patrick, G. (1980) The localisation of particles retained in the trachea of the rat, *Pathology* **131**:309–320.

Taulbee, D. B. and Yu, C. P. (1975) A theory of aerosol deposition in the human respiratory tract, *J. Appl. Physiol.* **38**:77–85.

Taulbee, D. B., Yu, C.P. and Heyder, J. (1978) Aerosol transport in the human lung from analysis of single breaths, *J. Appl. Physiol.* **44**:803–812.

Townsend, J. (1900) *Phil. Trans.* **193**:129.

Tsuda, A., Butler, J. P. and Fredberg, J. J. (1994) Effects of alveolated duct structure on aerosol kinetics II. Gravitational sedimentation and intertial impaction, *J. Appl. Physiol.* **76**:2510–2516.

Wang, C.-S. (1975) Gravitational deposition of particles from laminar flows in inclined channels, *J. Aerosol Sci.* **6**:191–204.

White, F. M. (1999) *Fluid Mechanics*, 4th edition, McGraw-Hill, Boston.

Willeke, K. and Baron, P. A. (1993) Gas and particle motion, in K. Willeke and P. A. Baron, eds, *Aerosol Measurement, Principles, Techniques and Applications*, Van Nostrand Reinhold, NY.

Xu, G. B. and Yu, C. P. (1986) Effects of age on deposition of inhaled aerosols in the human lung, *Aerosol Sci. Technol.* **5**:349–357.

Yeh, H.-C. and Schum, G. M. (1980) Models of human lung airways and their application to inhaled particle deposition, *Bull. Math. Biology* **42**:461–480.

Yu, C. P. and Cohen, B. S. (1994) Tracheobronchial airway deposition of ultrafine particles, *Ann. Occup. Hyg.* **38** (Supplement 1):83–99.

Yu, C. P., Zhang, L., Becquemin, M. H., Roy, M. and Bouchikhi, A. (1992) Algebraic modeling of total and regional deposition of inhaled particles in the human lung of various ages, *J. Aerosol Sci.* **23**:73–79.

# 8
# Jet Nebulizers

One of the principal difficulties in delivering a therapeutic agent to the lung is the development of an efficient way to break up bulk amounts of the compound into micron sized particles for inhalation. From a molecular point of view, it takes energy to break a bulk compound into particles because one must pull apart the molecules in the bulk phase, i.e. one must overcome the attractive forces between the molecules. The number of bonds being broken is proportional to the amount of new surface being created, so the amount of energy needed to break up a bulk compound into droplets or particles is proportional to the increase in surface area.

There are many ways to aerosolize compounds, and new methods continue to be developed. However, for compounds that can be dissolved or suspended colloidally in water, one of the oldest and simplest methods of creating an aerosol for inhalation is to blow air at high speed over a liquid surface. The kinetic energy of the air supplies the energy needed to break up the bulk liquid into droplets.

Jet nebulizers (sometimes called pneumatic nebulizers) are the most common type of inhalation device that uses this approach to droplet production. Jet nebulizers are a subset of the more general twin-fluid atomizers, and belong to the air-assist/airblast atomizer classes of atomizers discussed by Lefebvre (1989). However, traditional single-phase atomizers, in contrast to jet nebulizers, use a high pressure liquid feed supply to produce droplets from a high speed liquid jet. A pressurized liquid feed line is not normally used with nebulizers due to cost, safety, and portability issues. Thus it is useful to analyze nebulizers in their own right. This chapter is an introduction to the mechanics of jet nebulization. However, several concepts developed in this chapter are necessary in understanding droplet production with propellant metered dose inhalers, discussed in Chapter 10.

## 8.1 Basic nebulizer operation

Although there are many different designs of jet nebulizers, the basic geometry of a typical 'unvented' jet nebulizer is shown in Fig. 8.1.

The basic operating principle of an 'unvented' nebulizer is as follows. A pressurized air source (either from a pump/compressor or from a wall source) supplies high pressure air which flows through a nozzle (or orifice or venturi, depending on the design) where the air accelerates to high speed. The pressure near the nozzle is lower than in the reservoir and this draws liquid up the liquid feed tube.[1] The nozzle region is designed so that the high speed air here flows over a short section of liquid surface supplied by the liquid feed tube. This is the primary droplet production region. The droplets produced in this region

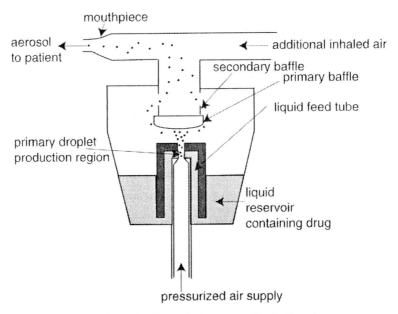

**Fig. 8.1** Schematic of a typical 'unvented' nebulizer design.

(via mechanisms we will soon discuss) then splash off primary baffles, producing smaller droplets, which then flow with the air out through secondary baffles (which filter out the larger particles) and into a mouthpiece (or mask) for patient inhalation of the aerosol. For most patients, the air supply from the pressurized source does not supply enough air to make up a typical inhalation flow rate, so additional ambient air is inhaled through the mouthpiece.

If the additional air that is brought in to make up the patient's full inhalation flow rate comes through the primary droplet production region, then the nebulizer is referred to as a 'vented' nebulizer (also called an 'active venturi' nebulizer, since the low pressure in the nozzle or 'venturi' region may actively draw air into the nebulizer even without a patient present.) By placing a one-way valve on the vent, additional air is entrained into the nebulizer only during inhalation, which lowers the flow rate of air through the primary production region during exhalation and reduces the amount of exhaled aerosol somewhat. A schematic of a basic vented nebulizer, with such a valve in place, is shown in Fig. 8.2.

For such a valved vented nebulizer, the valve opens to allow additional air to flow through the nebulizer during inhalation and closes on exhalation (to prevent aerosol exiting during exhalation). Valved vented nebulizers give somewhat higher efficiency than unvented or unvalved-vented nebulizers since they waste less aerosol on exhalation.

---

[1]The low pressure region near the nozzle has traditionally been explained as a 'venturi' effect, but this cannot be entirely correct, since at high enough air supply pressures the pressure at the nozzle would be higher than in the reservoir and nebulization would cease. The fact that nebulizers continue to operate at several atmospheres of air supply pressure runs counter to this explanation. Instead, entrainment of air in the jet downstream of the nozzle may be partly responsible for the slightly subatmospheric pressures in the primary droplet production region.

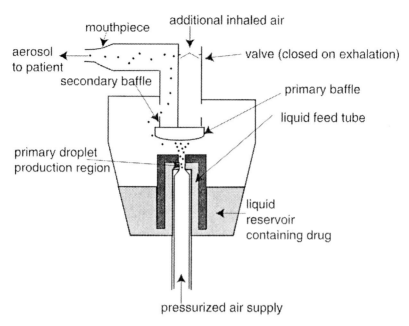

**Fig. 8.2** Schematic drawing of a valved 'vented' nebulizer design.

However, other differences are usually present between brands of vented and unvented nebulizers, so that the presence of a valved vent does not automatically ensure a nebulizer will give higher efficiency than other unvented or unvalved nebulizers.

It should be noted that for all inhalation aerosol devices requiring multiple breaths, there is a small 'connection' volume between the entrance to the respiratory tract (either the mouth or the nose) and the aerosol-containing volume of the device. After the first breath from the device, this connection volume will be filled on exhalation with exhaled air that does not contain significant amounts of aerosol. This exhaled air will then be rebreathed immediately on the next tidal breath through the device, causing the amount of aerosol inhaled to be smaller than would be expected if the connection volume was absent. For small tidal volumes (such as with toddlers and infants), this can cause a significant reduction in the amount of aerosol being inhaled from a nebulizer and is a reason for using as small a connection volume as is possible for such patients.

The constant supply of air through the pressurized air supply line with a jet nebulizer also introduces an age-dependence to the dose delivered to very young subjects (Collis *et al.* 1990). In particular, for young subjects, inhalation flow rates may be below the air flow rate supplied by the nebulizer (the excess air exits through the exhalation route of the nebulizer), so that even during inhalation there is aerosol exiting the nebulizer. Because this wastage does not occur until inhalation is below a certain flow rate (the value depends on the nebulizer, but is typically $4–8 \, l \, min^{-1}$), patients with inhalation flow rates above this flow rate will inhale the full dose, but patients inhaling below this flow rate will receive only a portion of this dose (with the inhaled dose depending on their flow rate). Weight/age correction of doses should bear this phenomenon in mind.

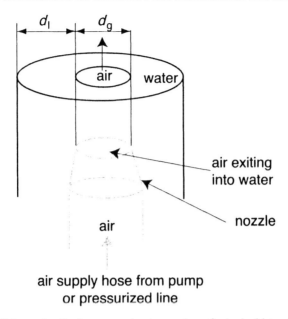

**Fig. 8.3** Schematic of primary production region of a typical jet nebulizer.

## 8.2 The governing parameters for primary droplet formation

The primary production of droplets in most nebulizers that are currently used in clinical practice occurs because of a high speed jet of air blasting up in a column through water, as shown in Fig. 8.3. Note that traditional airblast and air-assist atomizers have the air and water reversed from the usual jet nebulizer design (i.e. atomizers usually have the air forming a sheath on the outside of a central water jet). Note also that many nebulizer designs have been used that have primary droplet production regions that are different from the one shown in Fig. 8.3. However, our concern here is largely with the basic concepts involved, and these can be elucidated with the geometry shown in Fig. 8.3.

It must be realized that the droplets produced from the jet are not the final droplets that are inhaled, since they are usually too large for this purpose. Secondary processes (particularly impaction on baffles and aerodynamic breakup processes) occur that break up the primary droplets initially formed and filter out the larger particles. These secondary processes will be discussed later. However, before discussing the physical mechanisms that cause droplets to break up when high speed air flows over a liquid, let us first write down the dynamical nondimensional parameters that govern the process of primary droplet formation, aside from any geometrical parameters (such as nozzle shape etc.).

If we assume Newtonian behavior for the fluids involved, then the velocity **v** and pressure $p$ in both phases must obey the Navier–Stokes equations and continuity (mass conservation) equation. For a single-phase fluid of constant temperature, only two nondimensional parameters appear in the Navier–Stokes equations, and these are the Reynolds number $Re$, and Mach number $Ma$. If we use the subscript l to indicate the aqueous phase and g to indicate the air phase, then assuming incompressible flow for the liquid phase (meaning $Ma_l$ is not a parameter of importance), three nondimensional parameters are present in the Navier–Stokes equations. These parameters are the gas

Reynolds number $Re_g$, the liquid Reynolds number $Re_l$ and the gas Mach number $Ma_g$. More specifically,

$$Re_g = U d_g \rho_g / \mu_g$$

where $U$ is the velocity of the air–water interface (and can be approximated as the mean gas velocity in the air jet), and the symbols $\rho$ and $\mu$ indicate density and viscosity respectively. The liquid Reynolds number is

$$Re_l = U d_l \rho_l / \mu_l$$

and the Mach number of the gas jet is

$$Ma_g = U/c$$

Here, $c$ is the speed of sound in the gas.

We must also satisfy the dynamical boundary condition on the stress at the interface between the air and liquid. This condition states that the tangential stress at the interface is continuous. The only tangential stresses are due to viscous forces, and we have

$$(\sigma_{rz})_l = (\sigma_{rz})_g \tag{8.1}$$

$$(\sigma_{\theta z})_l = (\sigma_{\theta z})_g$$

where for a Newtonian fluid the stress tensor is proportional to the gradient of the velocity, e.g.

$$\sigma_{rz} = \mu \left( \frac{dv_r}{dz} + \frac{dv_z}{dr} \right)$$

If we nondimensionalize Eq. (8.1), the viscosity ratio $\mu_g/\mu_l$ and the diameter ratio $d_g/d_l$ appear. This can be seen by expanding Eq. (8.1) explicitly and nondimensionalizing to obtain

$$\mu_l \frac{\left( \dfrac{dv_r'}{dz'} + \dfrac{dv_z'}{dr'} \right)}{d_l} = \mu_g \frac{\left( \dfrac{dv_r'}{dz'} + \dfrac{dv_z'}{dr'} \right)}{d_g}$$

or

$$\left( \frac{dv_r'}{dz'} + \frac{dv_z'}{dr'} \right) \frac{\mu_l}{\mu_g} = \left( \frac{dv_r'}{dz'} + \frac{dv_z'}{dr'} \right) \frac{d_l}{d_g}$$

where primed quantities represent dimensionless quantities (e.g. $v' = v/U$, $r' = r/d$ where $d = d_g$ or $d_l$).

We must also consider the normal stress component at the interface. This component has the pressure appearing in it (since pressure is always normal to a surface, it does not appear when we consider the tangential components of force at the surface). In addition to the pressure and viscous stresses, surface tension also occurs at an interface. Including surface tension, then continuity of normal stress across an interface implies

$$(\sigma_{rr})_g = (\sigma_{rr})_l$$

i.e.

$$p_l - 2\mu_l \frac{dv_l}{dr} + \frac{4\sigma}{d_g} = p_g - 2\mu_g \frac{dv_g}{dr}$$

where $\sigma$ is the surface tension. The form of the surface tension term in this equation is derived in many standard textbooks in many different fields, including fluid dynamics (White 1998) and physical chemistry (Saunders 1966) and so is not given here. Nondimensionalizing this equation by dividing through by $\rho_g U^2$ gives

$$p_1' - 2\left(\frac{\rho_g}{\rho_1}\right) Re_1 \frac{dv_1'}{dr'} + \frac{4}{We_g} = p_g' - 2Re_g \frac{dv_g'}{dr'}$$

where the nondimensional Weber number $We_g$ is defined as

$$We_g = \frac{\rho_g U^2 d_g}{\sigma}$$

We could instead use a Weber number based on the liquid properties, e.g.

$$We_1 = \frac{\rho_1 U^2 d_1}{\sigma}$$

but this is not an independent parameter since

$$We_1 = We_g \left(\frac{\rho_g}{\rho_1}\right)\left(\frac{d_g}{d_1}\right)$$

Thus, this interface boundary conditions gives us only one more parameter, which is the Weber number.

We therefore have a total of seven nondimensional parameters. Only six of these are independent however, since

$$\frac{d_g}{d_1} = \frac{Re_g}{Re_1} \frac{\mu_g}{\mu_1} \bigg/ \frac{\rho_g}{\rho_1}$$

In all we then have six independent nondimensional parameters that govern the dynamics of the primary droplet formation in a nebulizer:

Reynolds number in the gas jet $Re_g = U d_g \rho_g / \mu_g$
Reynolds number in the liquid surrounding the gas jet $Re_1 = U d_1 \rho_1 / \mu_1$
Mach number in the gas jet $Ma_g = U/c$
gas Weber number $We_g = \rho_g U^2 d_g / \sigma$
viscosity ratio $\mu_g / \mu_1$
density ratio $\rho_g / \rho_1$

For a given nozzle geometry, primary droplet sizes are a function of only these parameters. However, some of these parameters may affect droplet sizes more strongly than others, as we shall see. Of course, changes to the geometry that holds the water around the air jet, as well as changes in nozzle geometry, can affect droplet production (and is known to be important for the case of a liquid jet spraying into air – see Reitz and Bracco 1982), but geometrical effects are not part of the dynamical equations governing the physics, so that dynamical dimensionless parameters are of little use in representing them. The physics of droplet formation will be the same though, and for a given geometry will be governed only by the above parameters.

Note that this choice of parameters is not unique. In fact, any of these parameters can be combined to make a new parameter that can be used instead of one of the above six parameters. For example, the Ohnesorge number $Oh$ is defined as

$$Oh = We^{1/2}/Re = \mu/(\rho d \sigma)^{1/2}$$

where we assume both $We$ and $Re$ use the same fluid properties, velocities and length scales. The Laplace number is defined as

$$Lp = 1/Oh^2 = Re^2/We$$

and the capillary number is defined as

$$Ca = We/Re$$

Each one of these is sometimes used instead of one of the six parameters we have defined.

## 8.3 Linear stability of air flowing across water

To understand how droplets are produced when air blasts up through the water in a nebulizer, it is useful to perform what is called a 'linear stability analysis' of a perfectly smooth air–water interface. Linear stability analysis is a standard procedure in both solid and fluid dynamics. In the present case, the idea behind such an analysis is to pretend that the air–water interface is perfectly smooth, except for a very small disturbance that is present on the interface. Then the equations governing what happens to this disturbance are written down and used to decide whether the disturbance will grow and lead to instability (and possibly droplet formation) or decay (with no droplets being produced). That droplets can result from nonlinear development of unstable waves on an air–water interface with large velocity difference between the air and water is quite apparent in the photos of Taylor and Hoyt (1983) shown in Fig. 8.4.

**Fig. 8.4** Instabilities can be seen developing on the surface of a water jet spraying into air in this photo from Taylor and Hoyt (1983), reprinted with permission.

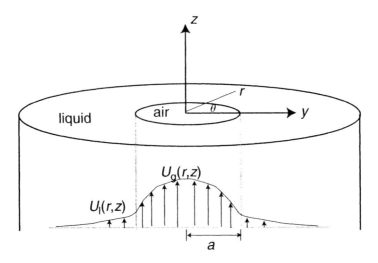

**Fig. 8.5** Basic state for linear stability analysis of an air jet in water.

Note, however, that if an instability is found, this does not mean we will have droplet production, since it is possible for the instability to lead to a new stable state that does not give droplets. For example, the smooth interface could be unstable but the instability could result simply in waves traveling on the surface that do not produce droplets (similar to the way wind produces waves on a lake that do not produce droplets).

To perform the stability analysis, we define the basic state of the fluid as that in which the air–water interface is completely smooth. In the standard analysis, the basic state is assumed to have velocity $U_g(r,z)$ in the air and $U_l(r,z)$ in the water as shown in Fig. 8.5.

Most previous work in this and similar geometries assumes the basic state is independent of $z$. This assumption must of course be incorrect, since the velocity of the liquid must vary with $z$ in order to have zero velocity at the nozzle tip ($z = 0$). However, inclusion of nonperiodic variation in $z$ makes the analysis more complicated (since it requires what is called 'spatial stability analysis'). Such variation should of course be included if any quantitative theoretical understanding is to be made, but to the author's knowledge such an analysis has not yet been made without neglecting the viscous forces, which are probably important as discussed below.

A small disturbance to the smooth interface at $r = a$ is assumed to be present on the surface (caused by disturbances that are always present in any real flow) so that the perturbed interface is at $r = a + \eta$, where in order for this to be a small disturbance we must have $\eta \ll a$. The velocity of the fluid in its perturbed state is now slightly different from the basic state, and so we write the velocity of the perturbed state as

$$\mathbf{v}_g = U_g + \mathbf{v}'_g$$

and similarly for the liquid state

$$\mathbf{v}_l = U_l + \mathbf{v}'_l$$

The pressures are also written as

$$p_g = P_g + p'_g \text{ and } p_l = P_l + p'_l$$

where the primed quantities indicate small disturbances and the capital quantities indicate the basic state (which is assumed known).

The goal of a linear stability analysis is to find out whether a given disturbance will grow and if so, the rate at which it grows. To achieve this goal, one must solve the equations governing the liquid and gas phases. However, because a small amplitude disturbance is assumed, all the equations governing the disturbance can be linearized, and so Fourier transforms in $z$ and $\theta$ and Laplace transforms in time can be used. As a result, one ends up analyzing what happens to disturbances that are of the form

$$\eta = \eta_0 f(r)\exp(ikz)\exp(il\theta)\exp(st)$$

where $\eta_0$ is the initial amplitude of the disturbance to the interface. Similar forms are assumed for the perturbations to the base velocity and pressure. Here, $k = 2\pi/\lambda$ and $l$ are wavenumbers that give the wavelength of the disturbance in the $z$ direction and $\theta$ direction, respectively. The parameter $s$ is a growth rate that governs how fast the disturbance grows in time. Note that in actual fact a disturbance would grow as it moved downstream in the $z$-direction (referred to as 'spatial instability', which is obtained by allowing $k$ and $l$ to be complex numbers), whereas the disturbance used in most linear stability analyses in this and similar geometries grows in time (leading to 'temporal instability', obtained by assuming $k$ and $l$ are real numbers). This difference has already been mentioned and spatial stability analysis is preferable, but is more complex, as mentioned above.

We can estimate the value of the wavelength $\lambda$ of disturbances on the interface by realizing that jet nebulizers are normally designed to produce primary droplets with diameters much smaller than the nozzle diameter $2a$ (incidentally, in the terminology of liquid jets, this means jet nebulizers are similar to liquid jets in the second wind-induced or atomization regimes – see Lin and Reitz 1998). Making the reasonable assumption that the primary droplets are of the same order in size as the wavelength of the unstable disturbances that created them, then we can assume that $\lambda \ll a$. Thus, for jet nebulizers it is reasonable to assume $ka = 2\pi a/\lambda \gg 1$. This is an important result, since it means that the curvature of the interface is unimportant in the stability analysis and allows us to instead consider a planar air–water interface. We can then make use of the vast quantity of literature on the flow of air over planar water surfaces. The work of Boomkamp and Miesen (1996) is particularly useful for this purpose. They perform a linear (temporal) stability analysis for the flow of two incompressible planar layers of fluid for a wide range of the relevant parameters. They include gravity in their analysis and allow the air–water interface to be inclined at an angle $\beta$, but this term is unimportant for jet nebulizers since the nondimensional parameter governing the importance of gravity (which is the inverse of the Froude number times $\cos \beta$) is small, indicating the unimportance of gravity in the physics of the primary droplet production process.

Although several interesting results can be deduced from the literature on air flow over a planar air–water interface, one of the most interesting results is an understanding of what causes the interface to be unstable (and potentially result in droplets, since if the interface is stable, droplet production is not possible with small disturbances to the interface). To understand the instability mechanism, we must first rewrite the condition (8.1) requiring continuity of shear stress at the interface for a planar surface. For the basic state of the undeformed air jet, this condition can be written as

$$\frac{\mu_g dU_g}{dy} = \frac{\mu_l dU_l}{dy} \tag{8.2}$$

Because $\mu_g$ and $\mu_l$ are not equal (e.g. for air $\mu_g = 1.8 \times 10^{-5}$ kg m$^{-1}$ s$^{-1}$ while for water $\mu_l = 10^{-3}$ kg m$^{-1}$ s$^{-1}$), we see that Eq. (8.2) implies that the slope of the basic state velocity profile is dramatically different between the two phases. For example, for air–water we can rewrite Eq. (8.2) as

$$\frac{dU_g}{dy} = 55 \frac{dU_l}{dy}$$

so that the slope of the basic state velocity profile in the air is 55 times that in the water.

Now consider a pertubation at the interface that moves the interface a small distance $\eta$ in the $y$-direction from its undisturbed position. This perturbation alters the velocity of the interface slightly, disturbing the basic state velocity at the interface by an amount $u'_g$ in the gas and $u'_l$ in the liquid. However, the basic state velocity also varies with distance $y$ from the interface, so that if a disturbance moves the interface a distance $\Delta y = \eta$ in the $y$-direction, this introduces an additional change in the $z$-component of the velocity at the interface by the amount $\eta \, dU/dy$ (since $\Delta U = \Delta y \, dU/dy$). However, we must have continuity of the velocity $\mathbf{v} = \mathbf{v}' + U$ across the interface. Thus, we must have

$$u'_g + \eta \frac{dU_g}{dy} = u'_l + \eta \frac{dU_l}{dy} \qquad \text{at the interface}$$

This can be rewritten as

$$u'_l = u'_g + \eta \left( \frac{dU_g}{dy} - \frac{dU_l}{dy} \right)$$

Thus, the component of the perturbation velocity is not continuous across the interface since $dU_g/dy \neq dU_l/dy$ because of the difference in viscosity across the interface (recall $dU_g/dy = 55 \, dU_l/dy$ for air–water). Thus, a disturbance that originates in the air jet that pushes the jet outward (i.e. gives $\eta > 0$) will result in a disturbance to the liquid velocity at the interface that is larger than the velocity of the disturbance in the air. For this to happen, energy must be transferred to the liquid from the air jet. It is this energy transfer from the air jet to the liquid jet because of the viscosity difference that drives the instability at the interface. Thus, *the discontinuity in viscosity at the interface is responsible for the instability of the interface.*

A more detailed analysis by Boomkamp and Miesen (1996) shows that the viscosity difference across the interface results in positive work being done on the interface and they coin the term 'viscosity-induced instability' to describe the instability that results from this effect.

Although many other explanations have been proposed for the instability at air–water interfaces (e.g. a Kelvin–Helmholtz instability), and indeed other mechanisms are responsible for this instability for different parameter regimes (e.g. surface tension can be an important mechanism for very low speed jets, while other mechanisms are important in other parameter regimes – see Boomkamp and Miesen 1996), the literature on instability of planar air–water flows under similar parameter regimes suggests that viscosity-induced instability is responsible for the instability that ultimately leads to droplet production with jet nebulizers.

## 8.4 Droplet sizes estimated from linear stability analysis

At this point it is useful to note that a commonly used rule of thumb in stability analysis proposes that the length scales developing from linear instability are the same as those of the most unstable waves in the linear stability analysis. This proposal does not have a rigorous basis, since initial conditions and nonlinearity are neglected. As a result, it works well in some situations but not in others, which limits its utility. However, if this rule of thumb applies in the present case, then we would expect the primary droplets formed in a jet nebulizer to be of the same size as the most unstable wavelength from a linear stability analysis.

The most unstable wavelength for the case when the wavelengths are much smaller than the jet diameter (as it is with existing jet nebulizers) was calculated by Taylor (1958). However, this analysis neglects the gas viscosity, which, given the basic viscosity-induced nature of the instability may not be reasonable. In addition, Taylor's analysis ignores the fact that the basic state changes with $z$, as well as the fact that disturbances to a jet grow in the $z$-direction ('spatial instability') rather than growing in time at a fixed $z$-location ('temporal instability'). However, if we are willing to accept these flaws, we can use the result from Taylor, who found the most unstable wavelength depends on the parameter

$$\Gamma = (\rho_l/\rho_g)(\sigma/\mu_l U)^2 = (\rho_l/\rho_g)(Re_l/We_g)^2$$

For $\Gamma > 100$ the most unstable wavelength is nearly independent of $\Gamma$ and given by

$$\lambda \approx \frac{1.26\sigma}{(\rho_g U^2)} \tag{8.3}$$

For an air–water interface, $\Gamma \gg 100$ for all subsonic values of the interface velocity $U$, so Eq. (8.3) can be used. Putting the properties for air and water into Eq. (8.3), we obtain

$$\lambda \approx \frac{0.0756}{U^2} \qquad (U \text{ in m s}^{-1}, \lambda \text{ in m}) \tag{8.4}$$

A typical velocity for the air–water interface in a jet nebulizer might be on the order of 50–150 m s$^{-1}$ or higher (since air velocities are often sonic, i.e. near 300 m s$^{-1}$, but boundary layers are expected next to the water surface so the interface velocity is less than this), which gives $\lambda \approx 3$–32 μm for droplet sizes from Eq. (8.4). Although this is in the right range for the primary droplet sizes of jet nebulizers, this result may be fortuitous given the number of fairly drastic assumptions we have made to this point (i.e. temporal instability, neglect of gas viscosity, and the assumption that the most unstable linear mode determines droplet sizes). Whether the inclusion of gas viscosity in determining the most unstable wavelength in a spatial linear stability analysis is important is not yet known (Lin and Reitz 1998). Even without resolving this issue, however, it is safe to suggest that nonlinear effects which are not considered by linear stability analysis are important in determining droplet sizes, so that a detailed quantitative understanding of droplet sizes cannot be obtained with temporal linear stability analysis alone.

## 8.5 Primary droplet formation

Linear stability analysis helps satisfy our scientific curiosity as to what causes the air–water interface to be unstable when an air jet is blasted through a liquid. However, from a practical viewpoint, the size of the droplets that result from this instability, and the amount of liquid in the droplets entrained in the air stream are what we normally would like to know for a jet nebulizer. Neither are supplied from linear stability analysis. Droplet size is of interest because of its importance in determining deposition in the respiratory tract, while entrainment is of interest because it directly affects the rate at which liquid can be nebulized. Unfortunately, linear stability analysis is based on the assumption that the jet interface is only slightly perturbed from a smooth state. The formation of a drop from this interface requires a large deformation of the interface, which cannot be described by a linear stability analysis. Since droplet formation is a highly nonlinear process, it would seem unlikely that linear stability analysis can quantitatively predict droplet sizes, as suggested above and by many other authors for liquid jets in air (Wu et al. 1991). Indeed, experimental descriptions of droplet formation on planar air–water interfaces (Woodmansee and Hanratty 1969) mention ripples forming on larger 'roll waves', with these ripples accelerating toward the front of a roll wave and breaking into droplets when they reach the crest of a roll wave. Such descriptions may indicate resonant interactions among different wavelength waves, a process that is quite common with other fluid dynamic instabilities (Craik 1985). Such processes are not describable by linear stability analysis.

The question then is, what can we use to predict primary droplet sizes? Unfortunately, the actual droplet formation process is not entirely understood at this time. From the literature on planar air–water interfaces, liquid jets in air, and annular two-phase flow (i.e. air flowing inside liquid-coated pipes), it is probable that waves forming on the air–water interface because of the above discussed viscosity-induced instability are important in the process (Woodmansee and Hanratty 1969, Hewitt and Hall-Taylor 1970, Wu et al. 1991, Chigier and Reitz 1996). However, the manner in which these waves yield droplets is not entirely understood. In fact, several mechanisms are suggested in the literature just cited and by other researchers. Some of these explanations include the shearing of roll waves, as mentioned above, via a secondary instability (Scardovelli and Zaleski 1999), as well as the creation of ligament-like 'fibers' that are pulled from the liquid by the air (Wu et al. 1991, Faragó and Chigier 1992) which then break up by the action of surface tension forces (similar to how a low velocity stream of water from a tap breaks up into drops, which is called 'Rayleigh' breakup for a liquid jet). Other explanations include the ejection of droplets due to high energy turbulence in the liquid (Wu et al. 1992), although turbulence levels in the low speed liquid feeds of jet nebulizers are probably not high enough for this latter mechanism to play a role with most jet nebulizers. Other mechanisms have also been suggested. The droplet formation process that actually occurs in jet nebulizers remains to be determined and may be different for different nebulizer designs and parameter ranges.

However, even if a model for the creation of the primary droplets in a jet nebulizer could be created, this model would be incomplete because two major processes cause these primary drops to break up into smaller droplets in jet nebulizers. The first of these processes can occur immediately upon formation of the primary droplet and is caused by the fact that at creation, the droplet may have an axial (z-component) velocity that is lower than that of the air in the high speed jet. Thus, there can be a large relative velocity between the droplet and its surrounding air. This results in strong aerodynamic forces on the droplet that can cause breakup.

The second cause of droplet breakup in jet nebulizers is impaction on baffles. Most jet nebulizers slam the primary droplet stream into a stationary plate (called a baffle). The high velocity impact of these droplets on the baffle can cause the droplets to splash, breaking them up into smaller droplets.

These two mechanisms are examined in the following sections.

## 8.6 Primary droplet breakup due to abrupt aerodynamic loading

It has long been known that drops can break up into multiple droplets if they are exposed to a sudden change in the speed of the air surrounding them, often referred to as sudden changes in 'aerodynamic loading'. In fact, there is a vast quantity of literature on this topic. A review of much of this work as it relates to sprays is given by Faeth *et al.* (1995), while a more general review is given by Gelfand (1996). Most of the work in this area focuses on the effect of shock waves in air as they move past droplets.

Unfortunately, the mechanics of secondary droplet formation due to sudden changes in aerodynamic loading is not entirely understood theoretically. Most of the present understanding comes from experimental observations. These experiments indicate that droplet breakup does not occur until certain levels of aerodynamic loading have been reached, i.e. no secondary droplet breakup occurs for slow enough rates of change in relative velocity between the droplet and the air. As the aerodynamic loading increases in strength, droplet breakup takes on different appearances in the experiments, giving rise to different droplet breakup regimes. For sudden aerodynamic loading, these regimes have been classified based on the values of two nondimensional parameters, which are often a Weber number and a Reynolds number based on the droplet diameter $d$:

$$We_d = \rho_g U^2 \, d/\sigma$$

$$Re_{dl} = U d \rho_l / \mu_l$$

Alternatively, the two parameters $We_d$ and an Ohnesorge number $Oh = \mu_l/(\rho_l \, d\sigma)^{1/2}$ are often used instead.

Based on experimental observations by many researchers, the following regimes have been suggested by various authors (Gelfand 1996, Shraiber *et al.* 1996, Faeth *et al.* 1995, Pilch and Erdman 1987):

$We_d < We_{co}$: no droplet breakup
$We_{co} < We_d < We_{c1}$: vibrational, bag, bag-stamen breakup
$We_{c1} < We_d < We_{c2}$: shear breakup (also called sheet stripping)
$We_d > We_{c2}$: wave crest stripping, catastrophic breakup

The events associated with these regimes are illustrated in Fig. 8.6.

These regimes must be interpreted broadly, since a complete consensus does not exist in the literature on what regimes are present or on the names used to identify the different regimes. The distinctions between them vary from experiment to experiment with various authors identifying additional regimes. However, it is generally agreed that there is a minimum value of $We_d = We_{c0}$ that is required, below which secondary droplet

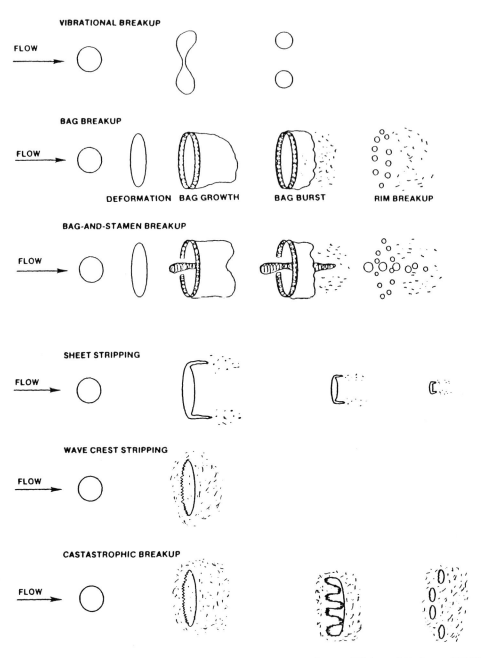

**Fig. 8.6** Schematic of secondary breakup mechanisms according to Pilch and Erdman (1987). Reprinted from Pilch and Erdman (1987) with permission.

breakup is not observed. Although the exact value of $We_{c0}$ varies from experiment to experiment (partly because it depends on the history of acceleration the drop is exposed to), a value near $We_{c0} \approx 12$ is typical (for $Oh < 0.1$).

As $We_d$ is increased, droplet breakup begins to occur near $We_d = We_{c0}$. Vibrational breakup (where drops oscillate at their frequency of natural oscillation before breaking up into several big fragments) is observed at the lowest $We_d$, followed at higher $We_d$ by 'bag' or 'parachute' fragmentation, and finally by 'bag-and-stamen' breakup (similar to 'bag' breakup except a central, re-entrant jet forms in the middle of the bag), and referred to as 'multimode' breakup by some authors.

At higher aerodynamic loading ($We_d > We_{c1}$), droplet breakup is observed to occur by stripping of droplets from sheets of liquid trailing behind the droplet. A typical value of $We_{c1}$ is 40–100 if $Oh < 0.1$.

At still higher aerodynamic loading where $We_d > We_{c2}$ (where $We_{c2} \approx 350$–$1000$ for $Oh < 0.1$), secondary droplets are stripped from the surface of the drop in a manner somewhat similar to the primary atomization of droplets from a planar air–water interface discussed above. This is followed at even higher $We_d$ by drops exploding into fragments, possibly due to a Rayleigh–Taylor instability at the droplet surface, which is the instability that occurs at a planar interface between two fluids when the fluids are accelerated from the lighter to the heavier fluid (Taylor 1950).

It is interesting to note that the values of $We$ demarcating the different droplet breakup regimes are found to be nearly independent of the viscosity ratio $\mu_g/\mu_l$ for measurements made in the range $10^{-4} < \mu_g/\mu_l < 10^{-2}$. Only for very viscous liquid drops ($\mu_g/\mu_l < 10^{-5}$) do experiments begin to show viscosity affecting these demarcations, moving each regime to higher values of $We_d$ (Faeth et al. 1995, Gelfand 1996). For droplets in room temperature air, this would imply that viscosities more than 180 times that of water are needed before the viscosity ratio affects the process, which is outside the range normally seen with pharmaceutical formulations. Note that if the gas viscosity is assumed to be nearly inviscid (i.e. its viscosity is assumed to be nearly zero), then the importance of liquid viscosity can be represented using the Ohnesorge number $Oh$:

$$Oh = \mu_l/(\rho_l d\sigma)^{1/2}$$

or the Laplace number $Lp = 1/Oh^2$, defined earlier, instead of the viscosity ratio. In this case, it is generally found that the $We$ numbers for the different breakup regimes are independent of liquid viscosity for $Oh < 0.1$ or so, with these $We$ numbers increasing with $Oh$ above 0.1 (Faeth et al. 1995).

Droplet size distributions after secondary breakup due to shock waves are polydisperse, but have been found to obey Simmons universal root-normal distribution with $MMD/SMD = 1.2$ (Faeth et al. 1995), where $SMD$ is the Sauter mean diameter, defined in Chapter 2. Thus only one moment of the size distribution needs to be specified if a two-parameter size distribution function is assumed. Hsiang and Faeth (1992) give the following empirical correlation for secondary droplet sizes:

$$SMD/d \approx 6.2\,(\rho_l/\rho_g)^{1/4}\,Re_{dl}^{-1/2}$$

where $d$ is the primary droplet size before breakup. This correlation was developed based on measurements with $We < We_{c2}$, $\rho_l/\rho_g > 500$ and $Oh < 0.1$, and may not be valid outside this range. More recently, Chou and Faeth (1998) suggest that in the bag breakup regime there are two distinct, nearly monodisperse, droplet populations that result, one from the breakup of the 'bag' and the other from Rayleigh breakup (i.e. surface tension pinching off) of the basal ring that occurs at the base of the bag. Chou and Faeth find these two droplet populations carry approximately equal amounts of mass, but differ dramatically in size, with

$$\text{bag droplet diameters} = 0.042d \qquad (8.5a)$$

$$\text{basal ring droplet diameters} = 0.36d \qquad (8.5b)$$

where $d$ is the initial droplet size.

One final interesting aspect of the data on secondary droplet breakup is the observation that the breakup time is essentially independent of $We_d$ or $Oh$ for $Oh < 0.1$. Droplet breakup time is given approximately by

$$\tau_b = C\tau_0 \qquad (8.6)$$

where $C$ is a constant ($C \approx 2.5$–$5$) and $\tau_0 \approx d(\rho_l/\rho_g)^{1/2}/U$ is the time scale for growth of the Rayleigh–Taylor instability. Breakup times are actually dependent on $We$ and $Oh$ (see Shraiber et al. (1996) for correlations giving $\tau_b(We,Oh)$), but vary little for $Oh \ll 1$ and $We \gg 1$. For the bag breakup regime, Chou and Faeth (1998) find the bag droplets in (8.5a) develop first and have $C = 3$–$4$, while the basal ring droplets in (8.5b) develop slightly later, with $C = 4$–$5$ or so.

## Example 8.1

Calculate the type of droplet breakup, breakup time and secondary droplet size that might be expected of the primary water droplets created in a nebulizer if the diameter of the primary droplets is 50 μm and the difference in speed between the droplets and the air is 150 m s$^{-1}$. Assume sudden aerodynamic loading.

### Solution

We need to calculate the Weber number

$$
\begin{aligned}
We_d &= \rho_g U^2\, d/\sigma \\
&= (1.2\ \text{kg m}^{-3} \times (150\ \text{m s}^{-1})^2\ 50 \times 10^{-6}\ \text{m})/(0.072\ \text{N m}^{-1}) \\
&= 19
\end{aligned}
$$

and Reynolds number

$$
\begin{aligned}
Re_{dl} &= Ud\rho_l/\mu_l \\
&= (150\ \text{m s}^{-1} \times 50 \times 10^{-5}\ \text{m} \times 1000\ \text{kg m}^{-3})/(0.001\ \text{kg m}^{-1}\,\text{s}^{-1}) \\
&= 7500
\end{aligned}
$$

Thus, we see that $We_d > We_{c0}$, but $We_d < We_{c1}$, so we would expect droplet breakup of a vibrational/bag/bag-stamen type.

Out of curiosity, since $Oh$ is often used, let us calculate $Oh$ as follows:

$$Oh = We_{dl}^{1/2}/Re_{dl} \text{ where } We_{dl} = \rho_l U^2\, d/\sigma$$

$$= \left(\frac{1000}{1.2} \times 19\right)^{\frac{1}{2}}/7500$$

$$= 0.017$$

This is much less than 0.1 and so the above values for $We_{c0}$ and $We_{c1}$ at low $Oh$ are reasonable.

Since $Oh \ll 1$ and $We \gg 1$, the time for droplet breakup is given by Eq. (8.6):

$$\tau_b \approx \frac{C\, d(\rho_l/\rho_g)^{1/2}}{U}$$

$$= (C \times 50 \times 10^{-6}\ \text{m} \times (1000/1.2)^{1/2})/(150\ \text{m s}^{-1})$$

$$= C \times 9.6 \times 10^{-6}\ \text{s}$$

where $C$ appearing in Eq. (8.6) is in the range 2.5–5, and so we obtain

$$\tau_b = 25\text{–}50\ \mu\text{s}$$

If the gas velocity is 200 m s$^{-1}$ then the droplet velocity is 50 m s$^{-1}$, and this would mean the droplet will travel between 1 and 2 mm before breaking up, although this is likely to be an underestimate, since droplet acceleration up to speeds nearer the gas velocity would be expected. If we assume the droplet travels at an air velocity of 300 m s$^{-1}$ instead (i.e. at the sonic velocity seen in most jet nebulizers), we obtain a distance of 6–12 mm before breaking up. Some nebulizers have at least this much distance between the droplet production area and any baffles, so that it would be possible for the primary droplets to break up before impacting on a baffle. However, some nebulizers do not have this much distance between the baffles and droplet production area, so that it would appear that aerodynamic loading may not be an important factor with some nebulizer designs.

To calculate secondary droplet sizes, we can use Eq. (8.5):

$$\text{bag droplet diameters} \approx 0.042d = 0.042(50\ \mu\text{m}) = 2.1\ \mu\text{m}$$

$$\text{basal ring droplet diameters} \approx 0.36d = 18\ \mu\text{m}$$

with each of these two droplet populations containing approximately half of the original droplet mass.

It should be noted that using data from the literature on droplet breakup from shock waves may not be quantitatively applicable to nebulizers because droplets in a nebulizer are not exposed to such a sudden change in velocity as that seen by a droplet as a shock wave passes over it. This brings us to our next topic.

## 8.7 Primary droplet breakup due to gradual aerodynamic loading

The data in the previous section are based mostly on experiments with abrupt acceleration of droplets caused by shock waves. In jet nebulizers, the primary droplets are not exposed to such shock waves. Although it is not clear what difference in velocity exists between the initially formed droplets and the air jet in a nebulizer (since the primary droplet formation process itself is not completely understood), it is reasonable to suggest that the droplets are already accelerating as they are forming and continue to be accelerated upon formation. Thus, primary droplets in jet nebulizers are probably exposed to a more gradual aerodynamic loading than occurs with shock waves.

Unfortunately, much less work has been done on breakup of droplets under such conditions. However, it is clear that the rate of change of relative velocity between the droplet and surrounding air (i.e. the acceleration) will affect droplet breakup. Shraiber *et* al. (1996) review the literature on this topic and propose the following empirical correlation based on their experimental data:

$$We_{c0} = 4 + (12 + \ln Oh)\exp[-(0.03 - 0.024 \ln Oh) H] \qquad (8.7)$$

where $We_{c0}$ is the Weber number at breakup, i.e. $We_{c0} = \rho_g d\left(U|_{t_c}\right)^2/\sigma$. The parameter $H$ ($1 \leq H \leq 12$ is the range over which Eq. (8.7) interpolates the experimental data) is an integrated Weber number accounting for the drop's history:

$$H = \frac{1}{\tau_n} \int_0^{t_c} We(t)\mathrm{d}t \qquad (8.8)$$

Here $t_c$ is the time taken to reach $We_{c0}$ (i.e. the time when breakup occurs) and $\tau_n$ is the natural fundamental frequency of oscillation of the drop, an empirical correlation for which is given by Shraiber et al. (1996) as

$$\tau_n = 0.83 \, \rho_l d^2 \, Oh/\mu_l \qquad (8.9)$$

They also found droplet sizes after breakup with gradual aerodynamic loading to be given by

$$SMD \approx 0.31d \qquad (8.10)$$

which, as they admit, does not account for any dependence on dynamical parameters, which is also true of the correlation given in Eq. (8.5) for breakup due to shock waves.

In order to use Eqs (8.7)–(8.9), the speed of the drop relative to the surrounding air as a function time, $U(t)$, must be known. If we neglect all forces on the droplet except that due to aerodynamic drag, then we know from Chapter 3 that the velocity of the drop will be given by

$$m\frac{\mathrm{d}U}{\mathrm{d}t} = F_{\mathrm{drag}}$$

where $m$ is the mass of the drop. Assuming a spherical drop, then $m = \rho_l \pi d^3/6$ and $F_{\mathrm{drag}} = C_d(\rho v^2/2) \times A$, where $A = \pi d^2/4$, and $C_d$ is the drag coefficient for air flow around the drop, and this simplifies to

$$\frac{\mathrm{d}U}{\mathrm{d}t} = -\frac{3C_d\rho_g U^2}{4\rho_l d} \qquad (8.11)$$

Since the drag coefficient is a function of the drop Reynolds number, i.e. $C_d = f(Re_{dl})$ as we discussed in Chapter 3, we must know $U(t)$ to determine the drag coefficient. For a typical nebulizer droplet with $U_0 = 50 \text{ m s}^{-1}$, $d = 50 \text{ μm}$, we obtain $Re = 500$, so we cannot use Stokes drag law to obtain $C_d$. In addition, the drop will deform significantly as it approaches breakup giving larger (and unknown) $C_d$ than for a sphere. However, if we assume a constant value of $C_d$ in (8.11), we can integrate this equation to give the droplet velocity

$$U(t) = U_0/[1 + 3C_d t(\rho_g/\rho_l)^{1/2}/(4\tau_0)] \qquad (8.12)$$

where

$$\tau_0 \approx d(\rho_l/\rho_g)^{1/2}/U_0 \qquad (8.13)$$

is the initial time scale for growth of the Rayleigh–Taylor instability mentioned earlier. Hsiang and Faeth (1993) suggest that a reasonable value of $C_d$ to use for the droplet breakups they observed is 5. If we use this value in Eq. (8.12), then we can substitute $U(t)$ from this equation into $We = \rho_g d U^2/\sigma$ and integrate Eq. (8.8) for $H$. Then, substituting this result into Eq. (8.7), we are left with a transcendental equation that we must solve to obtain the only remaining unknown, which is the breakup time $t_c$. Performing the above steps requires some tedious algebra as well as a numerical root-finding method, but with the help of a symbolic package such as *Mathematica* this is not too difficult to do. Once we have obtained $t_c$, we know the time at which droplet breakup occurs, as well as the critical Weber number at breakup, as exemplified in the example at the end of this section.

It should be pointed out that the inclusion of suspended particles in the liquid being nebulized can change the droplet sizes that are produced. To predict the aerodynamic breakup of primary droplets that contain suspended particles, the work of Shraiber *et al.* (1996) is useful. They measured the breakup of 2–6 mm drops that had 100–400 µm diameter quartz particles suspended in them (occupying volume fractions $\beta$ from 0–0.21). They found that the above Eqs (8.7)–(8.9) remain valid if the density and viscosity of the liquid in these equations is replaced with the viscosity and density of the suspension as a whole, where the viscosity of the suspension is estimated using

$$\mu_{susp} = \mu_l(1 + 2.5\beta)$$

and the density is given by the result

$$\rho_{susp} = \rho_l + \beta(\rho_s - \rho_l)$$

Since particles suspended in a nebulizer are usually much smaller than primary droplet sizes, as was the case in the experiments of Shraiber *et al.*, this approach may be reasonable for the prediction of droplet sizes before impaction on baffles. Note, however, that the effect of suspended particles on secondary droplet production due to impaction on baffles can be quite important, particularly if the suspended particles are of the same size order as droplets coming off the baffles (McCallion *et al.* 1996a, Finlay *et al.* 1997) so that prediction of final droplet sizes with nebulized suspensions requires consideration of the effect of the suspended particles on secondary breakup on baffles.

## Example 8.2

Assume that a 50 µm primary droplet produced by a nebulizer has an initial velocity at formation that is 150 m s$^{-1}$ below the velocity of the surrounding air. Estimate the breakup time $t_c$, the critical $We$ for this droplet at breakup, and the distance traveled before breakup if the air jet velocity is 300 m s$^{-1}$. Assume gradual aerodynamic loading.

## *Solution*

Using an air density of $\rho_g = 1.2$ kg m$^{-3}$, and droplet density of $\rho_l = 1000$ kg m$^{-3}$ we can evaluate the Rayleigh–Taylor time scale from Eq. (8.13):

$$\tau_0 \approx d(\rho_l/\rho_g)^{1/2}/U_0 = (50 \times 10^{-6}\,\text{m}(1000/1.2)^{1/2})/(150\,\text{m s}^{-1}) = 9.62 \times 10^{-6}\,\text{s}$$

Putting this into Eq. (8.12) for the droplet's relative velocity and assuming a drag coefficient of $C_d = 5$, as suggested by Hsiang and Faeth (1993), then we have

$$U(t) = U_0/[1 + 3C_d t(\rho_g/\rho_l)^{1/2}/(4\tau_0)]$$
$$= 150 \text{ m s}^{-1}/[1 + 3 \times 5 \times t(1.2/1000)^{1/2}/(4 \times 9.52 \times 10^{-6})]$$
$$= 150/(1 + 13\,500\,t) \text{ m s}^{-1}$$

Using the definition of the droplet natural frequency $\tau_n$ in (8.9) with $\sigma = 0.072 \text{ N m}^{-1}$, $\mu_l = 0.001 \text{ kg m}^{-1}\text{s}^{-1}$, $Oh = \mu_l/[\rho_l d\sigma]^{1/2} = 0.0167$, gives

$$\tau_n = 3.46 \times 10^{-5}\text{ s}$$

Putting this with our result above for $U(t)$ into (8.8) and integrating, then after a few steps we obtain

$$H = 40.1606\,t_c/(7.40741 \times 10^{-5} + t_c)$$

Putting this into (8.7) and simplifying we obtain the following equation for the time $t_c$ to breakup:

$$18.75/(1 + 13\,500\,t_c)^2 = 4 + 0.04579 \exp[3.816 \times 10^{-4}/(7.40741 \times 10^{-4} + t_c)]$$

This equation can be solved numerically for $t_c$ using Newton iteration with an initial guess obtained graphically after plotting this equation to see where the two sides are nearly equal. Doing so, we obtain

$$t_c = 7.57 \times 10^{-5}\text{ s}$$
$$= 75.7\text{ μs}$$

This is slightly longer than the 25–50 μs we found earlier for the same droplet using results from shock wave data.

The velocity of the drop relative to the air just prior to breakup will be

$$U(t_c) = 150/(1 + 13\,500 \times 7.57 \times 10^{-5}) = 74 \text{ m s}^{-1}$$

which is roughly half its initial relative velocity. (Note the actual velocity of the drop will be $v = v_{air} - U(t)$ where $v_{air}$ is the absolute velocity of the air in the jet and here is $300 \text{ m s}^{-1}$). The Weber number at breakup will be

$$We_{c0} = \rho_g dU(t_c)^2/\sigma$$
$$= 1.2 \times 50 \times 10^{-6} \times (74 \text{ m s}^{-1})^2/0.072 \text{ N m}^{-1}$$
$$= 4.6$$

This is much lower than the critical Weber number of 12 noted earlier for drops exposed to shock waves.

To obtain the distance, $s$, traveled by the drop prior to breakup we must integrate the equation

$$v = ds/dt$$

where $v = 300 - U(t) \text{ m s}^{-1} = 300 - 150/(1 + 13\,500\,t)$. Integrating, we have

$$s = 300\,t - 0.0111 \ln(1 + 13\,500t)$$

and putting in $t = 7.57 \times 10^{-5}$ gives $s = 0.015$ m or $s = 1.5$ cm.

This is larger than the values estimated earlier with the data from shock wave data. Indeed, a distance of 1.5 cm is large enough that droplet breakup due to aerodynamic loading might not occur before impaction on baffles, depending on how far the baffles are from where the droplet was formed. However, this distance is still within values that might be expected with some nebulizer designs, so it is possible that droplet breakup due to aerodynamic loading may occur with some nebulizers.

Note that if the observation of Shraiber *et al.* (1996) on droplet sizes is valid here, then we would expect a Sauter mean droplet diameter of $0.31 \times 50 \ \mu m = 15 \ \mu m$, and if the droplets obey Simmons universal root-normal distribution with $MMD = 1.2 \ SMD$, then we would expect $MMD = 1.2 \times 15 \ \mu m = 18 \ \mu m$, which is roughly the same size as the basal ring droplets predicted earlier when we used the equations from shock wave breakup.

## 8.8 Empirical correlations

Although there is a vast amount of literature on the production of droplets from air moving at high relative velocity across a liquid surface, much of our understanding of this process is limited to qualitative experimental observations. As a result, from the standpoint of predicting this process theoretically, our understanding remains relatively poor. For this reason, our ability to predict droplet sizes and entrainment rates is limited to using empirical correlations obtained from experiments. Lefebvre (1989) reviews many such correlations for atomizers, and the reader is referred there for an excellent summary of this body of work.

Unfortunately, the geometry of the spray generation region and surrounding regions can profoundly affect the droplet production process, so that such correlations are limited to the specific geometry they were developed for. This dramatically limits the utility of the many empirical correlations that have been presented in the literature on atomizers. In particular, since nebulizers differ in several aspects of their design from traditional geometries for which many such correlations have been developed, such correlations cannot generally be used directly with nebulizers.

Despite this lack of predictive correlations for the sizes of nebulized droplets before impaction on baffles, the form of the correlations that have been developed for atomizers that most closely resemble nebulizers (i.e. airblast atomizers) can be used to suggest a number of generalizations that are probably reasonable for nebulizers. In particular, correlations for airblast atomizers are often of the form

$$\frac{d}{L} = f\left(\frac{\dot{m}_{\mathrm{l}}}{\dot{m}_{\mathrm{g}}}, Oh, We\right) \tag{8.14}$$

where $d$ is droplet size (e.g. $MMD$ or $SMD$), and $L$ is some chosen length scale (usually some nozzle dimension such as the diameter). Here,

$$Oh = \mu_{\mathrm{l}}/(\rho_{\mathrm{l}} d_0 \sigma)^{1/2} \tag{8.15}$$

is an Ohnesorge number based on the liquid properties and $d_0$ is a characteristic nozzle dimension, while $We$ is a gas Weber number

$$We = \rho_{\mathrm{g}} U_{\mathrm{g}}^2 d_0 / \sigma \tag{8.16}$$

based on an average air jet velocity $U_{\mathrm{g}}$. The quantities $\dot{m}_{\mathrm{l}}$ and $\dot{m}_{\mathrm{g}}$ are the mass flow rates (e.g. in kg s$^{-1}$) of the liquid and gas in the atomizer, respectively.

Recall that at the start of this chapter we concluded that for the simple nebulizer design where a circular jet of air blasts through a column of water, the droplet sizes prior to impaction on baffles should obey

$$\frac{d}{L} = f\left( Re_g, Re_l, Ma_g, We_g, \frac{\mu_g}{\mu_l}, \frac{\rho_g}{\rho_l} \right) \quad (8.17)$$

where we recall the definitions of the various parameters as follows.

$Re_g = U d_g \rho_g / \mu_g$ is the Reynolds number in the gas jet and $U$ is the velocity of the air–liquid interface.

$Re_l = U d_l \rho_l / \mu_l$ is the Reynolds number in the liquid surrounding the gas jet.

$Ma_g = U/c$ is the Mach number in the gas jet.

$We_g = \rho_g U^2 d_g / \sigma$ is the gas Weber number.

If nebulizers obey a correlation like that in Eq. (8.14) for airblast atomizers, then comparing Eq. (8.14) and Eq. (8.17) we see the number of dynamical parameters governing droplet formation is less than suggested by Eq. (8.17). In particular, using the fact that the mass flow rate is given by

$$\dot{m} = \rho U A$$

where $U$ is an average velocity and $A$ is the cross-sectional area of the flow, it can be shown that the ratio of mass flow rates $\dot{m}_l/\dot{m}_g$ is a function of $Re_g$, $Re_l$, $\mu_l/\mu_g$ and $\rho_l/\rho_g$. Thus, if correlations for nebulizers are of the form given in Eq. (8.14) for airblast atomizers, we see that of the six parameters on the right-hand side of Eq. (8.17), only the Mach number does not appear in Eq. (8.14), since all the other parameters on the right-hand side of Eq. (8.17) appear in some form or another in Eq. (8.14) [recall that $Oh = We^{1/2}/Re$]. Thus, Eq. (8.14) suggests that it is reasonable to expect that droplet sizes prior to impaction on baffles in a nebulizer are not significantly affected by the Mach number in the gas jet, and we can then suggest that primary droplets in a nebulizer obey a relation of the form

$$\frac{d}{L} = f\left( Re_g, Re_l, We_g, \frac{\mu_g}{\mu_l}, \frac{\rho_g}{\rho_l} \right) \quad (8.18)$$

In addition, the form of Eq. (8.14) suggests that we do not need all five of the remaining independent dynamical parameters on the right-hand side of Eq. (8.8), since only three dynamical parameters appear in Eq. (8.14) for airblast atomizers. (Note, however, that each of the three parameters on the right-hand side of Eq. (8.14) can be obtained by some combination of the five parameters on the right-hand side of Eq. (8.18), so that Eq. (8.14) is a reduced form of Eq. (8.18)). Thus, if the empirical results for airblast atomizers are representative of what can be expected of jet nebulizers, then droplet sizes from nebulizers before impaction on baffles would obey an equation of the form

$$\frac{d}{L} = f\left( \frac{\dot{m}_l}{\dot{m}_g}, Oh, We \right) \quad (8.19)$$

Correlations of this form have been suggested for nebulizers (Mercer 1981) by the above analogy with atomizers, but whether such correlations are valid for jet nebulizers has not been examined in any detail to the author's knowledge.

If jet nebulizers do obey empirical equations analogous to those obeyed by airblast atomizers, then it is useful to know more specifically the form of the right-hand side of Eq. (8.14). Lefebvre (1989) gives an excellent summary of the many correlations that have been presented, and from this summary it can be seen that a number of airblast atomizers obey a correlation of the form

$$SMD \propto AWe^{-a}\left(1 + \frac{\dot{m}_l}{\dot{m}_g}\right)^{b} + BOh^{c}\left(1 + \frac{\dot{m}_l}{\dot{m}_g}\right)^{d} \qquad (8.20)$$

where $A$ and $B$ are numerical constants, and the exponents $a$, $b$, $c$ and $d$ are all generally greater than 0 and in the range 0–1.5 or so. Such forms do correctly predict that droplet sizes decrease with increasing air to liquid flow rate (e.g. by using a higher pressure air source, smaller droplets are expected). However, this form is not as predictive as it might appear, since in a nebulizer the liquid mass flow rate is not known *a priori* (in contrast to airblast atomizers where both liquid and air mass flow rates can be considered known). In fact, the liquid mass flow rate $\dot{m}_l$ in a nebulizer is a function of the gas flow rate $\dot{m}_g$ as well as the fluid mechanics of the fluid in the reservoir of the nebulizer, as we shall see in a later section. In addition, such correlations of course must be combined with some consideration of the effect that baffles have on the droplet size distribution, which is a major effect as we shall see.

Additional analogies on droplet production prior to impaction on baffles in jet nebulizers can be obtained by examining the well-studied geometry of a liquid jet spraying into air (i.e. plain jet atomizers). For this geometry, data suggest that the spray properties should be again a function of only three parameters: $We_g$, $Re_l$ and $\rho_g/\rho_l$ (Reitz and Bracco 1982, Wu *et al.* 1991, Chigier and Reitz 1996, Lin and Reitz 1998). However, nebulizers also differ enough from the plain jet atomizer geometry that such correlations are unlikely to be directly applicable to nebulizers.

Another well-studied geometry that more closely resembles the jet nebulizer geometry is annular multiphase flow, a cut-away of which is shown in Fig. 8.7.

The presence of a central column of air flowing inside a sheath of liquid makes this geometry quite similar to that of some jet nebulizers. Droplets are entrained at the gas–liquid interface in a manner that is probably very similar to that in a jet nebulizer as the high-speed air moves over the liquid. Empirical correlations for this geometry have been developed which predict the amount of liquid entrained in the gas. Experimentally it is found that below a certain critical gas flow rate, no entrainment occurs, i.e. no droplets are produced below a certain gas flow rate. A correlation to predict this critical gas flow rate is given in Ishii and Grolmes (1975) and Ishii and Mishima (1989). This condition

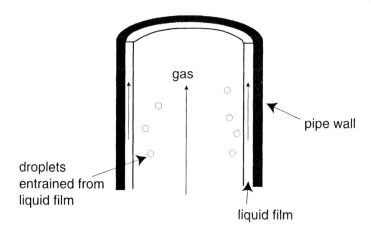

**Fig. 8.7** Schematic drawing of annular gas–liquid multiphase flow.

was developed based on a heuristic analysis of the shearing-off of roll wave crests, and is given by

$$\frac{\mu_1 U_g}{\sigma}\sqrt{\frac{\rho_g}{\rho_1}} \geq 11.78[N_\mu]^{0.8}\left(\frac{\rho_1 U_1 D}{\mu_1}\right)^{-1/3} \qquad \text{for } N_\mu \leq 15 \qquad (8.21)$$

where

$$N_\mu = \frac{\mu_1}{\left(\rho_1\sigma\sqrt{\dfrac{\sigma}{g(\rho_1 - \rho_g)}}\right)^{1/2}}$$

$U_g$ is the average velocity of the gas phase, $U_1$ is an average liquid velocity in the pure liquid film near the wall and $D$ is the outside diameter of the liquid film (i.e. the pipe diameter). Although this equation involves only dimensionless groups and might therefore be thought to be valid for gases and liquids with arbitrary physical properties, this is probably not the case. In fact, this correlation agrees well with data for air–water, but does not correlate for other more viscous liquids (Ishii and Grolmes 1975). Indeed, the presence of the gravitational acceleration, $g$, in $N_\mu$ above suggests that the angle of orientation of the tube should be important, which is not the case for nebulizers and makes one suspicious of the generality of this correlation. In fact, the presence of $g$ in this correlation arises because of the use of the wavelength

$$\sqrt{\frac{\sigma}{g(\rho_1 - \rho_g)}}$$

as a length scale, which is a characteristic lengthscale in gravity waves on horizontal air–water surfaces. However, such waves play no role in the physics of the droplet formation process as we have outlined earlier in this chapter. (It makes more sense to develop a correlation which instead uses the length scale $\sigma/(\rho_g U^2)$ from linear stability of wind-driven waves given earlier in Eq. (8.3).) Thus, the above correlation in Eq. (8.21) must be viewed as a curve fit that is validated only for the values of fluid properties (e.g. viscosity, surface tension and densities) used in the experiments (and might not be valid for the purpose of inferring the effects of changes in these fluid properties). Despite this fact, this correlation and others like it can be useful if the properties of the fluid to be nebulized differ little from that of water, which is often the case with nebulizers.

Although Eq. (8.21) might be used to predict what gas flow rate is needed in order to have any nebulization occur, of more interest is the amount of liquid that will be entrained as droplets in the gas flow. Correlations of this type have been developed for annular multiphase flow, and one such correlation that matches the experimental data is given in Ishii and Mishima (1989) as

$$E = (1 - e^{-10^{-5}\varsigma^2})\tanh(7.25 \times 10^{-7} We^{1.25} Re_1^{0.25}) \qquad (8.22)$$

where $E$ is the fraction of the water entering the pipe that is entrained up to a given distance $z$ along the pipe, where the location

$$\zeta = \frac{z}{D}\sqrt{\frac{\left[\dfrac{\sigma g(\rho_1 - \rho_g)}{\rho_g^2}\left(\dfrac{\rho_g}{\rho_1 - \rho_g}\right)^{2/3}\right]^{1/4} Re_1}{U_g}}$$

is the nondimensional axial distance from the start of the annular flow region. The Weber number used in this correlation is given by

$$We = \frac{\rho_g U_g^2 D}{\sigma}\left(\frac{\rho_1 - \rho_g}{\rho_g}\right)^{1/3}$$

while the Reynolds number is

$$Re_1 = \frac{\rho_1 U_1 D}{\mu_1}$$

As with Eq. (8.21), the use of the gravity wave length scale results in the gravitational acceleration $g$ appearing on the right-hand side, which reduces ones confidence in the validity of this correlation for predicting droplet production for fluids of arbitrary physical properties as already discussed above. However, for fluids with properties close to that of air–water (for which this correlation was developed) this correlation might be useful in predicting the entrainment efficiency (prior to impaction on baffles) for nebulizer designs whose droplet production region resembles the annular multiphase flow geometry.

It should be noted that the primary droplet production region in most jet nebulizers is not so simple as the annular multiphase flow geometry, e.g. having recirculating regions in the liquid next to the wall as shown in Fig. 8.8, resulting in the need to modify these correlations if quantitative predictions are desired, perhaps by reinterpreting the diameter of the pipe as an effective diameter such as the diameter over which the liquid is coflowing with gas flow.

Although entrainment efficiency will affect the time needed to nebulize a certain amount of drug (which is important, since too long a nebulization time will reduce

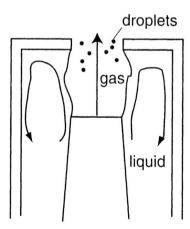

**Fig. 8.8** Recirculating regions in the liquid near the primary droplet production region may reduce the similarity between this geometry and annular multiphase flow.

patient compliance with a device), of perhaps even more interest is the size of droplets produced. Although we cannot predict droplet sizes ultimately released from a nebulizer without considering the effect of baffles, correlations that give the droplet sizes for annular multiphase flow may be of some use in predicting droplet sizes prior to baffle impaction. One such correlation that fits available experimental data reasonably well is given by Azzopardi (1985), which gives the Sauter mean diameter as

$$SMD = \sqrt{\frac{\sigma}{\rho_1 g}}\left(\frac{15.4}{We^{0.58}} + 3.5\frac{\dot{m}_{1_F}}{\rho_1 U_g}\right) \qquad (8.23)$$

where the Weber number in this correlation is

$$We = \frac{\rho_1 U_g^2}{\sigma}\sqrt{\frac{\sigma}{\rho_1 g}}$$

and $\dot{m}_{1_F}$ is the entrained liquid mass flux (i.e. the amount of liquid carried by the gas flow as droplets per unit area). Note that $\dot{m}_{1_F}/\rho_1$ has units of velocity and in a nebulizer can be interpreted as the average velocity of the liquid in the feed tube that supplies the liquid for the droplets. As with the other annular multiphase flow conditions above, this correlation was developed to match experimental data on air–water at room temperature, and the use of the gravity wave length scale makes its use for predicting the effect of using fluids with different physical properties questionable, as discussed above. Of particular interest in this equation is the lack of drop size dependence on the tube diameter. Note also that this equation lacks predictive ability just as the atomizer equations do, since the liquid mass flux in a nebulizer is not known *a priori* but must be determined by further analysis.

## Example 8.3

Use Azzopardi's correlation (Eq. 8.23) to estimate the droplet sizes before impaction that might be expected from the prototypical jet nebulizer primary droplet production region shown in Fig. 8.9. Assume the nebulizer can nebulize 2.5 ml of water in 5 s when driven by an air flow rate of $6\,l\,min^{-1}$. (Note that the actual time needed to nebulize

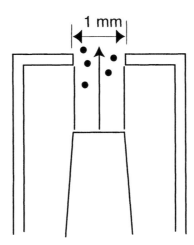

**Fig. 8.9** Schematic of primary droplet production region of a prototypical nebulizer.

2.5 ml will be much longer than 5 s since the nebulizer would include baffles placed directly in the line of travel of the droplets to reduce the droplet sizes. These baffles would dramatically reduce the rate at which droplets leave the nebulizer due to impaction on them, thus increasing nebulization time, typically to 5 or 10 minutes in many common jet nebulizers.)

## Solution

To use Azzopardi's correlation we must determine the gas velocity $U_g$. This can be obtained from the gas flow rate $Q = 6\,l\,min^{-1}$ and the area $A = \pi d_g^2/4$ (where $d_g = 0.001$ m) of the gas flow in the droplet production region using $Q = U_g A$. This gives us

$$U_g = Q/A$$

Converting $Q$ from $l\,min^{-1}$ to $m^3\,s^{-1}$ we have

$$Q = 6\,l\,min^{-1} \times 1\,min/60\,s \times 1\,m^3/1000\,l$$

$$or\ Q = 0.0001\ m^3\,s^{-1}$$

We then have $U_g = Q/(\pi d_g^2)/4 = 0.0001\ m^3\,s^{-1}/(\pi\,0.001^2\ m^2/4) = 127.3\ m\,s^{-1}$.

It should be noted that many nebulizers have orifice diameters, $d_g$, smaller than the 1 mm used here, in which case compressible flow considerations would be necessary to determine the flow velocity in the orifice (since the velocity in the orifice would reach Mach numbers $>0.3$ associated with compressibility). These considerations can be included if isentropic flow is assumed and are given in standard fluid mechanics texts (White 1998). Fortunately, in the present case, the gas velocity of $U_g = 127.3\ m\,s^{-1}$ that we have calculated is low enough that reasonable accuracy is possible without considering such compressible flow effects.

The next step in solving this problem is to calculate the Weber number in Azzopardi's correlation, defined as

$$We = \frac{\rho_l U_g^2}{\sigma}\sqrt{\frac{\sigma}{\rho_l g}}$$

$$= 1000\ kg\,m^{-3} \times (127.3\ m\,s^{-1})^2 \times (0.072\ N\,m^{-1}/(1000\ kg\,m^{-3} \times 9.81\ m\,s^{-2}))^{1/2}/$$
$$0.072\ N\,m^{-1}$$
$$= 6.10 \times 10^5$$

To use Azzopardi's correlation given in Eq. (8.23) as

$$SMD = \sqrt{\frac{\sigma}{\rho_l g}}\left(\frac{15.4}{We^{0.58}} + 3.5\frac{\dot{m}_{l_E}}{\rho_l U_g}\right)$$

we also need the entrained liquid mass flux $\dot{m}_{l_E}$. This can be inferred from the fact that 2.5 ml is nebulized in 5 s, since this means we have a mass of 2.5 g of water flowing through the cross-section $A$ in a time of 5 s. This gives a mass flux of

$$\dot{m}_{l_E} = 2.5 \times 10^{-3}\ kg/(5\ s)/(\pi\,0.001^2\ m^2)$$
$$= 636.6\ kg\,s^{-1}\,m^{-2}$$

Putting these numbers into Azzopardi's correlation then gives us

$$SMD = (0.072 \text{ N m}^{-1}/1000 \text{ kg m}^{-3}/9.81 \text{ m s}^{-2})^{1/2}$$
$$\times \ (15.4/(6.1 \times 10^5)^{0.58} + 3.5 \times 636.6/1000/127.3)$$
$$= 6.6 \times 10^{-5} \text{ m}$$
$$= 66 \ \mu\text{m}$$

If we assume Simmon's universal root-normal distribution is valid then we expect $MMD/SMD = 1.2$ as discussed in Chapter 2, and so we estimate that this nebulizer design will give us droplets with $MMD = 79.2 \ \mu\text{m}$. This is a typical size of droplet produced in a nebulizer without baffles (Nerbrink *et al.* 1994).

## 8.9 Droplet production by impaction on baffles

We have already mentioned that drops can break up into smaller droplets when they impact at high speed on a wall as a result of 'splashing'. An everyday occurrence of this type of phenomenon happens for example when raindrops splash upon landing in a puddle. In nebulizers, splashing will occur as droplets impact on baffles, as illustrated in Fig. 8.10.

Some work has been done on quantifying droplet splashing, much of which is reviewed in Rein (1993). For dry walls or walls with relatively thin layers of liquid on them (thin compared to the droplet diameter), splashing apparently results because the impacting drop forms a circular crown-like sheet coming out of the wall, which is unstable and results in droplets forming at the free rim of this sheet, as shown schematically in Fig. 8.11.

Recently, Yarin and Weiss (1995) have presented a theoretical framework that suggests that a shock-like kinematic discontinuity is responsible for the formation of the crownlike sheet. For smooth surfaces, these authors show that splashing can occur by this process only when

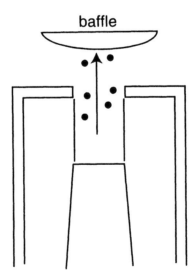

**baffle**

**Fig. 8.10** Droplets produced in the primary droplet production region will impact on primary baffles placed in the path of the particles.

**Fig. 8.11** Typical sequence of events occurring during droplet impaction and splash on a dry or thinly wetted wall.

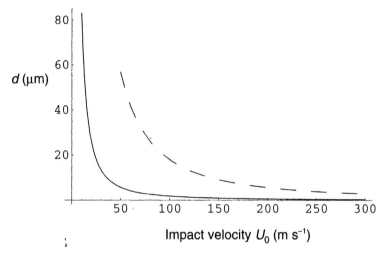

**Fig. 8.12** The critical diameter for splashing is shown for different droplet impact velocities for two different values of the constant $K_c$ in Eq. (8.24): solid line, $K_c = 57.7$; broken line $K_c = 324$. Droplet splashing can occur for diameters $d$ (in $\mu$m) or velocities $U_0$ (in m s$^{-1}$) above or to the right of these lines.

$$We_{dl}\, Re_{dl}^{1/2} = K_c^2 \text{ (for splashing to occur on a wall)} \qquad (8.24)$$

where $K_c$ is a constant and $We_{dl}$ is a droplet Weber number $We_{dl} = \rho_l d U_0^2/\sigma$ and the Reynolds number is $Re_{dl} = \rho_l d U_0/\mu_l$ where $U_0$ is the droplet velocity just prior to impaction. The value of $K_c$ depends on the properties of the surface (i.e. whether it is dry or covered by a liquid film) since Yarin and Weiss (1995) find $K_c = 324$ for drops impacting on a film that is approximately 1/6 their diameter in thickness, while Mundo *et al.* (1995) find $K_c = 57.7$ for drops impacting a smooth dry surface.

Of particular interest in our case is whether splashing of drops on baffles is likely to be an important mechanism of secondary droplet production in nebulizers. For this purpose, we can use Eq. (8.24), knowing that $K_c$ is likely to be somewhere in the range seen by the above authors[2].

Shown in Fig. 8.12 are two lines showing the critical diameter of water droplets that will 'splash' upon impaction at different velocities according to Eq. (8.24). Droplets smaller than this size (i.e. below the lines) will not splash, while those larger than this size will. The upper line corresponds to the value of $K_c = 324$ from Yarin and Weiss (1995) where droplets continually impacted on the same location with no active mechanism to

---

[2]Baffles in a nebulizer are unlikely to be completely dry because of droplets continually impacting on them, but the convection of air across the baffle as well as gravity will reduce the amount of liquid on the baffle so that it remains 'thin' in some sense. Each nebulizer design would be different in this sense, so an exact value of $K_c$ is difficult to decide on in general.

clear the resulting liquid film, while the lower line corresponds to the data of Mundo *et al.* (1995) for a dry surface. Nebulizer baffles are likely to have critical diameters for splashing that lie somewhere between these two lines.

Given our previous estimates of sizes of primary droplets produced in nebulizers, as well as data on the primary sizes of droplets from nebulizers (Nerbrink *et al.* 1994) and typical velocities of the primary droplet stream in nebulizers, it seems likely that a significant portion of the primary droplets in a nebulizer will splash. For example, it can be seen from Fig. 8.12 that with droplet velocities of 150 m s$^{-1}$ or higher, all droplets larger than about 9 μm in diameter will most certainly splash and thereby break up into smaller droplets, while even much smaller droplets will splash if the baffle is dry ($K_c = 57.7$). Thus, splashing on baffles is likely to be a significant mechanism of droplet breakup in nebulizers.

If droplet breakup on baffles due to splashing is important in nebulizers, then it is of considerable interest to know the sizes of droplets produced by this process. Unfortunately, data on droplet sizes from splashing is sparse. For droplets impacting on dry surfaces, Mundo *et al.* (1995) find that droplet sizes decrease with increasing values of the parameter

$$K = (We_{dl}\, Re_{dl}^{1/2})^{1/2} = [\rho_l^3 d^3 U_0^5/(\sigma^2 \mu_l)]^{1/4}$$

For example, they found the *SMD* of the secondary (splashed) droplets decreased from 0.6*d* at $K = 130$ (where *d* is droplet size before impact) to 0.1*d* at $K = 175$, becoming less dependent on *K* at the higher *K* values. Yarin and Weiss (1995) also find droplet sizes decrease with increasing *K* (their results are presented in terms of a dimensionless impact velocity defined as $u = K^{1/2}$), but the mode of the count distribution is nearly independent of *K* (and is near 0.06*d*) for *K* ranging from 324–1024. Mundo *et al.* (1995) also find the mode of the distribution nearly independent of *K* for $K \geq 162.5$, but find it strongly dependent on *K* for smaller *K*.

It should be noted that it is the normal component of velocity of the impacting droplets that leads to the splash mechanism, so that for oblique angles of impact the normal component of velocity should be used as the velocity in the above equations (as suggested by Mundo *et al.* 1995).

## Example 8.4

Let us reconsider the prototypical nebulizer from the example considered at the end of the previous section and now consider what happens if this droplet stream impacts a baffle. In the previous example, we estimated the *MMD* of the water droplets to be 66 μm using Azzopardi's correlation (Eq. (8.23)) for annular multiphase flow. A baffle is placed in this droplet stream such that the primary droplets can be approximated as having the same velocity as the air stream exiting the jet in the nebulizer, so that their velocity at impact will be 127 m s$^{-1}$. Estimate the size of the droplets produced after impaction.

### Solution

First, we note that the size estimate from the correlation of Azzopardi already includes the effect of possible secondary breakup due to aerodynamic loading (since this is implicitly included in the measurements of Azzopardi), so the droplet sizes impacting on

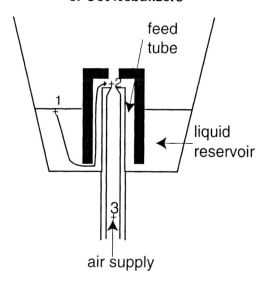

feed tube

liquid reservoir

air supply

**Fig. 8.13** Schematic of nebulizer for estimating the liquid feed rate.

the baffle are given by the result we obtained in the previous example, i.e. $MMD = 66$ μm.

Next, we must decide if splashing will occur. For this purpose, we must calculate the value of $K = (We_{dl}\ Re_{dl}^{1/2})^{1/2} = [\rho^3 d^3 U_0^5/(\sigma^2\mu)]^{1/4}$. Putting in the properties of water: $\rho = 1000$ kg m$^{-3}$, $\sigma = 0.072$ N m$^{-1}$, $\mu = 0.001$ kg m$^{-1}$ s$^{-1}$ with $d = 66$ μm, $U_0 = 127$ m s$^{-1}$, we obtain $K = 1163$. This is larger than either value of the critical values $K_c = 57.7$ or 324 for dry or wet surfaces mentioned in regard to Eq. (8.24), so we expect splashing.

To estimate the sizes of the droplets produced, we note that for such a large value of $K$, both Mundo *et al.* (1995) and Yarin and Weiss (1995) find the mode of the distribution for the secondary droplets is near $0.06d$, so a reasonable estimate for the $MMD$ of the droplets produced after impaction on the baffles is $0.06 \times 66$ μm $= 4$ μm.

Although this example is instructive, it is dependent on correlations like Eqs (8.20) or (8.23) to give the droplet sizes prior to impaction. However, we must know the liquid feed rate $\dot{m}_l$ to use Eq. (8.20) or the liquid feed velocity $\dot{m}_{lE}/\rho_l$ in order to use Eq. (8.23). This information is not generally known *a priori* for a nebulizer. However, we can estimate the liquid feed velocity or feed rate as follows by performing a relatively standard fluid dynamics analysis of the flow in the liquid feed tubes and reservoir. Let us consider the nebulizer geometry shown in Fig. 8.13, where only the primary droplet production region is shown. The liquid feed tube is assumed to be an annular region, but with minor modifications the following analysis could be applied to any feed tube geometry.

The liquid in the reservoir of the nebulizer flows from position 1 in the reservoir to position 2 (where it is turned into droplets) because of the difference in pressure between points 1 and 2. In particular, by considering the change in energy of the fluid as it moves from 1 to 2 (White 1998), we can write

$$\frac{p_1}{\rho_l g} + \frac{v_1^2}{2g} + z_1 = \frac{p_2}{\rho_l g} + \frac{v_2^2}{2g} + z_2 + h_f + \sum K \frac{v_2^2}{2g} \tag{8.25}$$

where $p$ is the pressure, $v$ is the velocity, $z$ is the vertical location (with gravity oriented vertically downwards), $h_f$ represents pressure losses due to viscous shear at the wall inside the feed tube region, $\Sigma K$ represents the sum of minor loss coefficients that account for pressure losses due to additional viscous effects that appear because the flow in the feed tube is not simply flow in an infinitely long straight tube (but has additional losses near the entrance of the feed tube, and possibly near the point 2). The subscripts 1 and 2 refer to values of quantities at points 1 and 2.

If we define the hydraulic diameter of the feed tube as

hydraulic diameter $D_h = (4 \times$ cross-sectional area of feed tube)/(perimeter of feed tube)

and assume the flow in the feed tube is laminar, then using standard fluid mechanics results for flow in tubes (White 1998) $h_f$ is given by

$$h_f = f \frac{L}{D_h} \frac{v^2}{2g} \qquad (8.26)$$

where $v = v_2$ is the average velocity in the feed tube, $L$ is the length the fluid must travel along the feed tube, and $f$ is a 'friction factor' given by

$$f = \frac{64v\zeta}{vD_h} \qquad (8.27)$$

The parameter $v$ is the kinematic viscosity of the liquid, and $\zeta$ is an empirical factor (with value between 0.6 and 1.5 or so) that depends on the cross-sectional shape of the feed tube (see White 1998). Making the reasonable assumption that $v_2 \gg v_1$, and substituting Eqs (8.26) and (8.27) into Eq. (8.25), we obtain the following quadratic equation for $v_2$:

$$\frac{v_2^2}{2g}(1 + \sum K) + \frac{32\zeta vL}{gD_h^2}v_2 + (z_2 - z_1) - \frac{p_1 - p_2}{\rho_1 g} = 0 \qquad (8.28)$$

Solving for $v_2$ we obtain

$$v_2 = \frac{\sqrt{2D_h^4(1 + \sum K)\left(\frac{\Delta p}{\rho_1} + g\Delta z\right) + 1024L^2v^2\zeta^2} - 32Lv\zeta}{D_h^2(1 + \sum K)} \qquad (8.29)$$

where we have defined $\Delta p = p_1 - p_2$ and $\Delta z = z_1 - z_2$.

To obtain $v_2$ from this equation we must know the pressure difference $\Delta p$ between points 1 and 2. Assuming the pressure supply line pressure, $p_3$, is known, the pressure $p_2$ would be known if we could estimate the pressure drop from point 3 to point 2. Unfortunately, this requires a compressible flow analysis that is beyond the scope of this book (the analysis requires more than a simple one-dimensional converging nozzle analysis since such an analysis cannot account for the previously mentioned fact that $p_2$ remains below $p_1$ for all values of $p_3$). However, once $p_2$ is known, we can use Eq. (8.29) to give an estimate for the average velocity $v_2 = \dot{m}_{lE}/\rho_1$ of the liquid in the liquid feed tube, and we can then use Azzopardi's empirical equation for droplet size without having to specify this value, allowing us to make a priori estimates of droplet sizes. We still must estimate the value of the minor loss coefficient term in Eq. (8.29) though. However, typical values for $\Sigma K$ would be expected to be near 1 for reasonable feed tube entrance geometries (White 1998).

## Example 8.5

Combine Azzopardi's Eq. (8.23) with Eq. (8.29) and the assumption that the primary baffles reduce the droplet sizes by a factor of 0.06 due to splashing, as indicated earlier, to predict the variation of droplet sizes as a function of the viscosity $\mu_l$ and surface tension $\sigma$ of the liquid in a jet nebulizer over the range of $\mu_l = 10^{-6} - 2 \times 10^{-5}\ m^2\ s^{-1}$, $\sigma = 0.03$–$0.072\ N\ m^{-1}$ (recall water has $\mu_l = 10^{-6}\ m^2\ s^{-1}$ and $\sigma = 0.072\ N\ m^{-1}$). In Eq. (8.29) assume an upstream pressure of $p_3 = 1.5\ p_{atm}$, a liquid feed tube hydraulic diameter of 1 mm and length $L = 2$ cm, a reservoir depth of $\Delta z = 1$ cm, a correction factor $\zeta = 1.5$ for the friction correction factor in Eq. (8.27), and minor loss coefficients $\Sigma K = 1$. Assume the velocity at the nozzle in the nebulizer is sonic. Assume a constant liquid density $\rho_l = 1000\ kg\ m^{-3}$ and assume $p_1 = 101$ kPa, $p_2 = 80$ kPa.

## Solution

This problem is a relatively simple matter of putting in the given numbers into Eq. (8.29)

$$v_2 = \frac{\sqrt{2D_h^4(1 + \Sigma K)\left(\dfrac{\Delta p}{\rho_1} + g\Delta z\right) + 1024L^2 v^2 \zeta^2 - 32Lv\zeta}}{D_h^2(1 + \Sigma K)} \tag{8.30}$$

The resulting values of $v_2$ can then be substituted directly into Eq. (8.23) in place of $\dot{m}_{lE}/\rho_l$, i.e. we obtain droplet sizes from

$$SMD = \sqrt{\frac{\sigma}{\rho_l g}}\left(\frac{15.4}{We^{0.58}} + 3.5\frac{v_2}{U_g}\right) \tag{8.31}$$

where

$$We = \frac{\rho_l U_g^2}{\sigma}\sqrt{\frac{\sigma}{\rho_l g}}$$

The only difficulty is posed by the fact that the gas velocity $U_g$ in the air jet at the nozzle is not known. However, we are told it is sonic, so we can use the equation for the speed of sound at the throat of a sonic converging nozzle (White 1998):

$$U_g = [2kRT_0/(k + 1)]^{1/2}$$

where $k = 1.4$ is the ratio of specific heats for air, $R = 287\ m^2\ s^{-2}\ K^{-1}$ is the ideal gas constant for air, and $T_0$ is the temperature in the low velocity air flow upstream of the nozzle, i.e. $T_0 = 293$ K. This gives $U_g = 313\ m\ s^{-1}$.

Putting in all the numbers to the above equations, we obtain the results shown in Fig. 8.14, where droplet diameter $d$ is in microns.

It can be seen that droplet sizes decrease with increases in liquid viscosity. This is a result of the decreasing liquid feed velocity $v_2$ in Eq. (8.31) that occurs because of increased viscous losses in the feed tube as the liquid becomes more viscous (i.e. more viscous liquids move more slowly through a tube). In addition, droplet sizes are seen to increase with increases in surface tension. Both of these results are in agreement with data obtained by McCallion et al. (1995). However, the above observed increase in droplet sizes with increases in surface tension is opposite to what is observed by McCallion et al. (1996b), who noted small increases in droplet sizes with decreases in

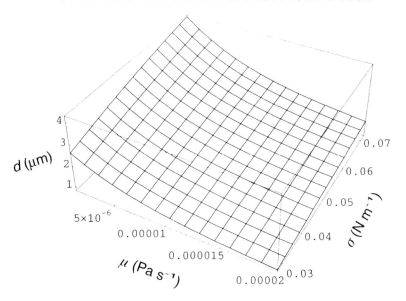

**Fig. 8.14** Predictions of droplet size (prior to impaction on secondary baffles) as a function of surface tension and viscosity for a prototypical nebulizer.

surface tension due to the addition of surfactants. However, there is a factor involved here that is not accounted for in our analysis, nor in the cited experimental measurements, and this is the effect of the vapor pressure of the liquid on droplet sizes. Indeed, droplets produced from the primary baffle will normally undergo some droplet shrinkage due to evaporation after splashing from the baffle as they make their way out of the nebulizer in order to come into equilibrium with the air that is carrying them. The addition of surface active agents can alter the vapor pressure and reduce this evaporation rate (since such agents may affect the ability of water molecules to leave the droplet surface as discussed in Chapter 4), an affect observed in other experiments (Otani and Wang 1984, Hickey *et al.* 1990). As a result, addition of surfactants may result in larger droplets with decreases in surface tension if the surfactant accumulates at the droplet surfaces and reduces hygroscopic shrinkage that would occur when these surfactants are not present.

Thus, the theory we have developed above assumes the vapor pressure remains constant while the viscosity and surface tension are varied. In previous experiments with nebulizers the surface tension or viscosity are usually varied by using different fluids, but this typically results in varying vapor pressures (Durox *et al.* 1999). For this reason, in order to predict the results of nebulizer droplet experiments done with different viscosity and surface tensions, we would need to include hygroscopic effects on the droplets as they travel from the primary production region to the exit of the nebulizer. This complicates matters, but could be done in principle, and may explain the often paradoxical and contradictory effects of changes in viscosity and surface tension on droplet sizes measured in experiments with nebulizers.

Note also that we have assumed that Eq. (8.23) is valid for arbitrary surface tension, whereas this equation has been validated only with water droplets in air, and its generality with other fluids is questionable as stated earlier. However, Eq. (8.20) also suggests that droplet sizes should increase with increasing surface tension, so that we cannot disregard this contradiction between theory and experiment. The above discussion based on vapor pressure variations may still explain this paradox.

It should also be noted that our calculations here do not account for the effects of secondary baffles, as described below, which will further alter the particle size distribution before the aerosol is measured.

## 8.10 Degradation of drug due to impaction on baffles

We have been treating impaction on primary baffles as being helpful to nebulization, since it enhances the production of smaller droplets via splash. However, splashing of a drop on a solid wall involves the propagation of a shock wave through the drop immediately after impact (Rein 1993) . This shock wave travels at the speed of sound of the liquid (which is very high in water, i.e. 1500 m s$^{-1}$, but still finite). Across the shock wave there is a discontinuity in the fluid state (i.e. velocity, pressure, density), which occurs over a length scale that is of the same size as the diameter of the solvent molecules (i.e. water). For large molecules or liposomes, which have sizes much larger than the solvent molecules, this discontinuity in pressure and velocity could result in forces that pull apart large molecules or liposomes that extend across the shock wave as it passes through the drop. Similar destructive forces may be present in the region of the kinematic discontinuity that is part of splashing as described by Yarin and Weiss (1995) which behaves like a shock wave (but with lower speed of propagation than sound waves in the liquid). Whether either of these shocks are partly responsible for the disruption of liposomes (Finlay and Wong 1998) and other large molecules (Niven *et al.* 1998) that has been observed with some jet nebulizers is a topic for future research.

## 8.11 Aerodynamic size selection of baffles

Not all droplets aimed at the baffle will impact on a baffle and splash, since smaller droplets may be able to go around the baffle with the air that goes around the baffle. In

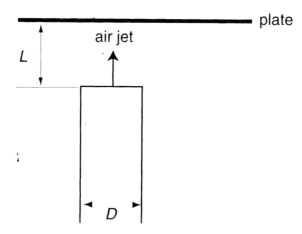

**Fig. 8.15** The geometry for impaction of a particle on a plate.

this manner, a baffle can cause a size reduction in the nebulized aerosol even without splashing, since the baffle can be thought of as a filter that prevents the large droplets from getting through.

To calculate the effectiveness of a baffle for this purpose, standard impactor theory (Willeke and Baron 1993) can be used. This theory considers the geometry shown in Fig. 8.15.

The fraction of particles $\eta$ of a given size that will impact on the plate for this geometry is known to be a function of three parameters. These parameters are the Stokes number

$$Stk = \rho_{\text{particle}}\, d^2 C_c/(18\mu_g D) \tag{8.32}$$

the jet Reynolds number

$$Re_g = \rho_g U D/\mu_g \tag{8.33}$$

and $L/D$ (the distance between the nozzle exit and the baffle divided by the jet diameter). The fraction of particles landing on the plate is then

$$\eta = f(Stk,\, Re_g,\, L/D) \tag{8.34}$$

For our purposes, it suffices to obtain an estimate of the 50% cut point diameter $d_{50}$, which is defined as the droplet diameter at which 50% of the incoming droplets will impact on the baffle. Droplets larger than $d_{50}$ have a greater than 50% chance of impacting, while smaller droplets have a less than 50% chance of depositing. From the literature on impactors, $d_{50}$ is known to be relatively insensitive to $Re_g$ and $L/D$ and can be obtained from the following empirical result:

$$d_{50} \approx \frac{[9\mu_g D/(\rho_1 C_c U)]^{1/2}}{4} \tag{8.35}$$

which is a reasonable approximation for $500 \leq Re_g \leq 10\,000$ and $1 \leq (L/D) \leq 5$. Here, $\mu_g$

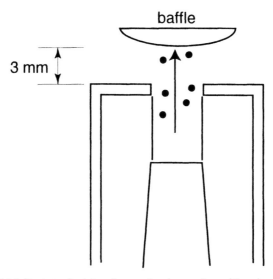

**Fig. 8.16** Prototypical droplet production region with primary baffle.

is the dynamic viscosity of the air jet, $D$ is the jet diameter, $C_c$ is the Cunningham slip factor, $U$ is the jet velocity and $\rho_l$ is the density of the liquid composing the droplets.

## Example 8.6

Consider the prototypical nebulizer geometry considered in the previous examples (and shown in Fig. 8.16). Determine the 50% cut point for the primary baffle of this nebulizer, i.e. determine $d_{50}$ for this nebulizer. Recall that the jet diameter was 1 mm, and the air velocity was 127 m s$^{-1}$. Assume the baffle is 3 mm from the nozzle exit.

### Solution

To calculate the diameter at which 50% of the droplets will impact on the baffle we use the empirical result in Eq. (8.35), i.e.

$$d_{50} \approx \frac{(9\mu_g D/(\rho_1 C_c U))^{1/2}}{4}$$

Putting in the viscosity of the air jet $\mu_g = 1.8 \times 10^{-5}$ kg m$^{-1}$ s$^{-1}$, the density of water droplets $\rho_1 = 1000$ kg m$^{-3}$, jet diameter $D = 0.001$ m, jet velocity $U = 127$ m s$^{-1}$ and using a Cunningham slip factor $C_c = 1 + 2.52\lambda/d$, we obtain

$$d_{50} \approx 0.2 \ \mu\text{m}$$

Thus, with this nebulizer only the very smallest droplets will make it past the baffle without impacting.

Nerbrink *et al.* (1994) calculate $d_{50}$ for the primary baffles of several nebulizers and obtain values that are approximately 10 times higher than the value in the above example. However, their values seem unreasonably high and may be erroneous due to an error in their use of the above empirical equation. Indeed, in order to obtain $d_{50} \approx 3 \ \mu\text{m}$ in the above example would require, for example, either increasing $D$ or decreasing $U$ by a factor of 100, or alternatively increasing $D$ to 1 cm while decreasing the jet velocity to 13 m s$^{-1}$, none of which is realistic. Thus, it would appear that primary baffles placed directly in the path of the air jet in a jet nebulizer serve largely to cause droplet breakup due to splashing on impact, and have too small a cut point to serve as a filter of the primary droplet stream since they remove virtually all of the incoming droplets of diameter $\geq 1 \ \mu\text{m}$ that are desired for inhaled pharmaceutical aerosols. Of course, once a droplet impacts, the splashing droplets associated with this impact will have a velocity away from the baffle and will be carried away from the baffle without further interaction with the baffle. This would appear to be the main route followed by droplets from a nebulizer design like that in Fig. 8.16.

It should be noted that some nebulizers have a secondary (or even tertiary) set of baffles that are designed to operate like an impactor stage to filter out the larger particles. These secondary baffles are placed in the air stream at some location downstream of the primary baffles when the jet velocity is much lower (due to entrainment as well as adiabatic expansion). Because of the much lower velocity of the air traveling past these baffles, droplet splash is not significant and the purpose of these baffles is truly aerodynamic size selection. Unfortunately, the flow in the region of the secondary baffles in nebulizers is often quite complex, partly due to the complex geometry of the baffles, but also due to the possibly turbulent, recirculating nature of the flow here, and so is not easily predicted due to the poor capabilities of turbulence models with such flows. As a result, accurate calculation of $d_{50}$ for these baffles may be difficult and the design of such baffles is largely empirical.

## 8.12 Cooling and concentration of nebulizer solutions

It has long been known that jet nebulizers become cooler than their surroundings and that the concentration of drug in solution increases during operation. These phenomena can be explained from energy and mass conservation considerations (Mercer *et al.* 1968). Cooling occurs because the droplets inside the nebulizer evaporate to come into equilibrium with the air entering the nebulizer (which is generally not saturated). This causes the droplets to cool, as we saw in Chapter 4. Most of these droplets impact on baffles and walls in the nebulizer and return to the liquid reservoir in the nebulizer, cooling the nebulizer and its contents. Because the droplets lose water to the air as they evaporate and humidify this air, water ends up leaving the nebulizer as water vapor, while drug is left behind with the droplets that impact before leaving. As a result, an amount of water leaves the nebulizer as water vapor, resulting in water leaving the nebulizer at a faster rate than the drug. This results in concentration of the drug in the nebulizer.

We can be more specific about cooling of the nebulizer with the following analysis. In particular, if we consider a control volume $V$ with surface $S$ that surrounds a nebulizer, the energy equation for this volume is given by

$$\int_S q \, dS - \int_S \rho h \mathbf{v} \cdot \hat{n} \, dS = \frac{d}{dt} \int_V \rho \hat{u} \, dV \qquad (8.36)$$

where we have neglected differences in kinetic and gravitational energy between the inlets and outlets compared to differences in enthalpy. Here $q$ is the rate of heat transfer through the nebulizer walls over the surface $S$ surrounding the nebulizer, $h$ is the specific enthalpy of the gas (air + water vapor) and droplets entering or exiting the nebulizer, and the right-hand side is the rate of change of internal ('thermal') energy of the nebulizer and its contents (including plastic walls, liquid in reservoir, droplets and air). This can be written more simply as

$$\dot{Q} + \sum h \dot{m}_{in} - \sum h \dot{m}_{out} = mc \frac{dT}{dt} \qquad (8.37)$$

where $\dot{Q}$ is the rate of heat transfer to the nebulizer from the ambient room, and the two summation terms give the rate at which enthalpy is convected in and out of the nebulizer where the air flows in or out of the nebulizer. The sums are to be done over the different components of the material entering or exiting the nebulizer, i.e. they give a term for air, for water vapor and for the droplets, where $\dot{m}_{in}$ or $\dot{m}_{out}$ are the mass flow rates of each of these components. On the right hand side, we have replaced the right hand side of Eq. (8.36) using a mass-averaged specific heat, $c$, for the nebulizer and its contents that we assume is temperature independent for the range of temperatures we expect, where $T$ is a mass-averaged temperature of the nebulizer and its contents and $m$ is its mass.

Note that the enthalpy of the air, the water vapor and the droplets entering or exiting the nebulizer can all be written as functions of temperature as

$$h = c_p T + \text{arbitrary constant}$$

Also, we must account for the fact that the mass flow rates of water vapor and droplets are different at the entrance and exit (since no droplets enter the nebulizer but some do leave, and since the droplets will humidify the air to some extent so that the water vapor concentration of the air is different on leaving than entering the nebulizer), while for air the mass flow rates going in and out of the nebulizer are the same. Including these considerations, we can write Eq. (8.37) as

$$\dot{Q} + c_{pa}(T_{in} - T_{out})\dot{m}_{air} + c_{pw}(c_{s\,in}T_{in} - c_{s\,out}T_{out})\frac{\dot{m}_{air}}{\rho_{air}} - c_l T_{out}\dot{m}_l = mc\frac{dT}{dt} \quad (8.38)$$

where $c_{pa}$ is the specific heat of air, $c_{pw}$ is the specific heat of water vapor, $c_l$ is the specific heat of liquid water, $T_{in}$ is the temperature of the air and water vapor entering the nebulizer, $T_{out}$ is the temperature of the air, water vapor and droplets exiting the nebulizer, $\dot{m}_{air}$ is the mass flow rate of air through the nebulizer, and $\dot{m}_l$ is the mass flow rate of liquid droplets leaving the nebulizer. The water vapor concentration in the air at the entrance is $c_{s\,in}$, while that at the exit is $c_{s\,out}$ (where both of these are a function of temperature via a Clausius–Claperyon equation like that given in Chapter 4 and note that they are to be evaluated at the same ambient pressure as $\rho_{air}$ since strictly speaking it is the mass fraction $c_s/\rho_{air}$ at the inlet or outlet that should appear in this equation). In Eq. (8.38), the inlet temperature $T_{in}$ can be considered known and equal to the ambient temperature $T_0$. In addition, the inlet water vapor concentration $c_{s\,in}$ can also be assumed known (based on the relative humidity and temperature in the air supply line). The air mass flow rate through the nebulizer $\dot{m}_{air}$ can also be considered known.

Equation (8.38) can be simplified by making two additional assumptions. In particular, if we assume the rate of heat transfer to the nebulizer $\dot{Q}$ can be obtained from a thermal resistance $h_{res}$ (Incropera and DeWitt 1990), i.e.

$$\dot{Q} = h_{res}(T_0 - T) \quad (8.39)$$

and if we also assume the outlet temperature $T_{out}$ is equal to the temperature of the nebulizer $T$, then Eq. (8.38) can be written as

$$\frac{d}{dt}(mcT) = T_0\left(h_{res} + c_{pa}\dot{m}_{air} + c_{pw}c_{s\,in}\frac{\dot{m}_{air}}{\rho_{air}}\right) - T\left(k + c_{pa}\dot{m}_{air} + c_{pw}c_{s\,out}\frac{\dot{m}_{air}}{\rho_{air}} + c_l\dot{m}_l\right) \quad (8.40)$$

Defining

$$a = T_0\left(h_{res} + c_{pa}\dot{m}_{air} + c_{pw}c_{s\,in}\frac{\dot{m}_{air}}{\rho_{air}}\right) \quad (8.41)$$

and

$$b = h_{res} + c_{pa}\dot{m}_{air} + c_{pw}c_{s\,out}\frac{\dot{m}_{air}}{\rho_{air}} + c_l\dot{m}_l \quad (8.42)$$

then this equation can be written in the form

$$\frac{d}{dt}(mcT) = a - bT \quad (8.43)$$

If we assume $m$, $c$, $a$ and $b$ are independent of $T$ or time $t$, Eq. (8.43) can be solved to show that the temperature obeys

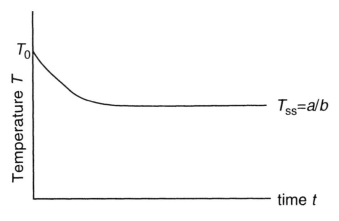

**Fig. 8.17** Nebulizer temperature plotted against time.

$$T = \frac{a}{b} + \left(T_0 - \frac{a}{b}\right)e^{-bt/mc} \qquad (8.44)$$

In this case the temperature will follow an exponential decay from initial temperature $T_0$ to an equilibrium temperature given by $T_{ss} = a/b$, where $T_{ss} < T_0$ (since $T_{ss}/T_0 = a/(bT_0) < 1$ from the definition of $a$ and $b$), as shown in Fig. 8.17.

The actual time dependence of the nebulizer temperature will not be given by Eq. (8.44), since the mass of the nebulizer and its contents, $m$, and the specific heat $c$ are not constant (e.g. $m$ decreases with time). In addition, the value of $b$ in Eq. (8.43) will depend on $T$ because of the dependence of the water vapor concentration $c_{s\,out}$ on $T$ (recall from Chapter 4 that water vapor concentration $c_s$ varies as $e^{-1/T}$), so that solution of Eq. (8.43) is not straightforward. These factors could be included and a more general solution to Eq. (8.43) could be sought, but the value of the thermal resistance, $h_{res}$, in Eq. (8.39) is not usually known since it will be affected by conductive heat transfer from the patient's hand holding the nebulizer, as well as convective motion of air next to the nebulizer and cannot be readily predicted (although Mercer *et al.* (1968) estimate values for $h_{res}$ for two nebulizers based on fitting Eq. (8.44) to experimentally measured temperature profiles). Despite these difficulties, it is clear from this analysis that nebulizer temperature can be expected to decay in an approximately exponential manner over time to a constant value, as has indeed been observed by many researchers – see Mercer *et al.* (1968), Stapleton and Finlay (1995), among many others.

A straightforward consideration of mass conservation can be used to detail how the concentration of solute (which includes drug and usually NaCl) in a jet nebulizer increases with time. In particular, the mass of solute in the nebulizer is related to the volume of liquid in the nebulizer by

$$m_s = CV_1 \qquad (8.45)$$

where $C$ is the concentration (e.g. in $kg\ m^{-3}$) of the solute in the liquid. The mass of solute in the nebulizer can only change due to solute being carried out of the nebulizer by liquid droplets exiting the nebulizer. However, we have said that the rate at which mass leaves the nebulizer as droplets is $\dot{m}_1$. The volume flow rate of these drops is $\dot{m}_1/\rho_1$, and their concentration is $C$ (since the concentration of the drops is nearly the same as the concentration of the liquid in the reservoir (Stapleton and Finlay 1995)) so the rate at which solute leaves the nebulizer will be

$$\frac{dm_s}{dt} = -\frac{\dot{m}_l}{\rho_l}C \tag{8.46}$$

Substituting Eq. (8.45) into Eq. (8.46), we obtain an equation for the rate of change of solute concentration:

$$C\frac{dV_1}{dt} + V_1\frac{dC}{dt} = -\frac{\dot{m}_l}{V_1}C \tag{8.47}$$

Notice that to solve this equation for $C(t)$ we must know the volume of liquid $V_1$ in the nebulizer. However, we can develop an equation for $V_1$ from mass conservation. In particular, the volume of liquid in the nebulizer will change because the air picks up water vapor as it travels through the nebulizer (since the drops in the nebulizer try to bring the water vapor concentration in the air up to the same level that is present at their surfaces, which is usually nearly saturated for the isotonic solutions that are used in nebulizers). In addition, liquid is lost as droplets exiting the nebulizer. More specifically, we can write

$$\rho_l \frac{dV_1}{dt} = \frac{\dot{m}_{air}}{\rho_{air}}(c_{s\ in} - c_{s\ out}) - \dot{m}_l \tag{8.48}$$

where $\rho_{air}$ and $c_s$ are to be evaluated at ambient pressure, and we have neglected any density changes in the liquid due to increases in solute concentration.

To determine how the concentration of solute $C$ changes with time we must solve Eq. (8.48), and put our solution for $V_1(t)$ into Eq. (8.47). We can then solve Eq. (8.47) to obtain $C(t)$. However, the right-hand side of Eq. (8.48) will depend on the temperature of the nebulizer $T$, since $c_{s\ out}$ is a function of the temperature. Thus, we must solve the equations we wrote down earlier for the nebulizer temperature $T$ before we can solve Eq. (8.48). However, we saw above that this cannot be done easily, so instead, to obtain an idea of how the concentration changes with time, let us assume that the right-hand side of Eq. (8.48) does not vary with time. This means we are assuming that the rate of water vapor transport and liquid droplet transport out of the nebulizer are constant in time. With this assumption, we can solve Eq. (8.47) to obtain

$$V_1 = V_{10} - \lambda t \quad \text{where } \lambda = \frac{1}{\rho_l}\left[\dot{m}_l + \frac{\dot{m}_{air}}{\rho_{air}}(c_{s\ out} - c_{s\ in})\right] \tag{8.49}$$

and $V_{10}$ is the initial volume of liquid in the nebulizer. This equation clearly cannot be valid for all times, since the liquid volume $V_1$ decreases linearly with time according to this equation and will become negative at some time. Thus, it is clear that our assumption of a constant right-hand side to Eq. (8.48) is incorrect. However, if we substitute this equation into Eq. (8.47), we obtain a linear equation for $C$ that can be solved to give

$$C(t) = C_0\left(\frac{V_{10}}{V_{10} - \lambda t}\right)^{\frac{\dot{m}_l}{\lambda \rho_l} - 1} \tag{8.50}$$

where $C_0$ is the initial concentration of solute in the nebulizer. This equation predicts an algebraic increase of concentration with time. The actual concentration variation with time will not obey this equation exactly, since the rate of change of volume in the nebulizer is not constant. However, Eq. (8.50) can be used to fit experimental data (Mercer *et al.* 1968), where concentration increases with time, as noted by many authors (Phipps and Gonda 1990).

## 8.13 Nebulizer efficiency and output rate

So far we have mostly been discussing the sizes of droplets that are produced from a nebulizer and the physics of droplet production. Although droplet size is an important parameter with nebulizers, of nearly equal importance is the efficiency of the nebulizer (i.e. the fraction of the drug put into the nebulizer that is delivered to the patient), as well as the rate at which a nebulizer can deliver a certain amount of drug (i.e. its output rate). Output rate is important since low output rates lead to long treatment times, which reduce patient compliance (compliance is a measure of what fraction of patients actually take the medications they are prescribed).

One of the primary determinants of both efficiency and output rate is the ability of the nebulizer to return droplets to the nebulizer reservoir for renebulization when they collect on primary and secondary baffles and nebulizer walls. However, this is not easy to predict since it involves thin film instability and dewetting phenonema, which are not readily predicted quantitatively for the complex geometries inside a nebulizer, and remain a topic for future work.

A major determinant of output rate in a jet nebulizer is the rate at which primary droplets are created, which will increase if the mass flow rate of liquid into the droplet production region increases with all other variables unchanged. Most jet nebulizers use narrow-diameter channels or tubes to supply the primary droplet production with liquid, so that increases in the viscosity of the nebulizer liquid will decrease the flow rate in these tubes (because of increased viscous friction at the walls of these tubes). As a result, the output rate of a jet nebulizer can be expected to decrease with increasingly viscous liquids. Indeed, because the viscosity of liquids increase with decreases in temperature, nebulizer liquid output rate can be expected to decrease as the temperature of a nebulizer is lowered. Because jet nebulizers drop significantly in temperature from the start to the finish of nebulization, this effect could cause nebulizer liquid output rates to decrease during nebulization with jet nebulizers. However, such an effect may be partially compensated for by increased drug concentration in the nebulizer, so that drug output rates may remain nearly constant (Smaldone *et al.* 1992). Whether such temperature-dependent changes in viscosity do indeed affect liquid output rates is a topic for future research.

## 8.14 Charge on droplets produced by jet nebulization

Several authors have measured the charge on droplets produced by atomization and have found that such droplets may have significant electrical charge (Chow and Mercer 1971), an effect which can be particularly important when a nebulizer is used to generate monodisperse aerosols by evaporation of nebulized water droplets that contain suspended monodisperse particles (Whitby and Liu 1968). Although different mechanisms for the charging of droplets have been proposed (Natanson 1949, Jonas and Mason 1968, Matteson 1971), it is likely that this phenomenon is caused by differences in the mobility of positive and negative ions, as suggested by Matteson (1971). In particular, ions in solution tend to move away from any newly created air–water surface, since they are attracted by intermolecular forces with the underlying water molecules. As newly created surface areas begin to form, such as occurs when primary droplets start to pull away from waves on the air–water surface, or as droplets pull away from ligaments that occur in the first stages of a droplet splashing from a baffle, the fresh surface will initially have the same composition as the bulk. However, rapid motion of ions away from this surface will occur. But the negative and positive ions will not move away from this surface at the same rate, since in general these ions have different sizes, and therefore feel different intermolecular forces with the underlying water molecules. As a result, if a drop pinches off from the bulk fluid quickly enough, it may carry away more ions of one charge than the other (e.g. if negative ions move more slowly than positive ions, then more negative than positive ions would be carried away by the drop), resulting in a tendency to have a net charge on the droplet.

This effect is particularly pronounced at low ion concentrations ($< 10^{-4}$ mole $l^{-1}$) with water, where differences in the mobility of $OH^-$ and $H^+$ ions are important (with the negative hydroxyl ions being less mobile, so that a net negative charge occurs when nebulizing distilled water). However, with the addition of significant amounts of solute, differences in the mobility between the negative and positive added solute ions begin to dominate. In fact, if the positive ions of the added solute are less mobile than the negative ions (as occurs with $K^+$ and $Cl^-$), then less charge occurs on the droplets as more solute is added, since the net motion of positive solute ions will counterbalance the net motion of hydroxyl ions into the droplets. At a certain solute concentration, this can counterbalance the effect of the reduced mobility of the hydroxyl ions, and droplets without charge can be produced (typically around $10^{-4}$ moles $l^{-1}$, Matteson 1971). With further increases in solute concentration, charged droplets are again produced. Note that for NaCl, the negative chloride ions are larger and less mobile than the sodium ions, so the addition of NaCl should initially serve to add to the net negative charging of the water droplets that occurs due to the hydroxyl ions. Thus, we expect to see nebulized saline droplets having a net negative charge at low NaCl concentrations. However, interionic interactions begin to occur at higher concentrations ($> 10^{-5}$ moles $l^{-1}$ or so) which reduce the effectiveness of the above charging mechanism, so that charge is a nonlinear function of solute concentration. In fact, for all electrolyte solutions (both those with more mobile positive ions and those with more mobile negative ions), much lower levels of charging are observed at concentrations above $10^{-4}$ moles $l^{-1}$ (Matteson 1971). Since most nebulized pharmaceutical aerosols use nearly isotonic amounts of NaCl (i.e. 0.15 moles $l^{-1}$) in order to avoid irritating isotonic airway surfaces, charging is much less important for such aerosols than for distilled water.

It should be noted that the addition of surface active ions (surfactants) will alter the above charging mechanism, and can result in much higher charging levels (Matteson 1971).

Regarding levels of charge, Chow and Mercer (1971) find charge levels of less than 20 elementary charges per droplet for nebulizer droplets up to 10 µm in diameter produced from 0.1% and 1.0% NaCl–uranine (9 : 1 by mass) aqueous solutions, with charge levels given empirically by $|n'| \approx 4.8\, d^{0.6}$ where $d$ is particle diameter in microns and $n'$ is the number of elementary charges on the droplet. As we saw in Chapter 3, these charge levels are not large enough to significantly affect deposition in the lung. However, Chow and

Mercer observed larger levels of charge at higher concentrations ($|n'| \approx 18.7\, d^{0.9}$), which may be large enough to cause significant effects on deposition in the lung if such charges are not neutralized in the respiratory tract by the hydrated ion clusters discussed in Chapter 3.

Note that the above mechanism of charging cannot operate for nonionic liquids. Thus, nebulization of methanol, for example, would not be expected to yield droplets with net charge. This is indeed observed to be the case in experiments (Matteson 1971).

## 8.15 Summary

From our discussion in this chapter it should be apparent that there are several major processes involved in producing droplets from a nebulizer. These include the primary production process due to the flow of high speed air across a water surface, the possible subsequent breakup of these primary droplets due to the lower speed of the droplets compared to the surrounding air (i.e. aerodynamic breakup), breakup of droplets due to splashing on primary baffles, and aerodynamic size selection by secondary baffles. In addition, interfacial phenomena are important in determining how much drug is wasted due to drops clinging to the sides of the nebulizer, and in determining the charge on nebulized droplets. Although we have seen that certain aspects of these processes may be predicted based on studies aimed at understanding each of these processes on their own, no generalized theory is available that allows detailed quantitative prediction of nebulizer behavior with inclusion of all of the above processes. Such an understanding awaits further research.

## References

Azzopardi, B. J. (1985) Drop sizes in annular two-phase flow, *Exp. Fluids* 3:53–59.

Boomkamp, P. A. M. and Miesen, R. H. M. (1996) Classification of instabilities in parallel two-phase flow, *Int. J. Multiphase Flow* 22:67–99.

Chigier, N. and Reitz, R. D. (1996) Regimes of jet breakup and breakup mechanisms (physical aspects), Chapter 4 of *Recent Advances in Spray Combustion: Spray Atomization and Drop Burning Phenomena*, ed. K. K. Kuo. AIAA, Reston, VA.

Chou, W.-H. and Faeth, G. M. (1998) Temporal properties of secondary drop breakup in the bag breakup regime, *Int. J. Multiphase Flow* 24:889–912.

Chow, H. Y. and Mercer, T. T. (1971) Charges on droplets produced by atomization of solutions, *Am. Ind. Hyg. Assoc. J.* April:247–255.

Craik, A. D. D. (1985) *Wave Interactions and Fluid Flows*, Cambridge University Press, Cambridge.

Collis, G. G., Cole, C. H. and Le Souëf, P. N. (1990) Dilution of nebulised aerosols by air entrainment in children, *Lancet* 336:341–343.

Durox, D., Ducruix, S. and Lacas F. (1999) Flow seeding with an air nebulizer, *Exp. Fluids* 27:408–413.

Faeth, G. M., Hsiang, L.-P. and Wu, P.-K. (1995) Structure and breakup properties of sprays, *Int. J. Multiphase Flow* 21:Suppl. 99–127.

Faragó, Z. and Chigier, N. (1992) Morpological classification of disintegration of round liquid jets in a coaxial air stream, *Atomization and Sprays* 2:137–153.

Finlay, W. H. and Wong, J. P. (1998) Regional lung deposition of nebulized liposome-encapsulated ciprofloxacin, *Int. J. Pharmaceutics* 167:121–127.

Finlay, W. H., Stapleton, K. W. and Zuberbuhler, P. (1997) Predicting lung dosages of a nebulized suspension: Pulmicort[R] (Budesonide), *Particulate Sci. Technol.* 15:243–251.

Gelfand, B. E. (1996) Droplet breakup phenomena in flows with velocity lag, *Proc. Energy Combust. Sci.* **22**:201–265.

Hewitt, G. F. and Hall-Taylor, H. S. (1970) *Annular Two-Phase Flow*, Chapter 8, Pergamon Press, New York.

Hickey, A. J., Gonda, I., Irwin W. J. and Fildes, F. T. J. (1990) The effect of hydrophobic coating upon the behavior of a hygroscopic aerosol powder in an environment of controlled temperature and relative humidity, *J. Pharm. Sci.* **79**:1009–1014.

Hsiang, L.-P. and Faeth, G. M. (1992) Near-limit drop formation and secondary breakup, *Int. J. Multiphase Flow* **18**:635–652.

Hsiang, L.-P. and Faeth, G. M. (1993) Drop properties after secondary breakup, *Int. J. Multiphase Flow* **19**:721–735.

Hsiang, L.-P. and Faeth, G. M. (1995) Drop deformation and breakup due to shock wave and steady disturbances, *Int. J. Multiphase Flow* **21**:545–560.

Incropera, F. P. and DeWitt, D. P. (1990) *Introduction to Heat Transfer*, Wiley, New York.

Ishii, M. and Grolmes, M. A. (1975) Inception criteria for droplet entrainment in two-phase concurrent film flow, *AIChE J.* **21**:308–318.

Ishii, M. and Mishima, K. (1989) Droplet entrainment correlation in annular two-phase flow, *Int. J. Heat Mass Transfer* **32**:1835–1846.

Jonas, P. R. and Mason, B. J. (1968) Systematic charging of water droplets produced by break-up of liquid jets and filaments, *Trans. Faraday Soc.* **64**:1971–1992.

Lefebvre, A. H. (1989) *Atomization and Sprays*, Taylor & Francis, Bristol, PA.

Lin, S. P. and Reitz, R. D. (1998) Drop and spray formation from a liquid jet, *Ann. Rev. Fluid Mech.* **30**:85–105.

Matteson, M. J. (1971) The separation of charge at the gas–liquid interface by dispersion of various electrolyte solutions, *J. Colloid Interface Sci.* **37**:879–890.

McCallion O. N. M., Taylor, K. M. G., Thomas, M. and Taylor, A. J. (1995) Nebulization of fluids of different physicochemical properties with air-jet and ultrasonic nebulizers, *Pharm. Res.* **12**:1682–1688.

McCallion, O. N. M., Taylor, K. M. G., Thomas, M. and Taylor, A. J. (1996a) Nebulization of monodisperse latex sphere suspensions in air-jet and ultrasonic nebulisers, *Int. J. Pharm.* **133**:203–214.

McCallion, O. N. M., Taylor, K. M. G., Thomas, M. and Taylor, A. J. (1996b) The influence of surface tension on aerosols produced by medical nebulisers, *Int. J. Pharmaceutics* **129**:123–136.

Mercer, T. T. (1981) Production of therapeutic aerosols: principles and techniques, *Chest* **80**:813–818.

Mercer, T. T., Tillery, M. I. and Chow, H. Y. (1968) Operating characteristics of some compressed-air nebulizers, *Am. Ind. Hyg. J.* **29**:66–78.

Mundo, C. H. R., Sommerfeld, M. and Tropea, C. (1995) Droplet-wall collisions: experimental studies of the deformation and breakup process, *Int. J. Multiphase Flow* **21**:151–173.

Natanson, G. L. (1949) The electrification of drops during atomization as a result of fluctuations in the ion distribution, *Zh. Fiz. Khim.* **23**:304–314.

Nerbrink, O. Dahlback, M. and Hansson, H.-C. (1994) Why do nebulizers differ in their output and particle size characteristics, *J. Aerosol Med.* **7**:259–276.

Niven, R. W., Wedeking, T. and Smith, J. G. (1998) Effects of formulation on gene delivery to the lung, in *Respiratory Drug Delivery VI*, Interpharm Press, Buffalo Grove, IL, pp. 177–185.

Otani, Y. and Wang, C. S. (1984) Growth and deposition of saline droplets covered with a monolayer of surfactant, *Aerosol Sci. Technol.* **3**:155–166.

Phipps, P. R. and Gonda, I. (1990) Droplets produced by medical nebulizers, *Chest* **97**:1327–1332.

Pilch, M. and Erdman, C. A. (1987) Use of breakup time data and velocity history data to predict the maximum size of stable fragments for acceleration-induced breakup of a liquid drop, *Int. J. Multiphase Flow* **13**:741–757.

Rein, M. (1993) Phenomena of liquid drop impact on solid and liquid surfaces, *Fluid Dynam. Res.* **12**:61–93.

Reitz, R. D. and Bracco, F. V. (1982) Mechanism of atomization of a liquid jet, *Phys. Fluids* **25**:1730–1742.

Saunders, L. (1966) *Principles of Physical Chemistry for Biology and Pharmacy*, Oxford University Press, New York.

Scardovelli, R. and Zaleski, S. (1999) Direct numerical simulation of free-surface and interfacial flow, *Ann. Rev. Fluid Mech.* **31**:567–603.

Shraiber, A. A., Podvysotsky, A. M. and Dubrovsky. V. V. (1996) Deformation and breakup of drops by aerodynamic forces, *Atomization and Sprays* **6**:667–692.

Smaldone, G. C., Dickinson, G., Marcial, E., Young, E. and Seymour, J. (1992) Deposition of aerosolized pentamidine and failure of pneumocystis prophylaxis, *Chest* **101**:32–87.

Stapleton, K. W. and Finlay, W. H. (1995) Determining solution concentration within aerosol droplets output by jet nebulizers, *J. Aerosol Sci.* **26**:137–145.

Taylor, G. I. (1950) *Proc. Roy. Soc.* **201**:192–196.

Taylor, G. I. (1958) Generation of ripples by wind blowing over a viscous fluid, in *Collected Works of G.I. Taylor*, ed. G. K. Batchelor, Cambridge University Press, Cambridge.

Taylor, J. J. and Hoyt, J. W. (1983) Water jet photography – techniques and methods, *Exp. Fluids* **1**:113–120.

Whitby, K. T. and Liu, B. Y. H. (1968) Polystyrene aerosols – electrical charge and residue size distribution, *Atmosph. Environ.* **2**:103–116.

White, F. M. (1999) *Fluid Mechanics*, 4th edition, McGraw-Hill, Boston.

Willeke, K. A. and Baron, P. A. (1993) *Aerosol Measurement, Principles, Techniques and Applications*, Van Nostrand Reinhold, New York.

Woodmansee, D. and Hanratty, R. J. (1969) Mechanism for the removal of droplets from a liquid surface by a parallel air flow, *Chem. Eng. Sci.* **24**:299–307.

Wu, P.-K., Ruff, G. A. and Faeth, G.M. (1991) Primary breakup in liquid–gas mixing layers, *Atomization and Sprays* **1**:421–440.

Wu, P.-K., Tseng, L.-K. and Faeth, G. M. (1992) Primary breakup in liquid–gas mixing layers for turbulent liquids, *Atomization and Sprays* **2**:295–317.

Yarin, A. L. and Weiss, D. A. (1995) Impact of drops on solid surfaces: self-similar capillary waves, and splashing as a new type of kinematic discontinuity, *J. Fluid Mech.* **283**:141–173.

# 9
# Dry Powder Inhalers

The concept of using a device to disperse small powder particles for inhalation into the lung to give therapeutic effect is easy to grasp in a general sense. However, of the three most commonly used methods for delivering therapeutic agents to the lung in broad clinical use today (i.e. nebulizers, powder inhalers and pressurized metered dose inhalers), powder aerosols are the latest of the three to be developed, partly due to the difficulty in manufacturing and reproducibly dispersing small, controlled amounts of fine particles.

In common with the other delivery methods, the mechanics of dry powder inhalers is not well understood. For dry powder inhalers (or 'DPIs' as they are commonly referred to), this lack of understanding is largely a result of our current inability to predict the adhesive and aerodynamic forces on the irregularly shaped and rough-surfaced, small particles that are normally used in dry powder inhalers. Despite this, a partial understanding of the mechanics involved can be obtained by considering simplified systems, a theme we have pursued in previous chapters, and to which we again turn.

## 9.1 Basic aspects of dry powder inhalers

Before delving into the mechanics of dry powders, let us first consider some basic aspects of these inhalers.

The purpose of a dry powder inhaler is to insert a prescribed dose of powder aerosol into the air inhaled by a patient during a single breath. This is shown schematically in Fig. 9.1. A dose of powder is presented for inhalation in the device (this dose is usually prevented from being exposed to ambient air until the patient is ready to inhale, since condensation from humid air onto the powder can interfere with powder dispersal, as we shall later see in more detail). During inhalation, the patient's air entrains powder as it flows through the device. Since coughing can be induced by inhalation of large amounts of powder, total amounts of inhaled powder (including active drug and any excipients) are usually less than 10–20 mg. In 'passive' dry powder inhalers, the motion of the inhaled air supplies all of the energy associated with pick-up (i.e. entrainment) and breakup (i.e. deaggregation) of the powder particles, whereas in 'active' dry powder inhalers, an external source of energy (such as a battery or stored mechanical energy) releases energy during inhalation to help disperse the powder (e.g. by spinning an impeller, or blasting the powder with compressed air). Normally the powder is delivered during a single, large breath. For descriptions of the basic operation of many different dry powder inhaler designs see Dunbar *et al.* (1998).

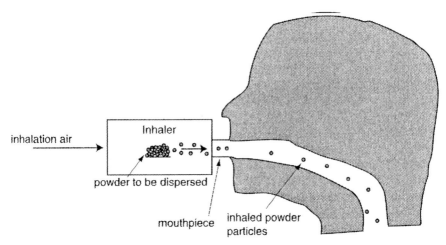

**Fig. 9.1** Schematic of the basic operation of a dry powder inhaler.

## 9.2 The origin of adhesion: van der Waals forces

One of the primary obstacles that a dry powder inhaler must overcome is the tendency of powder particles to adhere, prior to their dispersal, to any surfaces (particularly the surfaces of fellow particles) with which they are in contact. We have seen in Chapter 7 that inhaled particles that are larger than several microns in diameter typically have a high probability of depositing in the mouth–throat, so that clumps of aggregated powder particles have a greater chance of depositing in the mouth–throat than if the particles are not aggregated. Thus, if the lung is the intended delivery target, it is normally desirable that inhaled powder particles disperse into individual particles. For this to happen, the adhesive forces between particles in the bulk powder must be overcome. Before examining how this can be made to happen, let us examine the origin of these forces. Since this topic is an entire field in itself, our treatment here is necessarily brief (the reader is referred to Israelachvili (1992) for a more complete discussion).

Adhesion due to adhesive forces (between objects of different material) or auto-adhesive[1] forces (between separate particles of the same material) are simply the result of electromagnetic forces acting between the electrons and protons of the individual molecules making up the objects. Unfortunately, the quantum mechanical nature of the electron 'clouds' that surround molecules yields a set of equations for the distribution of these clouds that cannot be solved analytically, and which are overwhelmingly too demanding to solve numerically for the relatively large molecules and large number of such molecules contained in particles seen in pharmaceutic applications. However, it is possible to calculate the forces between simple objects, such as a plane and a spherical particle, or two spherical particles, if some additional information is introduced. Because such calculations are strongly dependent on the detailed shape of the particle where it contacts the surface, such simplified cases actually have little predictive value for typical pharmaceutic particles due to the irregular (rough) shape of such particles. In fact, it is not currently possible to predict the force of adhesion with real (rough) pharmaceutic

---

[1]The term cohesion refers to the intermolecular forces within a single solid body.

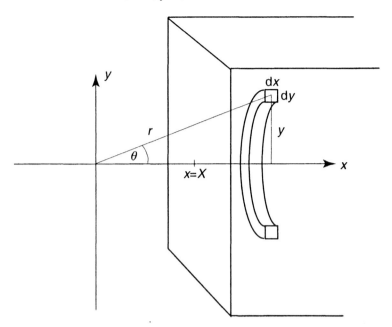

**Fig. 9.2** Schematic geometry for calculating the force of attraction between a molecule at the origin and a wall that occupies the region $x \geq X$.

dry powder particles. Even so, it remains instructive to consider the case of ideally smooth particles, if only to demonstrate the molecular origin of such adhesive forces.

To this end, let us perform a simplified calculation of the force of molecular attraction of a spherical particle next to a flat wall. We begin by first examining the force on a single molecule near a wall, as shown in Fig. 9.2.

Consider a single molecule placed at the origin, with a wall at distance $X$ from the origin in the $x$-direction. To determine the total force the wall exerts on the molecule, we can add up the force each ring of radius $y$, like the one shown in Fig. 9.2, exerts on the molecule, since all molecules in this ring are essentially at the same distance $r$ from our molecule at the origin. If the wall contains $n_m$ molecules per unit volume, then each ring contains a total of $2\pi y \, n_m \, dx \, dy$ molecules. If we knew the force, $f$, exerted by each molecule in this ring on the molecule at the origin, then we could simply add up these individual molecular forces to obtain the total force of molecular attraction of the ring due on the molecule at the origin[2]. Then, to obtain the total force $f_x$ (which is in the $x$-direction) that the wall exerts on the molecule at the origin, we would add up all the forces the rings making up the wall exert on the molecule as follows:

$$f_x = \int_{y=0}^{\infty} \int_{x=X}^{\infty} (f\cos\theta) 2\pi y n_m dx \, dy \tag{9.1}$$

where $f$ is the force of a single molecule in the wall on the molecule at the origin, and

[2]This is actually a simplification because molecular forces are not purely additive since the presence of one molecule affects the forces that neighboring molecules exert on other molecules – we ignore such 'nonadditivity' in our simple derivation here, but its presence is one of several factors that interfere with the rigor and accuracy of our present simplified approach.

$\cos \theta = x/r$ resolves the intermolecular force into the $x$-direction (the forces in the $y$ and $z$-direction cancelling due to axisymmetry).

However, the force of attraction between individual molecules is difficult to determine, as mentioned above. Its rigorous determination requires quantum mechanical considerations that are beyond the scope of this text. For our purposes, it is enough to know that such complex considerations result in intermolecular forces of the form

$$f = \frac{C}{6r^7} \qquad (9.2)$$

where $r$ is the distance between molecules and $C$ is a constant that depends on the properties of the molecules involved. Equation (9.2) may be more familiar when it is realized that it arises from differentiating the intermolecular potential energy, $w$, where $w = -Cr^{-6}$ is the attractive potential energy term occurring in the Lennard–Jones potential commonly encountered in introductory chemistry. As noted above, the origin of the intermolecular potential energy lies in the quantum mechanical electromagnetic interactions between the electrons and protons of the molecules, referred to generically as van der Waals forces. Such interactions involve Coulombic forces between charges, dipoles and induced dipoles (where one molecule induces a dipole in a nearby molecule, which can be time-dependent as the electrons in one molecule move about and cause time-dependent dipoles giving so-called 'dispersion' intermolecular forces). For separations larger than a few nanometers, it is also necessary to include the fact that electromagnetic signals travel at the speed of light, so that if, for example, electrons in a molecule in the wall are inducing a dipole in our molecule at the origin in Fig. 9.2, there is a lag (or 'retardation') before our molecule 'feels' this dipole and responds (because of the finite time it takes for electromagnetic signals to travel between the molecules). However, by this time the electrons in the wall molecule may have moved to another position and are now inducing a different dipole strength. Such considerations involve so-called 'retardation' effects, first included by the Russian physicist E. M. Lifschitz in the 1950s, and result in Eq. (9.2) having a $1/r^8$ dependence when $r$ is larger than a few nanometers.

Without worrying about how such complex considerations give rise to Eq. (9.2) or their effect on this equation, we can substitute Eq. (9.2) into Eq. (9.1) and integrate using $r = (x^2 + y^2)^{1/2}$ to obtain the total force exerted on our molecule at the origin by the wall (and which is in the $x$-direction):

$$f_x = \frac{\pi C n_m}{2X^4} \qquad (9.3)$$

Having calculated the force exerted on a single molecule due to the wall, we can proceed to calculate the total force on a spherical particle due to this force acting on all the individual molecules in the particle. Figure 9.3 shows the relevant geometrical considerations.

The distance, $D$, between the wall and the particle is not zero because of repulsive molecular forces that set in when molecules are close together (which give rise to the familiar $r^{-12}$ term in the Lennard–Jones intermolecular potential $w = -Cr^{-6} + Br^{-12}$). Typical values of separation $D$ are a few tenths of a nanometer, and are substance dependent, but an often used value is 0.4 nm (Krupp 1967, Rietema 1991). When the attractive and repulsive molecular forces are balanced, the point of closest

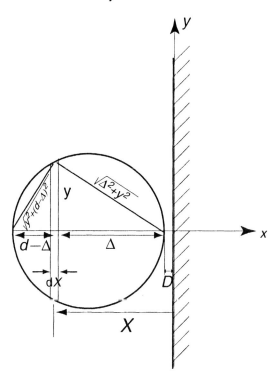

**Fig. 9.3** A spherical particle of diameter $d$ situated a distance $D$ from a plane wall.

approach of the sphere to the wall will be referred to as the particle contact point (with the corresponding point on the wall being the wall contact point).

Since we have already worked out the force, $f_x$, exerted by the wall on one molecule situated a distance $X$ from the wall, the force on a slice of the particle situated a distance $X$ from the wall will simply be the force $f_x$ times the number of molecules contained in this slice, i.e. the van der Waals force, $dF_{vdW}$, on the slice of thickness $dX$ shown in Fig. 9.3 is given by

$$dF_{vdW} = f_x\, n_{mp}\, \pi y^2\, dX \tag{9.4}$$

where $n_{mp}$ is the molecular number density of the particle. The total force on the particle due to molecular forces from the wall is then

$$F_{vdW} = \int_{X=D}^{X=D+d} f_x(X) n_{mp} \pi y^2 dX \tag{9.5}$$

Using Pythagoras' theorem, we can relate the distance $y$ in Fig. 9.3 to $\Delta$ and the particle diameter $d$ as follows:

$$d^2 = [y^2 + (d - \Delta)^2] + (\Delta^2 + y^2) \tag{9.6}$$

which can be simplified to

$$y^2 = (d - \Delta)\Delta \tag{9.7}$$

Substituting Eq. (9.7) into Eq. (9.5), realizing that

$$X = D + \Delta \tag{9.8}$$

and using Eq. (9.3) for $f_x$, we can rewrite Eq. (9.5) as

$$F_{vdW} = \frac{\pi^2 C n_{mp} n_m}{2} \int_{\Delta=0}^{d} \frac{(d-\Delta)\Delta}{(\Delta+D)^4} d\Delta \tag{9.9}$$

Integrating Eq. (9.9), we obtain our result

$$F_{vdW} = \frac{\pi^2 C n_{mp} n_m}{12} \frac{d^3}{D^2(d+D)^2} \tag{9.10}$$

Since, for pharmaceutical inhalation particles, $d$ is in the micrometer range, while as mentioned above $D$ is a few tenths of a nanometer, we have $D \ll d$ and Eq. (9.10) reduces to the commonly used form

$$F_{vdW} = A \frac{d}{12D^2} \tag{9.11}$$

where $A = \pi^2 C n_{mp} n_m$ is the Hamaker constant, after Hamaker (1937).

Equation (9.11) can be derived more rigorously, including nonadditivity, by instead using a macroscopic approach (rather than the microscopic approach we have just used), as was first done by Lifshitz (1956). Such a macroscopic approach treats the electric and magnetic fields at a continuum level, but includes the effects of spontaneous quantum fluctuations in the locations of the electrons. This macroscopic theory is beyond the scope of this text, but the interested reader is referred to Israelachvili (1992) or Krupp (1967) for a description. Such a macroscopic approach results in Eq. (9.11) but the Hamaker constant $A$ is replaced by $A = 3\hbar\bar{\omega}/4\pi$ where $\hbar\bar{\omega}$ is sometimes called the 'Lifshitz–van der Waals constant'. Here $\hbar = h/2\pi$ where $h = 6.63 \times 10^{-34}$ J s is Planck's constant, and $\bar{\omega}$ is an integral, over all electromagnetic frequencies, of an integrand involving the imaginary components of the dielectric constants of the medium involved in the adhesive interaction (the imaginary component of the dielectric constant results in absorption and conversion of electromagnetic waves to heat).

Values of Hamaker constants can be determined for different materials from measurements of optical properties at different electromagnetic frequencies (Israelachvili 1992). Values of $A$ and $\hbar\bar{\omega}$ are known for the interaction of various pure substances consisting of simple molecules (Visser 1972, Israelachvili 1992), and typical values are near $A = 10^{-19}$ J, but vary by more than four orders of magnitude about this value (from $0.001 \times 10^{-19}$ J $- 20 \times 10^{-19}$ J) for different materials and different intervening materials between the particle and the wall (here we have assumed a vacuum between the wall and particle, but the presence of a different medium, such as water, will result in a different Hamaker constant).

The derivation of Eq. (9.11) can be modified to determine equations for the force of molecular attraction between various shaped simple bodies, such as two spherical particles of different diameters $d_1$ and $d_2$:

$$F_{vdW} = \frac{A}{6D^2} \frac{d_1 d_2}{d_1 + d_2} \tag{9.12}$$

## 9.3 van der Waals forces between actual pharmaceutical particles

Equations such as Eqs (9.11) or (9.12) is useful in understanding the origins of adhesive forces. However, their validity for pharmaceutical inhalation applications is limited by the fact that the intermolecular force $F_{vdw}$ is strongly dependent on the shape of the particles in the immediate vicinity of the contact point. This can be seen if the following function is plotted:

$$G(\Delta) = \frac{\int_0^{\Delta} \frac{(d-\Delta)\Delta}{(\Delta+D)^4} \, d\Delta}{\int_0^{D} \frac{(d-\Delta)\Delta}{(\Delta+D)^4} \, d\Delta} \qquad (9.13)$$

The function $G(\Delta)$ represents the relative contribution to the force $F_{vdw}$ in Eq. (9.9) exerted by the wall on the portion of the spherical particle that is within the distance $\Delta$ of the particle contact point in Fig. 9.3. The value of $G$ varies from 0 to 1, with $G(d) = 1$, since the entire particle is within the distance $d$ of the particle contact point, while the value of $G(0) = 0$ since none of the spherical particle is closer to the wall than the particle contact point. Values of $G$ are shown in Fig. 9.4 for a particle separation of $D = 0.00008d$ (which occurs for a particle-wall separation of $D = 0.4$ nm and a particle diameter of $d = 5$ μm).

For the particle–wall separation of $0.00008d$ shown in Fig. 9.4 it can be seen that 95% of the total force on the particle in this case is due to the force of the wall on the part of the particle that lies within $0.0005d$ of the contact point, where $d$ is the particle diameter. By calculating the mass of this part of the particle, it can be shown that this represents

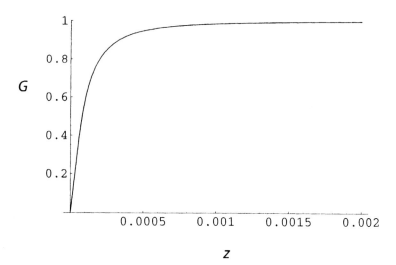

**Fig. 9.4** Values of the function $G$ in Eq. (9.13) are shown plotted against $z$, the relative distance from the point closest to the wall, i.e. $z = \Delta/d$, where $d$ is particle diameter. The function $G$ represents the relative amount of the attractive force of the wall on the particle due to those parts of the particle that are within the distance $z$ of the particle contact point. A value of particle–wall separation of $D/d = 0.00008$ has been used in this plot.

0.00008% of the particle's mass, i.e. 0.00008% of the particle mass that is closest to the wall gives rise to 95% of the adhesive force on the particle. For a 2 μm diameter particle, similar calculations result in 95% of the adhesive force being determined by the 0.0005% of the particle's mass that lies within $0.0013d$ of the contact point.

Because of the sensitivity of the adhesion force to the particle shape in the immediate neighborhood of the contact region, it is necessary that the shape of any particle adhering to a wall be known very accurately in the region of the contact point. However, actual powder particles generally have unknown, irregular shapes at this level of detail due to surface roughness (Podczeck 1997), so that Eqs (9.11) or (9.12) have little predictive power for powder particles normally occurring in inhaled pharmaceutical aerosols and as a result these equations are largely of pedagogical interest.

The effect of particle surface roughness on adhesion is complex (Zimon 1982) and no generally applicable method is available to predict its affect. Protrusions and indentations on a body that are much larger than intermolecular spacings, but still relatively small compared to the particle size, are called asperities. Powder particles are normally well covered by asperities, and so adhesive forces in the contact regions are determined by the shape of these asperities. One approach to estimating adhesive forces in the presence of this roughness is to use an effective diameter of the asperities in Eqs (9.11) or (9.12) instead of the particle diameter, since we have already seen that it is the surface shapes in the immediate diameter of the contact region that determine the total van der Waals force. With this approach, the attractive force $F_{vdW}$ resulting at a single contact point from Eqs (9.11) or (9.12) is much smaller than for an ideally smooth particle since the diameters of asperities are typically much smaller than particle diameter. In this sense, roughness reduces van der Waals adhesive forces, and this reduction is clearly seen when the idealized case of a hemispherical asperity is considered (Rumpf 1977). In general though, two rough surfaces mutually adhering to each other will have many points of contact where their asperities meet, so that this reduction in adhesive force is counterbalanced somewhat by an increase in the number of contact points. Determining the net effect of these two opposing factors from considerations like those we have used above for a smooth particle would require characterizing the three-dimensional detailed shapes of the surface asperities over the many regions where the particles are in contact, a task which is too demanding at present given the knowledge that no two powder particles have the same detailed surface shape.

Even if we did know the detailed shape of the surfaces in contact, there is an additional factor that we have neglected in our analysis to this point, and this is the elastic distortion of the surfaces that occurs when they press into each other due to their mutual intermolecular attraction. Since such distortions will change the surface shapes in the region of contact (tending to flatten the particle or its asperities), they can influence the van der Waals adhesive force. Theories are available for predicting this effect for an ideally smooth sphere contacting an ideally smooth plane wall (see the next section). However, such theories do not apply to particles with irregular surface roughness (Podczeck et al. 1996a). Indeed, prediction of van der Waals adhesion for typical powder particles (i.e. with surface roughness) would require solving the equations governing adhesion and elasticity with coupling between the two (since particle deformation affects adhesion, while the adhesive forces affect the deformation). Such a self-consistent approach would have to be applied over the complex three-dimensional surface shape associated with the two rough surfaces in contact. The demanding nature of such a task, combined with the paucity of knowledge on actual surface shapes for real powder particles, and the knowledge that no two powder particles have the same surface shape, has prevented much progress being made in predicting adhesive forces between typical powder particles from their intermolecular forces.

## Example 9.1

Particles made from lactose monohydrate are commonly incorporated into dry powder inhalation formulations as 'carrier' particles. These lactose particles are usually much larger than the drug particles in the mixture (typically being 50–100 μm in diameter, Larhrib et al. 1999). Their large size increases the aerodynamic forces on them, as we shall see later in this chapter, allowing them to be entrained by the inhalation air flowing past them, and carrying drug particles with them. (Without the carrier particles, the drug particles are often poorly entrained since they are too small to feel strong enough aerodynamic forces to overcome their autoadhesion.) Podczeck et al. (1996a) used a centrifuge technique to measure the adhesion force between lactose monohydrate particles and a flat lactose surface, finding an average value of $2.5 \times 10^{-8}$ N for this force. The lactose particles had a mean diameter of 62.3 μm. Assuming a value of the Hamaker constant $A = 7 \times 10^{-19}$ J and a particle–wall separation of 0.4 nm, calculate the adhesive force predicted and comment on the results using

(a) Eq. (9.11) with lactose particle diameter $d = 62$ μm,
(b) Eq. (9.12) instead using the diameter of the surface roughness asperities $d_1 = d_2 = 56$ nm, as implied by Podczeck et al. (1996a).

## Solution

(a) This is a simple matter of substituting values into Eq. (9.11):

$$F_{vdW} = A \frac{d}{12 D^2}$$

Using $A = 7 \times 10^{-19}$ J, $D = 0.4 \times 10^{-9}$ m and $d = 62$ μm, we obtain $F_{vdW} = 2 \times 10^{-5}$ N. This force is nearly three orders of magnitude larger than the measured value, which is to be expected since the lactose particles used by Podczeck et al. (1996a) are not smooth spheres, but have roughness which will reduce autoadhesion as mentioned above.

(b) If we include the effects of roughness by using Eq. (9.12) assuming that contact occurs between two asperities with diameter $d = 56$ nm, we obtain the result $F_{dvw} = 2.0 \times 10^{-8}$ N for the force due to contact of the asperities. We do not know how many asperities are in contact between the particle and the wall, but given the particle size of 62 μm, and the asperity diameter of 56 nm, it is likely to be a number much larger than one so that $F_{dvw}$ will be many times larger than $2.0 \times 10^{-8}$ N. Thus, Eq. (9.12) also gives a value that appears to be many times larger than that measured experimentally, a conclusion similar to that reached by Podczeck et al. (1996) who also included elastic deformation (via JKR and DMT theories discussed in the following section) in their theoretical prediction of adhesion. Thus, we see that theoretical prediction of adhesive van der Waals forces for actual pharmaceutic inhalation aerosol particles with surface roughness remains an elusive goal at present.

## 9.4 Surface energy: a macroscopic view of adhesion

The above discussion was focused on determining adhesion by taking a molecular viewpoint. However, it is useful to instead define a macroscopic quantity of a surface called its surface energy, $\gamma$, defined as the change in free energy when the surface area is increased by unit area. (Surface energy is also sometimes referred to as interfacial energy.) Since the surface energy will depend on what medium the new surface is exposed to, $\gamma$ is defined as the change in free energy when the surface area is increased by unit area in a vacuum.

If instead, new surface is created in the presence of some other medium (e.g. air instead of a vacuum), then the presence of vapor molecules (e.g. water vapor) can dramatically change the surface energy (particularly for solid surfaces) due to the adsorption of a very thin layer of vapor molecules onto the surface. Thus, to be precise, subscripts are commonly used to indicate the mediums that are interacting at the surface. For example, $\gamma_S$ is often used to indicate the surface energy of a solid surface in a vacuum, $\gamma_{SV}$ indicates the surface energy of a solid surface (indicated by the subscript 's') exposed to air with water vapor (v) present, while $\gamma_{SL}$ indicates the surface energy of a solid surface exposed to a particular liquid (such as water). The surface energy of water in air involves a liquid exposed to a gas with vapor in it, so the symbol $\gamma_{LV}$ is often used, where in this case $\gamma_{LV}$ is simply the surface tension of the water–air interface and has a value of 0.072 J m$^{-2}$ at room temperature. Alternatively, numeric subscripts are sometimes convenient, so that for example, $\gamma_1$ indicates the surface energy of medium 1 in a vacuum, while $\gamma_{12}$ indicates the surface energy of a surface separating medium 1 and medium 2.

A parameter directly related to the surface energy is the work of adhesion $\Gamma$, defined as the reversible work done when planar unit areas of material 1 and material 2 are separated from contact and moved infinitely far apart. In this case, two surfaces of unit area end up being created, so there is a factor of two between $\Gamma$ and $\gamma$, i.e.

$$\Gamma = 2\gamma \tag{9.14}$$

If unit area of material 1 and 2 are in contact and separated to infinity in a vacuum, the work of adhesion is given the symbol $\Gamma_{12}$. A relation between work of adhesion and surface energy in this case can be obtained by realizing that there are two ways we can create surfaces 1 and 2 at infinity. One way would be to create a unit area of surface where 1 and 2 are in contact (requiring energy $\gamma_{12}$), then separate medium 1 and 2 at this interface (requiring energy $\Gamma_{12}$), for a total amount of energy expended equal to $\gamma_{12} + \Gamma_{12}$. Alternatively, we could simply create unit area of medium 1 in a vacuum (requiring energy $\gamma_1$), and similarly for medium 2 (requiring energy $\gamma_2$). Since the two approaches are equivalent, we must have

$$\gamma_{12} + \Gamma_{12} = \gamma_1 + \gamma_2$$

or, as is more commonly written

$$\gamma_{12} = \gamma_1 + \gamma_2 - \Gamma_{12} \tag{9.15}$$

Equation (9.15) relates the surface energies to the work of adhesion.

If material 1 and 2 are separated in the presence of a third medium 3 (rather than in a vacuum), then the work of adhesion is given the symbol $\Gamma_{132}$. Using the same argument that resulted in Eq. (9.15), but replacing the vacuum with medium 3, we must also have

$$\Gamma_{132} = \gamma_{13} + \gamma_{23} - \gamma_{12} \tag{9.16}$$

Since $\gamma'_{11} = 0$, if medium 1 and 2 are the same Eq. (9.16) reduces to

$$\Gamma_{121} = 2\gamma'_{12} \qquad (9.17)$$

and the symbol 2 is used instead of 3 since there are only two media.

Values of surface energies are available for many materials from standard physical chemistry references and handbooks, but can also be obtained by measuring the shapes of liquid drops on surfaces, although traditional methods involving, for example, contact angles cannot usually be accurately applied to powders due to the lack of a flat, smooth material surface that is needed for these methods. Instead, for powders, surface energies can be estimated using measurements of the rate of capillary rise of a liquid in a column of powder dipped in liquid (Grundke *et al.* 1996, Desai *et al.* 2001).

For our purposes, one of the main reasons for introducing the concept of surface energy and work of adhesion is their connection to van der Waals forces. This can be seen by considering again the force between the spherical particle and a plane wall shown in Fig. 9.3, but instead considering the force of attraction between two cylindrical shells, as shown in Fig. 9.5.

If we define $f_{vdW}$ as the force of attraction per unit area (due of course to molecular van der Waals forces) between two flat walls located a distance $X$ apart, then the force between the two shells in Fig. 9.5 can be approximated as $2\pi y dy f_{vdW}(X)$, assuming the particle's diameter is much greater than the particle–wall separation (i.e. $d \gg D$). The total force between the sphere and the wall is then simply obtained by integration as

$$F_{vdW} = \int_{X=D}^{X=D+(d/2)} 2\pi y f_{vdW}(X)dy \qquad (9.18)$$

(where this is an approximation, since we are ignoring the force of attraction exerted by

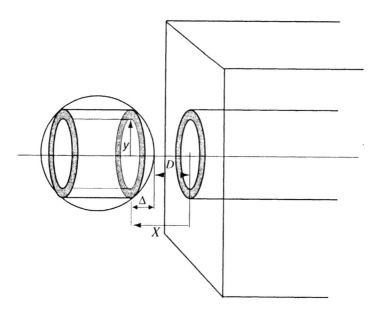

**Fig. 9.5** The geometry of a semi-infinite cylindrical shell of radius $y$ in a wall and a corresponding co-axial cylindrical shell in a neighboring spherical particle.

those parts of the flat wall that are outside the cylinder of diameter $d$ centered at the contact point, as well as the finite length of the cylindrical shells in the sphere – these are very reasonable approximations since we already know that almost all of the van der Waals force is due to the small region around the contact point). From Eq. (9.7), we see that for $d \gg D$ we can write $y^2 \approx \Delta d$. Using this result and Eq. (9.8), we rewrite Eq. (9.18) as

$$F_{vdW} = \pi d \int_{X=D}^{X=D+(d/2)} f_{vdW}(X)dX \qquad (9.19)$$

However, since force times distance is work, the integral in Eq. (9.19) is simply the total work done in taking two unit area flat surfaces from a separation of $D$ (which is the separation at contact) out to a separation of $d/2$ (which is essentially infinity, since we have already assumed $d \gg D$). As a result, the integrand is simply the work of adhesion $\Gamma$, and Eq. (9.19) can instead be written as

$$F_{vdW} = \pi d \Gamma \qquad (9.20)$$

which is a special case of the so-called Derjaguin approximation that relates the force of adhesion for spherical surfaces to the work of adhesion for flat surfaces (Israelachvili 1992).

From Eq. (9.14), $\Gamma = 2\gamma$, so Eq. (9.20) can be written

$$F_{vdW} = 2\pi d\gamma \qquad (9.21)$$

where $\gamma$ is the surface free energy of the particle and wall material. Thus, we have the result that the force of adhesion is directly proportional to the surface free energy $\gamma$, with particles made from material with high surface free energy 'sticking' together more adhesively than particles with low surface energy.

When the particle and the wall are made from different materials (1 and 2) and there is a third substance (medium 3) filling the space outside the particle and wall, Eq. (9.20) can be written more generally as

$$F_{vdW} = \pi d \Gamma_{132} \qquad (9.22)$$

where $\Gamma_{132}$ is the work of adhesion in the presence of these three media (from Eq. (9.16)).

By considering two spheres instead, one can use the same approach to derive the force of adhesion of two smooth spheres of diameter $d_1$ and $d_2$, with the result

$$F_{vdW} = \pi \frac{d_1 d_2}{d_1 + d_2} \Gamma_{132} \qquad (9.23)$$

One advantage of introducing the surface energy is that elastic deformation of the particle adhering to the wall is readily included (Johnson et al. 1971, Derjaguin et al. 1975), and this results in the two limiting cases:

$$F_{vdW} = 2\pi d\gamma \quad \text{(DMT limit)} \qquad (9.24)$$

$$F_{vdW} = \frac{3}{2}\pi d\gamma \quad \text{(JKR limit)} \qquad (9.25)$$

(the asymptotic JKR theory of Johnson et al. (1971) being valid for small, hard or weakly adhering particles, and the DMT theory of Derjaguin et al. (1975) being valid for large, soft or strongly adhering particles, and deformations in between requiring numerical solution of the governing equations – see Muller *et al.* (1983). Equation (9.24) results in the surprising conclusion that elastic deformation either has no effect on the adhesive force, or reduces adhesion, results that are in contrast to experimental data on powders which have surface roughness present (Zimon 1982, Rietema 1991).

The van der Waals force of adhesion given by Eqs (9.20)–(9.23) is the same as that appearing in Eqs (9.11) and (9.12). However, Eqs (9.20)–(9.23) involve the macroscopic quantity $\gamma$ or $\Gamma_{132}$, which can be obtained from the surface free energies (Eq. (9.16)) of the materials involved, rather than requiring determination of a Hamaker constant as in Eqs (9.11) or (9.12). The use of equations like (9.20)–(9.25) is particularly advantageous when surface molecules are adsorbed onto the surfaces, such as commonly occurs in air with adsorbed water vapor molecules and many powder surfaces, since characterization of the adsorbed layer is not needed, in contrast to a Hamaker constant approach (Xie 1997). Although this advantage makes the use of surface energies more readily realistic, particularly since elastic deformations of the particles can be included via Eqs (9.24) or (9.25), it does nothing to allow inclusion of surface roughness effects, since Eqs (9.20)–(9.25) all assume ideally smooth surfaces. The same difficulties we discussed earlier regarding inclusion of the effect surface asperities on Eqs (9.11) and (9.12) apply to Eqs (9.20)–(9.25), so that these equations too have little practical use in estimating adhesive forces between actual pharmaceutical inhalation aerosol particles, as discussed in the previous section.

## Example 9.2

We calculated the adhesive force of attraction between lactose particles in a previous example using Eq. (9.11), where a Hamaker constant $A = 7 \times 10^{-19}$ J and a particle separation $D = 0.4$ nm were used. If the surface free energy $\gamma_{12}$ of lactose in air is 58 mJ m$^{-2}$ (Podczeck 1999), combine Eq. (9.22) and Eq. (9.11) to determine if $A = 7 \times 10^{-19}$ J is a reasonable value for the Hamaker constant.

## Solution

From Eq. (9.11), we have the following expression for the adhesive force between a lactose particle and wall:

$$F_{\text{vdW}} = A \frac{d}{12D^2} \tag{9.11}$$

However, Eq. (9.22) gives us the same force instead with the work of adhesion appearing. Equating these two we have

$$A \frac{d}{12D^2} = \pi d \Gamma_{123}$$

where mediums 1 and 3 are the same (lactose), and medium 2 is air, so Eq. (9.17) gives $\Gamma_{121} = 2\gamma_{12}$ and we can solve for the Hamaker constant to find

$$A = 24\pi D^2 \gamma_{12}$$

Putting in $D = 0.4$ nm and $\gamma_{12} = 58$ mJ m$^{-2}$, this gives $A = 7 \times 10^{-19}$ J, so that the value we used in our previous example is in agreement with measured surface energies.

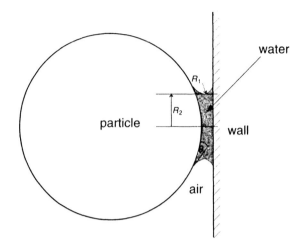

**Fig. 9.6** Water condensation ('capillary condensation') occurs between a particle and a wall in humid air due to the Kelvin effect (discussed in Chapter 4) that affects the vapor pressure at curved liquid surfaces. $R_1$ and $R_2$ are the two radii of curvature of the water–air surface, while $\theta$ is defined as the 'contact angle' and is a measurable property of the materials involved.

## 9.5 Effect of water capillary condensation on adhesion

So far we have been discussing adhesive forces without consideration of the fact that water can condense in the contact region between two adhering surfaces due to the Kelvin effect that we examined in Chapter 4, and as shown in Fig. 9.6. This is referred to as capillary condensation, and the resultant increase in adhesion is often referred to as the capillary force (although of course it is a manifestation of intermolecular forces just as the van der Waals force derived above, with the added complexity of hydrogen bonding interactions and other intermolecular effects associated with the strongly polar nature of water molecules).

Equations (4.65) and (4.66) show that water will condense between the particle and the wall, and will be in equilibrium with the water vapor in the air outside the particle and wall, when the relative humidity ($RH$) in the air satisfies

$$RH = \exp\left(-\frac{2\sigma M}{R_u \rho R_1 T}\right) \qquad (9.26)$$

where $\sigma = \gamma_{LV}$ is the surface energy (i.e. surface tension) of water in air, $M$ is the molar mass of water ($M = 0.018$ kg mol$^{-1}$), $\rho$ is the density of water ($\rho = 998$ kg m$^{-3}$), $R_u$ is the universal gas constant ($R_u = 8.314$ kg mol$^{-1}$ m$^2$ s$^{-2}$ K$^{-1}$), $T$ is the temperature in Kelvin, and we have assumed that $R_1 \ll R_2$ (otherwise $R_1 R_2/(R_1 + R_2)$ should be used instead of $R_1$ in Eq. (9.26)). Equation (9.26) differs from the equation given in Chapter 4 by a negative sign in the exponent, since here the direction of the radius of curvature is opposite to the water droplet considered in Chapter 4. Solving Eq. (9.26) for the menisus radius $R_1$ in Fig. 9.6 gives

$$R_1 = -\frac{2\sigma M}{\rho R_u T \ln(RH)} \qquad (9.27)$$

Values of $R_1$ are plotted in Fig. 9.7 for various $RH$.

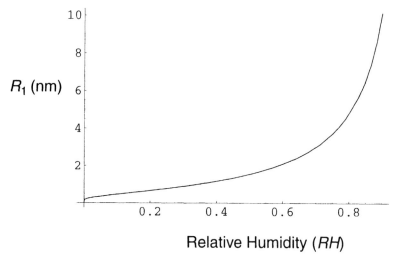

Relative Humidity (*RH*)

**Fig. 9.7** Values of the radius of curvature $R_1$ in Fig. 9.6 are shown for water at room temperature in air of various relative humidities ($RH$) up to 0.9 (i.e. 90%). Note that $R_1 \to \infty$ as $RH \to 1$.

Recalling that a typical value of wall–particle separation in air is 0.4 nm, Fig. 9.7 demonstrates that the contact region between particles and walls in room temperature air is likely to be water-filled over a reasonably wide range of relative humidities, as is also demonstrated in the next example.

## Example 9.3

Make a simple estimate of the range of humidities where water in the contact region will completely dominate the adhesive force. Assume the contact angle $\theta$ (shown in Fig. 9.6) is small, so that such an estimate can be made by determining the value of the relative humidity at which the diameter, $2R_1$, of the curved air–water interface in Fig. 9.6 is the same as distance $X_c$ from the wall within which 95% of the van der Waals force is determined when no water is present, i.e. $X_c = D + \Delta_c$, where $G(\Delta_c) = 0.95$ in Eq. (9.13). In this case, the region between the particle and the wall that gives 95% of the adhesive force when water is not present will be filled with water, as shown in Fig. 9.8. Assume a 5 μm diameter particle and a particle–wall separation of $D = 0.4$ nm.

## *Solution*

For a 5 μm diameter particle and $D = 0.4$ nm, we can use Eq. (9.13) or Fig. 9.4 to determine $\Delta_c = 0.00051d$, so that when the water is not present, 95% of the adhesive force is determined by that part of the particle within this distance of the particle contact point. Setting $2R_1 = D + \Delta_c$ with this $D$ and $\Delta_c$ and solving for $R_1$ gives $R_1 = 2.9$ nm. With this value of $R_1$, Eq. (9.25) (or Fig. 9.7) gives $RH = 0.7$, so that for humidities above $RH = 70\%$, adhesion is dominated by the presence of water in the contact region. This value is similar to the often quoted $RH$ of 65% above which capillary condensation of water negatively affects the dispersion of inhaled pharmaceutical powders (which, recall, are normally less than 5 μm in diameter). This agreement is coincidental though since we have not included any effects of surface roughness in our analysis.

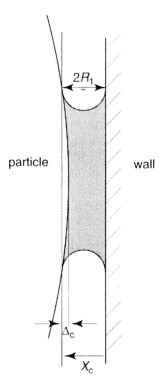

**Fig. 9.8** The relevant geometry for capillary condensation of a particle considered in the example. The water is assumed to wet the surface at nearly zero contact angle $\theta$ (i.e. the air–water surface is nearly tangential to the particle and wall surfaces).

The above example and discussion suggests that capillary condensation plays an important role in adhesion of powder particles for moderately high humidities (typically $> 65\%$ $RH$ at room temperature). For this reason, it is useful to develop an expression for the force of adhesion between a spherical particle and a wall when capillary condensation is present. For this purpose, let us define medium 1 as the particle and wall substance (i.e. we are assuming the wall and particle are made of the same material), medium 2 as water, and medium 3 as air. From a macroscopic viewpoint, the following three forces are present: (1) the gauge pressure, $p$, of the water which acts over the area $\pi R_2^2$ in Fig. 9.6; (2) the air–water surface tension ($\gamma_{23}$), which acts over the circumference $2\pi R_2$ at an angle $\theta$ shown in Fig. 9.6; and (3) the van der Waals force of adhesion of the wall on the particle, which acts across the water and so is given by Eq. (9.22) with the work of adhesion given by $2\gamma_{12}$. Assuming no deformation of the particle (as we did in our derivation of the van der Waals force of adhesion earlier), the force of adhesion in the presence of capillary condensation, $F_{cap}$, must then be in the $x$-direction and given by

$$F_{cap} = p\pi R_2^2 + 2\pi R_2\gamma_{23}\sin\theta + 2\pi d\gamma_{12} \tag{9.28}$$

The pressure in the water is lower than ambient pressure by an amount given by the well-known result for the pressure difference across a cylindrical interface (White 1999)

$$p = \frac{\gamma_{23}}{R_1} \tag{9.29}$$

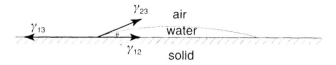

**Fig. 9.9** The interfacial surface tensions exhibited by a water drop on a solid surface in air result in a contact angle $\theta$.

Substituting this into Eq. (9.28), we obtain

$$F_{\text{cap}} = \pi \gamma_{23} R_2 \left( \frac{R_2}{R_1} + 2 \sin \theta \right) + 2\pi d \gamma_{12} \tag{9.30}$$

Since capillary condensation will occur most readily for small contact angles (otherwise $R_1$ is large and capillary effects occur only at very high humidities), let us assume small contact angles, so that $R_2/R_1 \gg 1$ and we can neglect the $2 \sin \theta$ term in Eq. (9.30). Also, making use of Eq. (9.7) (so that $R_2^2 \approx d\Delta_c$ where $\Delta_c$ is shown in Fig. 9.8) and assuming $R_2 \ll d$ and $R_1 \gg D$ (which allows us to approximate $\Delta_c$ as $2R_1 \cos \theta$) we finally obtain

$$F_{\text{cap}} = 2\pi d (\gamma_{23} \cos \theta + \gamma_{12}) \tag{9.31}$$

Equation (9.31) can be simplified by using the following result

$$\gamma_{13} = \gamma_{23} \cos \theta + \gamma_{12} \tag{9.32}$$

where Eq. (9.32) can be derived by considering a drop of water (medium 2) on a flat surface (medium 1) in air (medium 3), as shown in Fig. 9.9.

Assuming that any elastic deformation of the solid does not supply any horizontal force in the region where the interfacial tensions meet, then the horizontal components of the interfacial tensions must balance, which leads to Eq. (9.32).

Substituting Eq. (9.32) into Eq. (9.31) we obtain our final result for the adhesive force of a wall on a sphere of diameter $d$ (where both wall and particle are made of the same material) in the presence of capillary condensation:

$$F_{\text{cap}} = 2\pi d \gamma_{13} \tag{9.33}$$

It can be shown that for two particles with capillary condensation between them, the capillary force of adhesion between them is given by Eq. (9.33) with $d$ replaced by $d_1 d_2/(d_1 + d_2)$.

Equation (9.33) is a surprising result, since it is identical to the equation we derived for the van der Waals force without considering capillary condensation (Eq. 9.21), which at first thought seems to suggest that capillary condensation has no effect on adhesion. This is not true, of course, and the reason is somewhat subtle. In particular, here $\gamma_{13}$ is the surface energy of the wall–air surface in the presence of water vapor of a given relative humidity, which includes the effect of water molecules adsorbed on the surfaces. Changes in relative humidity can affect the surface energy $\gamma_{13}$ due to changes in the amount of water on the surface, with increases in $\gamma_{13}$ resulting in increased adhesion at higher humidities. This effect is clearly seen in Fig. 9.10 where the median adhesion force for micronized lactose monohydrate particles is shown after storage at various relative humidities and 20°C (from Podczeck *et al.* (1997), measured using a centrifuge

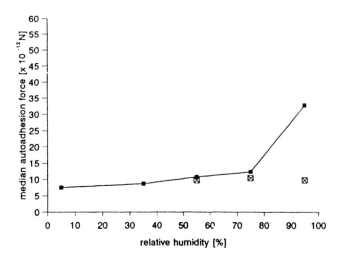

**Fig. 9.10** The median adhesion force between finely milled (to $<5\,\mu m$ in diameter) lactose monohydrate adhering to walls made from compacted lactose monohydrate particles (of much larger size before compacting) and stored at various relative humidities (solid squares and line). For the open square symbols, the samples were restored at 5% relative humidity after the first storage period. Reprinted from Podczeck *et al.* (1997) with permission.

technique). Although the particles used in Fig. 9.10 have surface roughness, which we have not included in deriving Eq. (9.33), capillary condensation like that described by Eq. (9.33) is responsible for the increased adhesion at high humidity.

Figure 9.10 also shows that capillary condensation may be removed by drying (seen by the open symbols in Fig. 9.10, where the samples were put in 5% *RH* after storage at the humidities shown), at least for the lactose monohydrate particles examined by Podczeck *et al.* (1997). Note, however, that this is not always the case, since fusing of adhering surfaces with solid bridges can occur (Rumpf 1977, Padmadisastra *et al.* 1994), which is not reversible. Such solid bridges form when molecules of the adhering surfaces dissolve into the water in the liquid capillary water to such an extent that when the capillary water is removed by drying, the dissolved material forms a solid bridge between the two surfaces, dramatically (and irreversibly) increasing the adhesive force.

It should be remembered that Eq. (9.33) was derived under the assumption of small contact angle. Solution of the more general case of arbitrary contact angle requires solving the Laplace–Young equation governing the shape of the air–water interface, which can be derived using the calculus of variations by minimizing a functional consisting of the free energy of the surface, or from geometrical and mechanical considerations (Batchelor 1967), and is solved by Orr *et al.* (1975).

Hydrophilic substances have small contact angles, while those with large contact angles are hydrophobic. This can be understood by realizing that hydrophilic molecules, by definition, have stronger intermolecular attractions to water molecules and so have higher work of adhesion $\Gamma_{12}$ in Eq. (9.15), resulting in a lower $\gamma_{12}$ in Fig. 9.9. All else being equal, a decrease in $\gamma_{12}$ in Fig. 9.9 must result in a decrease in contact angle $\theta$ in order for the interfacial surface tensions to balance.

For materials with large enough contact angles, capillary condensation will not occur (indeed if $\theta > 90°$, supersaturated humidities are needed since the Kelvin effect is now

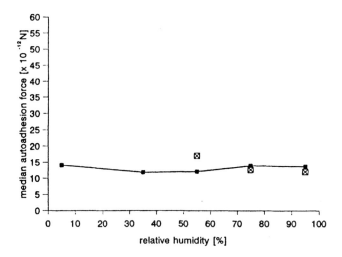

**Fig. 9.11** The median adhesion force between finely milled (to $< 5 \, \mu m$ in diameter) salmeterol xinafoate particles adhering to walls made from salmeterol xinafoate particles (of much larger size before compacting) and stored at various relative humdities (solid squares and line). For the open square symbols, the samples were restored at 5% relative humidity after the first storage period. Reprinted from Podczeck *et al.* (1997) with permission.

the same as it is for the droplets considered in Chapter 4 and results in increases in vapor pressure at the capillary surface). Thus, particles made from hydrophobic materials do not exhibit increased adhesion due to capillary condensation, as is seen in Fig. 9.11 for salmeterol xinafoate (a bronchodilator used in asthma treatment), which has a contact angle of 68°.

As with all the idealized considerations of adhesion of smooth particles in this chapter, surface roughness complicates matters. For surface asperities that are much larger than the meniscus radii $R_1$ and $R_2$, capillary condensation at each contact point may be viewed qualitatively as being similar to that considered above using the asperity diameter instead of particle diameter. However, for small surface asperities or at higher humidities, capillary condensation may fill the regions between asperities, so that using the particle diameter in Eq. (9.33) is more appropriate from a qualitative viewpoint in this case. Since particle diameters are normally much larger than asperity diameters, adhesion can become much larger as a result of this effect. Note, however, that quantitative estimation of adhesion forces for particles with surface roughness is not possible with smooth-particle theories (Podczeck *et al.* 1996b).

## 9.6 Electrostatic forces

### 9.6.1 Excess charge

We have discussed in Chapter 3 the effect of electostatic forces on the motion of charged particles. However, electrostatic forces can also contribute to the adhesion of particles. In particular, two charged particles of diameter $d_1$ and $d_2$ in contact having net excess charges $q_1$ and $q_2$ will be held together (or repelled if the charge is of the same sign) with the Coulomb force that we saw in Chapter 3:

$$F_{coul} = \frac{q_1 q_2}{4\pi \left(\dfrac{d_1 + d_2}{2}\right)^2 \varepsilon_0} \tag{9.34}$$

where $\varepsilon_0 = 8.85 \times 10^{-12}\,C^2\,N^{-1}\,m^{-2}$ is the permittivity of free space. Alternatively, if we treat the excess charge as being distributed over the surface of the particles (as it would be for conducting particles, rather than appearing as an idealized point charge at the center of the particle as in Eq. 9.34) with surface charge density $\sigma$, one obtains (Rumpf 1977)

$$F_{coul} = \frac{\pi d^2 \sigma_2 \sigma_1}{4\varepsilon_0} \tag{9.35}$$

Krupp (1967) supplies an equation for the case (of nonconductors) where the charge is distributed exponentially over a shell of thickness $\delta$ next to the outer surface of the particle near a wall.

For typical powders in dry powder inhalers, very high charge levels are needed in order for forces associated with excess charge to contribute significantly to adhesion, as demonstrated by the next example.

## Example 9.4

Calculate the charge to mass ratio necessary in order for the excess charge force in Eqs (9.34) and (9.35) to contribute more than 10% of the total measured median auto-adhesion force for

(a) lactose monohydrate particles (density $1530\,kg\,m^{-3}$) considered in a previous example (where the median autoadhesion force was $2.5 \times 10^{-8}\,N$ and median particle diameter was 62.3 μm);
(b) micronized lactose monohydrate particles shown in Fig. 9.10, assuming a particle diameter of 3 μm.

## *Solution*

Giving the symbol $F_{ad}$ to the measured median autoadhesive force, and setting $F_{coul} \geq 0.1 F_{ad}$, Eq. (9.34) for two particles of the same diameter $d$ but opposite charge $q$ gives us

$$0.1 F_{ad} \leq \frac{q^2}{4\pi\varepsilon_0 d^2}$$

Solving for $q$, we obtain

$$q \geq 2d\sqrt{0.1\pi\varepsilon_0 F_{ad}}$$

Estimating the mass of a particle as $m = \rho\pi d^3/6$, then this equation implies a charge to mass ratio

$$\frac{q}{m} \geq \frac{12}{\rho d^2}\sqrt{\frac{0.1\varepsilon_0 F_{ad}}{\pi}} \qquad (9.36)$$

Beginning instead with Eq. (9.35), we set the adhesive force to 10% of the measured value

$$0.1 F_{ad} \leq \frac{\pi d^2 \sigma^2}{4\varepsilon_0}$$

From this equation we can determine the surface charge, which is related to the charge on a particle by multiplying by the surface area $\pi d^2$. The result is

$$\frac{q}{m} \geq \frac{12}{\rho d^2}\sqrt{\frac{0.1 F_{ad}\varepsilon_0}{\pi}} \qquad (9.37)$$

which is identical to Eq. (9.36). Thus, Eqs (9.34) and (9.35) give identical predictions of the charge to mass ratio here.

Equation (9.36) (or Eq. (9.37)) gives us an estimate of the amount of excess charge needed for Coulomb forces to account for more than 10% of adhesion forces. For (a), we substitute $F_{ad} = 2.5 \times 10^{-8}$ and $d = 62.3 \times 10^{-6}$ m, to obtain $q/m = 1.7 \times 10^{-4}$ C kg$^{-1}$.

For part (b), referring to Fig. 9.10 a value of $F_{ad} = 15 \times 10^{-12}$ is reasonable. Using this value and $d = 3 \times 10^{-6}$ m in Eqs (9.36) or (9.37), we obtain $q/m \geq 1.8 \times 10^{-3}$ C kg$^{-1}$.

Measured values of the charge on lactose powders are lower than these values, where charge-to-mass ratios are typically less than $10^{-4}$ C kg$^{-1}$ (Carter *et al.* 1998, Bennett *et al.* 1999), so that we do not expect Coulomb forces due to excess charge to play a significant role in autoadhesion of lactose particles.

Although the actual charges and adhesive forces between particles in inhalation powders are usually unknown, charge levels of powders are generally well below $10^{-4}$ C kg$^{-1}$ (Bailey 1993, Byron *et al.* 1997), while adhesive forces are probably not that different from the above example, indicating that excess charge forces probably play only a minor role in adhesion of typical DPI powders. Of course, this conclusion may not be valid for powders with much lower adhesive forces (such as porous powder particles) and/or higher charge levels, so that comparison of excess charge forces to the total adhesive force for the specific powder under consideration is needed before entirely dismissing excess charge forces. Note that comparing the electrostatic force to the van der Waals force given by Eq. (9.11), (9.12) or Eqs (9.20)–(9.25) as a means of estimating the importance of electrostatic forces is of little use for most pharmaceutical inhalation powders since the van der Waals force given by these equations does not include the effect of surface roughness, an effect already discussed as being of paramount importance (and currently not amenable to prediction).

## 9.6.2 Contact and patch charges

In addition to excess charge forces, it is also possible for particles to be attracted to each other due to charges on their surface that can arise when particles made from different materials come into contact (giving rise to 'contact charges', see Bailey 1993; also referred to as an 'electrostatic double layer') or due to differences in the energy levels at different parts of the surface due to surface irregularities (giving rise to 'patch charges', Burnham *et al.* 1992, Pollock *et al.* 1995). Contact charging is thought to involve (Krupp 1967, Anderson 1996) the transfer of charged species (i.e. ions or electrons) from one surface to the other (with the charged species moving from the surface with higher chemical potential to the surface with lower chemical potential). Equilibrium is reached when this transfer of charge results in an electrostatic potential difference, the so-called contact potential, between the two surfaces. This contact potential counters the difference in chemical potential. Contact charges readily develop between particles or surfaces of different materials (but can occur between surfaces of the same material if the two surfaces have different contaminants or adsorbates), while patch charges operate between particles of either the same or different materials.

Since a predictive understanding of these surface charge mechanisms, particularly for nonconducting materials, is not available at present, it is difficult to examine their importance in adhesion of inhalation powders. However, the presence of capillary condensation eliminates both contact and patch charges, since the potential difference across the contact region is eliminated due to the conducting properties of water. At high relative humidities, we thus expect contact and patch charges to have a negligible effect on adhesion for hydrophilic powders. At low humidities, if contact or patch charges are much larger than van der Waals forces for such powders, then we would expect adhesive forces to increase as the relative humidity is lowered as the contact and patch charges start to appear with the removal of capillary condensation. From Fig. 9.10, we see that this is not the case with micronized lactose monohydrate, suggesting that patch and contact charge forces are not an important component of autoadhesion for this material (recall that contact charge mechanisms do not operate between two surfaces of the same material, but the different treatments of the particles and the surface that gave rise to Fig. 9.10 could have led to contact charges between them). Similar decreases in adhesive forces between lactose monohydrate and salmeterol xinoafoate (Podczeck *et al.* 1997) with decreases in relative humidity indicate that neither are contact charge forces comparable to van der Waals forces for these two materials.

In general, determining the importance of contact charges requires knowledge of the potential difference, $\Delta\phi_{con}$, that exists between the two contacting surfaces, while for patch charges, the potential difference between patches, $\Delta\phi_{pat}$, must be known. This information is generally not available for pharmaceutic inhalation powders. For a particle on a flat surface, Bowling (1988) gives the expression

$$F_{con} = \frac{\pi\varepsilon_0 d\Delta\phi_{con}}{2D} \tag{9.38}$$

for the contact charge contribution to the adhesive force, where $d$ is of course particle diameter and $D$ is the particle–wall separation. An order of magnitude estimate for the patch charge force between two particles in contact is given by Pollock *et al.* (1995) as

$$F_{patch} = 4\pi\varepsilon_0(\Delta\phi_{pat})^2 \tag{9.39}$$

Note that Eq. (9.39) suggests the patch charge force is independent of particle diameter and particle separation $D$, which is strictly true only for low surface curvature (Pollock *et al.* 1995).

**Example 9.5**

Estimate the potential difference $\Delta\phi$ needed for surface charge forces to contribute 10% of the total measured adhesion force of $2.5 \times 10^{-8}$ N for lactose monohydrate particles of diameter 62.3 μm adhering to a wall (Podczeck *et al.* 1996a) using

(a) Eq. (9.38) for contact charge forces,
(b) Eq. (9.39) for patch charge forces.

*Solution*

(a) Substituting $\varepsilon_0 = 8.85 \times 10^{-12}$ $C^2 N^{-2} m^{-2}$, $d = 62.3 \times 10^{-6}$ m, and particle–wall separation of $D = 0.4$ nm, with $F_{con} = 0.1 \times 2.5 \times 10^{-8}$ N and solving for $\Delta\phi_{con}$ in Eq. (9.38) gives a value of the contact potential of $\Delta\phi_{con} = 0.01$ V in order that contact charge forces contribute 10% to the total adhesion force. The contact potential between the lactose particles and lactose wall in Podczeck *et al.* (1996a) is unknown so that we cannot comment on whether this indicates contact charges are important, although as we have already noted, contact charge potentials do not exist between surfaces having the same chemical potential, so that we might expect contact potentials to be below this value if we think the surface of the compacted lactose particles making up the wall has similar chemical potential to the adhering lactose particles.

(b) Substituting $\varepsilon_0 = 8.85 \times 10^{-12}$ $C^2 N^{-2} m^{-2}$, $F_{pat} = 0.1 \times 2.5 \times 10^{-8}$ N into Eq. (9.39) and solving for $\Delta\phi_{pat}$ gives $\Delta\phi_{pat} = 4.7$ V. This is an unreasonably high patch charge potential, suggesting that patch charge likely does not play an important role in adhesion of these lactose particles.

Although it is difficult to say in general whether contact and patch charges are important since contact potentials are usually unknown and we lack theoretical methods for predicting surface charges associated with patch and contact charging, electrostatic charge (whether excess, contact or patch charge forces) has been thought to contribute little to adhesion in manufactured powders that have not been intentionally charged (Rietama 1991). This conclusion, however, has often been based on the comparison to van der Waals forces between ideal, smooth spheres. We have already seen that roughness dramatically reduces van der Waals forces, so that electrostatic forces can become important in powder adhesion in some cases (Zimon 1982).

Contact charging is certainly important, however, in determining the charge of powders after deaggregation. Indeed, the transfer of charge associated with contact charging when particles bounce off the containing walls inside inhalers, and possibly when drug particles are torn from their carrier particles, is apparently responsible for the $\mu C$ $kg^{-1}$ charges seen with some dry powder formulations after aerosolization (Byron *et al.* 1997). The effect of particle charge on deposition of particles in the lung was discussed in Chapter 3.

## 9.7 Powder entrainment by shear fluidization

In order for powder particles to be inhaled from a dry powder inhaler, the powder must be swept up (i.e. entrained) by the air being inhaled, a process sometimes referred to as fluidization. The geometry in which this entrainment occurs is different for each inhaler,

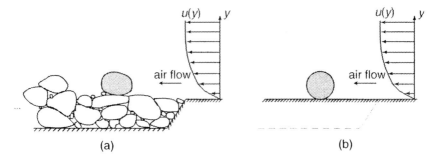

**Fig. 9.12** Entrainment of the shaded powder particle due to air flow as in (a) can be viewed simplistically using the geometry shown in (b), where the powder surface below the particle is treated as a plane wall (i.e. 'flat plate') and the particle is treated as a sphere. The streamwise velocity ($u$) of the air is zero at the wall, but increases with distance $y$ from the wall through a boundary layer.

so that the mechanics (both fluid and powder mechanics) of entrainment are device specific. However, a partial understanding of the mechanics involved in many inhaler designs can be had by considering the simplified problem of a spherical particle attached (by adhesion) to a plane wall with air sweeping along the wall, as shown in Fig. 9.12(b), which can be viewed as an idealization of the geometry shown in Fig. 9.12(a). In both Figs 9.12(a) and 9.12(b), for the particle to be entrained in the air flow and released for inhalation, the fluid exerts a lift (vertical) force on the particle that overcomes the adhesive force (for typical pharmaceutical inhalation powders, the weight of the particle is much smaller than either of these forces and can be neglected, as we shall soon see). Because the air in the boundary layer next to the wall undergoes strong shear $\tau = \mu \, du/dy$ due to the thin nature of boundary layers, the process of entrainment like that shown in Fig. 9.12 is sometimes called 'shear fluidization'. Of the various types of fluidization used in dry powder inhalers, shear fluidization is one of the more commonly used methods (Dunbar *et al.* 1998).

### 9.7.1 Laminar vs. turbulent shear fluidization

Before examining the mechanics of entrainment of a sphere on a wall as shown in Fig. 9.12, it is necessary to decide if the air flow is likely to be laminar or turbulent in the boundary layer flow approaching the particle. This is an important issue, since particle entrainment in a turbulent boundary layer is likely to be very different from that in a laminar boundary layer (Ziskind *et al.* 1995). This is because turbulent boundary layer structures ('eddies') are known to cause large, temporary increases in the instantaneous forces on particles adhering to walls, which can lead to dislodgement of the particle directly, or can cause the particle to vibrate itself off the wall over time due to elastic oscillations induced by the turbulent structures when their frequency is close to the natural frequency of elastic particle–wall distortions (Reeks *et al.* 1988, Lazaridis *et al.* 1998).

In general, as we saw in Chapter 6, if the Reynolds number is low enough, the flow is laminar, while at higher Reynolds numbers, there is a transition to turbulence. The transition to turbulence in wall boundary layers has been well studied. Unfortunately, general theoretical prediction of transition is not possible because of its dependence on external disturbances whose effect is not easily characterized (such as turbulence convected in from upstream flow patterns, acoustical disturbances, and the influence of wall roughness – see White 1991). However, we can avail ourselves of experimental data on this subject, where transition in flat plate boundary layers has been measured for a wide variety of external disturbances. These data suggest that the transition to turbulence depends on a Reynolds number, $Re_x$, defined as

$$Re_x = \frac{U_\infty x}{\nu} \tag{9.40}$$

Here, $U_\infty$ is the value of the fluid velocity outside the boundary layer (the so-called 'freestream' velocity) and $x$ is development length, i.e. the length of flat wall upstream along which the boundary layer develops. For $Re_x > Re_{x,tr}$, turbulence occurs. Values of $Re_{x,tr}$ are dependent on the extent of external disturbances, with values greater than $10^6$ being typical unless large external disturbances are present. Even in the presence of large amplitude disturbances (including freestream turbulence, acoustical disturbances and large wall roughness, the latter needing to satisfy $Uk/\nu > 120$ before any difference is seen from a smooth wall for 'sandpaper' type roughness, where $k$ is the roughness height and $\nu = \mu/\rho$ is the kinematic viscosity of air) flat plate boundary layers are normally laminar for $Re_x < 10^5$ (White 1991). Because of their limited size and relatively low flow velocities, $Re_x$ in wall boundary layers of the particle entrainment region of dry powder inhalers are typically not in the turbulent regime, as seen in the following example.

## Example 9.6

Estimate the cross-sectional area $A$ above which laminar wall boundary layers can be expected in the entrainment region of a prototypical dry powder inhaler like that shown in Fig. 9.13 where the air flows at $Q$ l min$^{-1}$ through a duct over a length $x$ prior to flowing over an indentation in the wall that is filled level to the wall with powder. Find a value for $A$ if $Q = 60$ l min$^{-1}$ and $x = 1$ cm.

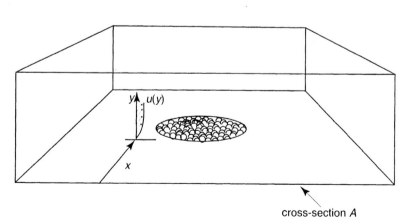

**Fig. 9.13** Air flows through a rectangular duct over a distance $x$ prior to reaching the powder contained in an indent in the wall. The duct cross-sectional area is $A$.

*Solution*

The Reynolds number $Re_x$ is given by Eq. (9.40), where we can estimate the freestream velocity from

$$U_\infty = Q/A$$

Thus, we have

$$Re_x = Qx/(\nu A)$$

Assuming laminar boundary layers will occur if $Re_x < Re_{x,tr}$, then we must have

$$A > Qx/(\nu Re_{x,tr})$$

to be assured of laminar boundary layers. For a flow rate of $Q = 60 \, \mathrm{l \, min^{-1}} = 10^{-3} \, \mathrm{m^3 \, s^{-1}}$ and $x = 0.01$ m, with the air viscosity and density being $\mu = 1.8 \times 10^{-5} \, \mathrm{kg \, m^{-1} \, s^{-1}}$, $\rho = 1.2 \, \mathrm{kg \, m^{-3}}$ and using $Re_{x,tr} = 100\,000$ as a very conservative estimate for the transition Reynolds number, we obtain

$$A > 5.6 \times 10^{-6} \, \mathrm{m^2}, \quad \text{i.e. } A > 5.6 \, \mathrm{mm^2}$$

Thus, for a flow passageway cross-sectional area larger than 5.6 mm², we do not expect turbulent wall boundary layers in the entrainment region in this case. This is smaller than the cross-sectional areas seen in the entrainment regions of most dry powder inhalers. The length, $x$, of flat wall upstream of the entrainment region is often shorter than 1 cm in typical dry powder inhalers, while the transition Reynolds number is probably larger than 100 000. As a result, this is probably a conservative estimate, so that smaller cross-sectional areas would probably be needed before an assumption of laminar boundary layers becomes questionable in many inhalers, although calculations are of course needed for each specific inhaler geometry to determine whether laminar or turbulent boundary layers may be expected.

From the above example, it is reasonable to suggest that for powder inhalers where entrainment of powder particles occurs by air flowing over the powder, as in Fig. 9.12(a), a laminar wall boundary is a good assumption in many cases.

It should be noted that the presence of the powder can 'trip' the boundary layer into being turbulent, but as mentioned above, roughness elements with height $Uk/\nu < 120$ have no effect on the transition Reynolds number $Re_{x,tr}$ in previous experiments, and the transition Reynolds number $Re_{x,tr} > 10^5$ for $Uk/\nu < 325$ (Schlichting 1979). For flow velocities less than 50 m s$^{-1}$ (which would occur in a 5 mm diameter circular passageway at 60 l min$^{-1}$), this implies that roughness heights $> 116$ µm are needed before a turbulent boundary layer begins to be induced below $Re_{x,tr} = 10^5$. Thus, even with large carrier particles of up to 100 µm in diameter, the transition Reynolds number is still above our conservative value of 100 000 mentioned above, so that the 'roughness' presented to the flow by the powder probably does not trip the boundary layer into becoming turbulent for most inhalers.

### 9.7.2 Particle entrainment in a laminar wall boundary layer

For the reasons just discussed, laminar boundary layers probably occur in many inhalers that use shear fluidization. Let us thus examine the aerodynamic forces on a particle attached to a flat wall in a laminar boundary layer, shown in Fig. 9.14.

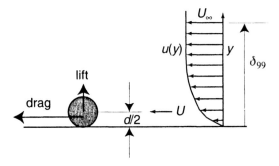

**Fig. 9.14** Drag and lift forces on a particle in a flat plate boundary layer.

Referring to Fig. 9.14, we define the drag and lift coefficients in terms of the lift and drag forces as

$$C_L = \frac{lift}{\frac{1}{2}\rho_{fluid}A_f U^2} \tag{9.41}$$

$$C_D = \frac{drag}{\frac{1}{2}\rho_{fluid}A_f U^2} \tag{9.42}$$

where $A_f = \pi d^2/4$ is the frontal cross-sectional area of the particle and $U = u|_{y=d/2}$ is the velocity of the fluid in the boundary layer at a distance $d/2$ from the wall.

The dependence of the lift force on the square of particle diameter (actually it is of even higher power than quadratic since $U$ itself increases with particle diameter for particles inside a boundary layer), while adhesive forces vary only linearly with diameter in our previous idealized equations ((9.11), (9.12), (9.20)–(9.25) and (9.33)). This is of course partly why large carrier particles (typically 50–100 μm in diameter) are often used to entrain the small 1–5 μm particles desired for lung deposition, since the large particles are exposed to large aerodynamic lift forces that pick the particles up and entrain them into the air flow. If properly designed, the small drug particles will adhere to the larger carrier particles and also be entrained. (Subsequent detachment of the small drug particles from the carrier particles occurs by turbulent deaggregation and impaction on walls, which we shall examine in subsequent sections.)

From dimensional analysis (Panton 1996), one can show that the drag and lift coefficients in Eqs (9.41) and (9.42) can be written as functions of the particle Reynolds number $Re_p$ and the ratio of particle diameter, $d$, to boundary layer thickness $\delta_{99}$, i.e. $d/\delta_{99}$. These parameters are defined as

$$Re_p = Ud/v \tag{9.43}$$

where $U = u|_{y=d/2}$ is the velocity in the boundary layer a distance $d/2$ from the wall, and $\delta_{99}$ ('delta 99') is defined as the value of $y$ at which $u = 0.99U_\infty$ in the upstream boundary layer, i.e.

$$u(\delta_{99}) = 0.99U_\infty \tag{9.44}$$

We thus write

$$C_L = f\left(Re_p, \frac{d}{\delta_{99}}\right) \tag{9.45}$$

$$C_D = g\left(Re_p, \frac{d}{\delta_{99}}\right) \tag{9.46}$$

The functions $f$ and $g$ in Eqs (9.45) and (9.46) are known in closed form only for the case $Re_p \ll 1$ and $d/\delta_{99} \ll 1$, in which case we can make use of the results for creeping flow in a linear shear layer (where $u(y) = 2Uy/d$), since for $y \ll \delta_{99}$ the boundary layer can be approximated as a linear profile (see Eq. (9.50) below). Various parts of the problem of a sphere attached to a wall in a linear shear flow have been solved for very low Reynolds numbers by different authors (Goldman *et al.* 1967, O'Neill 1968, Leighton and Acrivos 1985, Krishnan and Leighton 1995, Cichocki and Jones 1998) and verified experimentally (King and Leighton 1997). Total neglect of fluid inertia gives front-to-back flow symmetry on the sphere, resulting in drag but no lift, so that it is necessary to include terms to first order in Reynolds number in the flow field to obtain nonzero lift in solving this problem, which gives

$$C_L = 5.87 \tag{9.47}$$

$$C_D = 1.7 \times \frac{24}{Re_p} \tag{9.48}$$

for a particle on a wall in linear shear flow with $Re_p \ll 1$.

Notice that $C_L = 0$ for a sphere in a uniform flow (and the drag given by Eq. (9.48) is 1.7 times the value $24/Re_p$ for a sphere placed in a uniform flow at low $Re_p$ from Chapter 3). Thus, the presence of shear in the fluid motion upstream of the particle is necessary to produce lift on the particle, which is of course why the term 'shear fluidization' is used to describe entrainment of powder due to this lift (note the word shear in 'shear fluidization' does not refer to the shear force exerted on the particle, but rather the shear in the upstream boundary layer, since it is a combination of shear and pressure that exerts lift on the particle).

At this point it is worth examining whether the requirements that $Re_p \ll 1$ and $d/\delta_{99} \ll 1$ are likely to be satisfied in dry powder inhalers. For this purpose, it is necessary to estimate the boundary layer thickness $\delta_{99}$, as well as the velocity $U = u|_{y=d/2}$ appearing in Eq. (9.43). For a laminar flat plate boundary layer (White 1991, Panton 1996),

$$\delta_{99} = \frac{5.0x}{\sqrt{Re_x}} \tag{9.49}$$

$$\frac{u}{U_\infty} \approx \frac{3}{2}\eta - \frac{1}{2}\eta^3 \quad \text{where} \quad \eta \approx \frac{y}{0.918\delta_{99}} \tag{9.50}$$

Thus, the particle Reynolds number $Re_p$ and $d/\delta_{99}$ will vary with distance from the leading edge of the boundary layer, in addition to varying with freestream velocity $U_\alpha$ and must be calculated for each inhaler under consideration.

## Example 9.7

A prototype of a dry powder inhaler has a powder entrainment region as shown in Fig. 9.13, with $x = 0.5$ cm being the distance from the leading edge of the flat wall to the entrainment region. The cross-sectional area is 20 mm$^2$ and the inhaler flow rate is designed to be 60 l min$^{-1}$ (0.001 m$^3$ s$^{-1}$). Determine what size powder particles will meet the requirements $Re_p \ll 1$ and $d/\delta_{99} \ll 1$, which are needed for the drag and lift force on a spherical particle on the wall in the entrainment region to be given by Eqs (9.47) and (9.48). Also ensure that a laminar boundary layer, which has been assumed for Eqs (9.47) and (9.48), is expected.

## Solution

We can estimate the freestream velocity from the flow rate $Q$ and cross-section $A$ using

$$U_\infty = Q/A \tag{9.51}$$

which gives

$$U_\infty = 0.001 \text{ m}^3 \text{ s}^{-1}/20 \times 10^{-6} \text{ m}^2$$
$$= 50 \text{ m s}^{-1}$$

Substituting this in the definition of $Re_x$ in Eq. (9.40)

$$Re_x = \frac{U_\infty x}{\nu}$$

with $x = 0.005$ m, as given, and the kinematic viscosity of room temperature air being $\nu = 1.5 \times 10^{-5}$ m$^2$ s$^{-1}$, gives us $Re_x = 16\,667$. This is well below the conservative estimate of $Re_{x.tr} = 100\,000$ discussed earlier for the transition to turbulence, so we expect a laminar boundary layer.

Knowing that the boundary layer is laminar, we can now examine the particle sizes for which it is reasonable to use the low particle Reynolds number linear shear layer results in Eqs (9.47) and (9.48). First, let us examine the requirement that the particle lies in the approximately linear shear region of the boundary layer, i.e. $d \ll \delta_{99}$. Evaluating $\delta_{99}$ from Eq. (9.49) with $x = 0.005$ m and $Re_x = 16\,667$, we obtain $\delta_{99} = 193$ μm. Some inhalation powders will have diameters much less than this, i.e. $d \ll \delta_{99} = 193$ μm, but some (particularly those with carrier particles) will not. Recall that the reason for this assumption is that Eqs (9.47) and (9.48) are valid for a linear velocity profile, which is only valid in the inner region of the boundary layer very close to the wall. In the present example, the assumption of a linear velocity profile, as used to obtain Eqs (9.47) and (9.48), is a marginal assumption for particles larger than 20 μm or so.

Let us now examine the Reynolds number criterion $Re_p \ll 1$. To satisfy this criterion we must have

$$\frac{Ud}{\nu} \ll 1 \tag{9.52}$$

which we can write as

$$d \ll \frac{\nu}{U} \tag{9.53}$$

Substituting Eq. (9.50) for $U$ into Eq. (9.53), we have

$$d \ll \cfrac{v}{U_\infty \left[ \cfrac{3}{2} \left( \cfrac{d/2}{0.918\delta_{99}} \right) - \cfrac{1}{2} \left( \cfrac{d/2}{0.918\delta_{99}} \right)^3 \right]} \qquad (9.54)$$

Putting in Eq. (9.49) for $\delta_{99}$, Eq. (9.54) becomes

$$d \ll \cfrac{v}{U_\infty \left[ \cfrac{3}{2} \left( \cfrac{\sqrt{Re_x}(d/2)}{0.918 \times 5.0x} \right) - \cfrac{1}{2} \left( \cfrac{\sqrt{Re_x}(d/2)}{0.918 \times 5.0x} \right)^3 \right]} \qquad (9.55)$$

This is a cubic equation in $d$. Exact solutions for cubic equations are known, and are given in most symbolic computation packages, from which we obtain, after substitution of $x = 0.05$ m, $U_\infty = 50$ m s$^{-1}$, $Re_x = 16\,667$,

$$d \ll 8.4 \ \mu m \qquad (9.56)$$

This condition is not likely to be met for most powders considered to be useful for inhalation therapy. Decreasing the inhalation flow rate $Q$, increasing the cross-sectional area $A$, or increasing $x$ all result in larger values on the right-hand side of Eq. (9.56), but even decreasing $Q$ to 30 l min$^{-1}$ while simultaneously increasing $A$ to 100 mm$^2$ and $x$ to 5 cm still results in the requirement that $d \ll 84$ μm, which will not generally be met if typical carrier particles, such as lactose monohydrate, are used (carrier particles are usually in the 50–100 μm range). As a result, it is likely that most dry powder inhalers will have particle Reynolds numbers that are too high for Eqs (9.47) and (9.48) to be valid, although of course this must be ascertained on an individual basis. As an example of the values of typical particle Reynolds numbers encountered, for $x = 0.005$ m, $U_\infty = 50$ m s$^{-1}$, $Re_x = 16\,667$, a 30 μm diameter particle has $Re_p = 12.6$, while a 5 μm particle has $Re_p = 0.35$ under the same conditions.

As can be seen in the previous example, for many powder inhalers, particle Reynolds numbers are probably higher than the range of validity of Eqs (9.47) and (9.48). This is unfortunate, since no analytical results and few experimental results are available at higher particle Reynolds numbers in the range $Re_p < 100$ or so expected in powder inhalers. Willetts and Naddeh (1986) measured lift coefficient values for a sphere attached to a wall in laminar boundary layers, finding $C_L \approx 0.4$ for unspecified Reynolds numbers in the range $Re = 43$–100 and for spheres having $d/\delta_{99} \approx 1/4$, while $C_L \approx 0.05$ for $Re = 83$–140 and $d/\delta_{99} \approx 1/2$. Thus, with increasing Reynolds number and $d/\delta_{99}$, lift coefficients appear to decrease from the value of 5.87 given in Eq. (9.47). Of course, as Eqs (9.45) and (9.46) indicate, lift and drag coefficients are a function of both $Re_p$ and $d/\delta_{99}$ and more data are needed to suggest accurate values expected during laminar shear fluidization in powder inhalers. However, given the present data, lift coefficients of order 1 are to be expected. The resulting lift forces are large enough to overcome adhesive forces, as seen in the next example.

## Example 9.8

Assuming a lift coefficient $C_L = 1$ and a laminar boundary layer, estimate the lift force $F_L$ on a spherical particle adhered to a flat wall in a duct with 20 mm$^2$ cross-section with flow rate 60 l min$^{-1}$. Assume the particle is a distance $x = 3$ mm from the leading edge of the boundary layer. Obtain values and compare to the weight of the particle and the measured adhesion force for

(a) lactose monohydrate particles of diameter 62 μm where the measured median adhesion force is $2.5 \times 10^{-8}$ N (Podczeck et al. 1996),
(b) lactose monohydrate particles of diameter 3 μm where the measured median adhesion force is greater than $7 \times 10^{-12}$ N at all humidities (Podczeck et al. 1997).

Use a density for solid lactose monohydrate of 1530 kg m$^{-3}$.

## Solution

From Eq. (9.41), the lift force is given by

$$F_L = C_L \frac{1}{2} \rho_{fluid} U^2 \frac{\pi d^2}{4} \tag{9.57}$$

We must determine the velocity $U$ in Eq. (9.57), which, recall, is the velocity in the boundary layer a distance $d/2$ from the wall. For this we can use Eqs (9.49) and (9.50). First, we obtain

$$U_\infty = Q/A \tag{9.58}$$

which gives

$$U_\infty = 0.001 \text{ m}^3 \text{ s}^{-1}/20 \times 10^{-6} \text{ m}^2$$
$$= 50 \text{ m s}^{-1}$$

Substituting this into Eq. (9.40) we obtain

$$Re_x = \frac{U_\infty x}{\nu}$$
$$= 50 \text{ m s}^{-1}(0.003)/1.5 \times 10^{-5} \text{ m}^2 \text{ s}^{-1}$$
$$= 10\,000$$

Using this $Re_x = 10\,000$ and $x = 0.003$ m in Eq. (9.49), we obtain $\delta_{99} = 150$ μm. Putting $\delta_{99} = 150$ μm into Eq. (9.50) with $y = d/2$ we obtain the velocity at the distance $d/2$ from the wall, which for the two parts of the problem gives

(a) $U = 16.6$ m s$^{-1}$
(b) $U = 0.82$ m s$^{-1}$

We can now substitute these values into Eq. (9.57) with the two different particle diameters in (a) and (b) to obtain the following lift forces:

(a) $F_L = 4.9 \times 10^{-7}$ N
(b) $F_L = 2.8 \times 10^{-12}$ N

By comparison, the weight of the particles is $W = \rho g \pi d^3/6$, which gives

(a) weight $W = 1.9 \times 10^{-9}$ N
(b) weight $W = 2.1 \times 10^{-13}$ N

For part (a) the lift force is 20 times greater than the median adhesion force, and is 258 times the weight of the particle. For (b), the median adhesion force is 2.5 times greater than the lift force, and is 33 times the weight of the particle.

From the above numbers, we expect powder (a), consisting of large lactose particles, to be entrained by shear fluidization (since the lift force greatly exceeds the adhesion and gravity forces), but expect powder (b) to remain largely undispersed (since the lift force is well below the adhesion force of $7 \times 10^{-12}$ N).

Although the lift forces involved must be calculated for the specific inhaler under consideration, the previous example shows that the shear fluidization lift force on a particle in a laminar boundary layer is large enough to overcome typical adhesion forces of the carrier particles common to many inhaler formulations, but is likely insufficient to directly fluidize powders consisting solely of 1–5 μm particles typically desired for inhalation. This is part of the reason why carrier particles are often used in powder inhalers (another major reason being that they make it easier to portion out small doses of drug, since accurately metering a few micrograms of drug powder by itself is difficult, but is made easier by the presence of several milligrams of carrier particles).

After particle lift-off from the powder bed, the particle experiences a strong drag force (which is now given by the Stokes drag or other higher Reynolds number corrections discussed in Chapter 3) which accelerates the particle in the streamwise direction. The particle will also experience a torque that tends to increase its angular velocity (Crowe *et al.* 1998) since its angular velocity is initially lower than that of the fluid in the boundary layer (note that the resulting spin can induce centrifugal forces that may pull apart particles that are entrained as agglomerates – see the following example). However, lift forces also continue to be exerted on the particle due to its presence in the shear of the boundary layer (the 'Saffman lift force', which occurs on a particle in an unbounded linear shear layer), as well as possibly negative lift forces due to the difference in angular velocity of the particle and fluid (the 'Magnus lift force', which occurs when a particle has different angular velocity than the fluid). These forces are discussed in Crowe *et al.* (1998) in the absence of wall effects, while Wang *et al.* (1997) give a review of work including the presence of a wall, although most of the results in the presence of a wall are for particle Reynolds numbers and shear rates that are lower than expected in the entrainment region of dry powder inhalers. It should be noted that corrections to the Saffman lift force that account for the presence of the wall result in a negative lift force (i.e. a force that is directed towards the wall) once the particle is several particle diameters from the wall (Willetts and Naddeh 1986, Wang *et al.* 1997). However, by the time the particle reaches this distance from the wall, the above mentioned torque on the particle may have resulted in angular velocities that give rise to positive Magnus forces that counter this negative lift in the low shear regions of the outer boundary layer. The dynamics of particles immediately after entrainment are thus complex, and few studies or models are applicable in the range of parameters expected in dry powder inhalers.

## Example 9.9

Perform a simple analysis to crudely estimate the centrifugal force on a 5 μm lactose sphere attached to a 75 μm lactose carrier particle that has been entrained in a laminar boundary layer with freestream velocity $U_z = 50$ m s$^{-1}$. For simplicity, neglect the presence of the 5 μm sphere in estimating the rotational speed of the 75 μm particle. Also neglect the presence of the wall in calculating the lift so that the correlation of Mei (1992) can be used:

$$F_L = F_{saff}\left[(1 - 0.3314\sqrt{\beta})\exp\left(-\frac{Re_p}{10}\right) + 0.3314\sqrt{\beta}\right] \qquad Re_p \leq 40$$

$$= F_{saff}[0.0524\sqrt{\beta Re_p}] \qquad\qquad\qquad Re_p > 40 \qquad (9.59)$$

where $Re_p = dv_{rel}/v$ (here $v_{rel}$ is the speed of the fluid relative to the particle). Assume a value of $v_{rel}$ equal to $U_\infty$. The parameter $\beta$ is given by

$$\beta = \frac{d}{2v_{rel}}\frac{1}{2}\frac{du}{dy} \qquad (9.60)$$

and $F_{saff}$ is the Saffman lift expression valid for very low Reynolds number,

$$F_{saff} = 1.61d^2(\mu\rho_{fluid})^{1/2}\sqrt{\frac{1}{2}\frac{du}{dy}}\,v_{rel} \qquad (9.61)$$

Estimate the velocity gradient in the above expressions as $du/dy = U_\infty/\delta_{99}$.

Estimate the torque on the particle using the low Reynolds number result (Crowe *et al.* 1998)

$$T = \pi\mu d^3\left(\frac{1}{2}\frac{du}{dy} - \omega_p\right) \qquad (9.62)$$

where $\omega_p$ is the angular velocity of the particle (assumed to be positive and in the same direction as a rolling particle). For simplicity, estimate the torque assuming $\omega_p = 0$.

## Solution

We need to estimate the boundary layer thickness in order to estimate $du/dy$. Using Eq. (9.49) with $x = 0.01$ m and $U_\infty = 50$ m s$^{-1}$, we obtain $\delta_{99} = 274$ μm.

Estimating $du/dy$ as $U_\infty/\delta_{99}$, we obtain

$$du/dy = 182\,707\ s^{-1}$$

Using this value of $du/dy$ in Eq. (9.61) with $\rho_f = 1.2$ kg m$^{-3}$ for the density of air, we obtain

$$F_{saff} = 9 \times 10^{-7}\ N$$

Substituting $v_{rel} = 50$ m s$^{-1}$, $du/dy = 182\,707$ s$^{-1}$ into Eq. (9.60) gives $\beta = 0.07$. Using this value of $\beta$ with $F_{saff} = 9 \times 10^{-7}$ N, and $Re_p$ estimated as $U_\infty d/v = 250$, Eq. (9.59) gives

$$F_L = 2 \times 10^{-7}\ N \text{ for the lift force}$$

Assuming a constant particle acceleration of $a = F_L/m$ where $m$ is the mass of the particle, we can then estimate the time taken for the particle to traverse the boundary layer (at which time the torque on the particle will reverse direction as the particle enters the uniform stream) by solving $d^2y/dt^2 = a$ with zero initial conditions and constant acceleration, to obtain

$$\frac{at^2}{2} = \delta_{99} \tag{9.63}$$

Using $a = F_L/m$, where $m = \rho\pi d^3/6$ and $\rho = 1530 \text{ kg m}^{-3}$ for lactose, we obtain $t = 982 \text{ }\mu s$.

The angular velocity of the particle can now be estimated using the angular momentum equation

$$I = \frac{d\omega_p}{dt} = T \tag{9.64}$$

where $I$ is the moment of inertia and

$$I = \frac{2}{5}m\left(\frac{d}{2}\right)^2$$

for a sphere. Assuming a constant torque given by Eq. (9.62) with $\omega_p = 0$, we obtain $T = 2.2 \times 10^{-12} \text{ N m}$. The angular velocity of the particle after a time $t$ is given by

$$\omega_p = Tt/I \tag{9.65}$$

which gives $\omega_p = 11\,259 \text{ rad s}^{-1}$. If the 75 µm particle executes this angular velocity, the 5 µm particle would feel a centrifugal force

$$F_{cent} = m'\omega_p^2(d'/2) \tag{9.66}$$

where $m'$ is the mass of the 5 mm particle and $d' = 30 \text{ µm} + 5 \text{ µm}$. Substituting in the numbers, we obtain

$$F_{cent} = 4 \times 10^{-13} \text{ N}$$

This is much smaller than the median adhesion forces for fine lactose particles adhering to lactose substrates (Podczeck et al. 1997), so that it seems unlikely that the centrifugal force resulting from particle spin will lead to particle detachment. However, our estimate here is very crude, and more detailed calculations are not readily performed, since more exact equations for the torque in a shear layer and for the Magnus force (which we have neglected here, but should be included in a more detailed analysis) are not available for the parameter range considered here.

It should be noted that the drag force during shear fluidization is typically much larger than the lift force (this is apparent from Eqs 9.47 and 9.48 for low Reynolds numbers) so that even if the particles are not lifted and entrained, they may roll and/or slip in response to the large drag force (Krishnan and Leighton 1995, King and Leighton 1997), although the powder bed presents irregular roughness elements as large as the particle itself that partly inhibit such rolling and sliding. Still, rolling or sliding may result in some motion of particles over the irregular powder bed. However, rolling and sliding do not themselves lead to the desired entrainment into the flow stream (simply resulting in translation of the particle along the powder bed or possibly the downstream wall). Although rolling is the major mechanism resulting in motion of smooth spherical particles on flat surfaces (Wang 1990), the large entrainment rates desired for inhalation powders are better achieved by having the lift forces exceed adhesion forces. However, more work is needed to fully elucidate the entrainment mechanisms associated with pharmaceutic inhalation powders.

### 9.7.3 Particle entrainment in a turbulent wall boundary layer

We have already seen that laminar boundary layers are probably more common along walls in dry powder inhalers than are turbulent boundary layers, largely due to the relatively compact size of such inhalers and the short development lengths involved in any wall shear layers. However, turbulent boundary layers may be present in some inhaler designs. For this reason, it is useful to briefly examine shear fluidization in turbulent boundary layers.

One of the principal parameters for detachment of particles that are smaller than the boundary layer thickness in turbulent boundary layers is the nondimensional particle diameter $d^+$, defined as

$$d^+ = \frac{u_\tau d}{\nu} \tag{9.67}$$

where $u_\tau$ is the so-called 'friction velocity' defined as

$$u_\tau = \sqrt{\frac{\tau_w}{\rho_{fluid}}} \tag{9.68}$$

where $\rho^{fluid}$ is the fluid density and $\tau_w$ is the mean shear stress at the wall, which for a turbulent wall boundary layer can be approximated as (White 1999)

$$\tau_w \approx \frac{0.0135 \mu^{1/7} \rho_{fluid}^{6/7} U_\infty^{13/7}}{x^{1/7}} \tag{9.69}$$

Combining Eqs (9.67)–(9.69), we can write

$$d^+ \approx \frac{0.1162 \rho_{fluid}^{13/14} U_\infty^{13/14} d}{x^{1/14} \mu^{13/14}} \tag{9.70}$$

Typical values of $d^+$ for powder inhalers are less than 100.

Values of the mean (i.e. time-averaged) lift force on particles attached to a wall and embedded in a turbulent boundary layer have been measured by several authors. For values in the range $3 < d^+ < 100$, the empirical correlation of Hall (1988), is useful:

$$F_L = 20.9 \rho_{fluid} \nu^2 \left(\frac{d^+}{2}\right)^{2.31} \tag{9.71}$$

Hall (1988) also measured the lift force on a spherical particle where the flow was disturbed by a regular array of rods lying on the wall (perpendicular to the flow direction) upstream and downstream of the particle, with rod diameters varying from 0.67 to 2.5 times the particle diameter. The spacing between the rods was equal to their diameter, and the particle was placed midway between two of the rods. The lift force was reduced in magnitude by roughly a factor of 5 from the smooth wall case of Eq. (9.71). This latter result is interesting because it suggests that a particle does not need to be fully exposed to the fluid, as in Fig. 12(a), to have a lift force exerted on it, and that even particles that are 'shielded' from the flow by upstream particles may be entrained by this lift force, albeit with a reduced magnitude to the lift. However, lift forces in a turbulent boundary layer can greatly exceed typical adhesion forces, so that even 'shielded' particles can probably be readily entrained.

For values of $0.3 < d^+ < 2$, the empirical correlation of Mollinger and Nieuwstadt (1996) can be used:

$$F_L = 56.9 \rho_{\text{fluid}} v^2 \left(\frac{d^+}{2}\right)^{1.87} \tag{9.72}$$

It should be noted that the lift force on a particle will fluctuate considerably about the mean values given in Eqs (9.71) and (9.72) because of the presence of turbulence in the boundary layer. Indeed, Mollinger and Nieuwstadt (1996) find r.m.s. values of the lift force that are approximately 2.8 times the mean value in Eq. (9.72).

Various researchers have developed models for turbulent detachment of particles from walls (see Soltani and Ahmadi 1994, Ziskind *et al.* 1995 for reviews), some of which include models for the fluctuating lift forces (with mean given by Eq. (9.63) or (9.72)), as well as simplified models of the effect of surface roughness (Soltani and Ahmadi 1995, Ziskind *et al.* 1997). These models include detachment due to rolling or sliding associated with the drag force and its resultant moment, with rolling and sliding rather than lift thought to be the dominant mechanism of removal for spheres on flat walls. It should be noted that the terms 'detachment' or 'removal' in these studies do not refer to entrainment into the flow, but simply mean the particle moves from its original position (often by translating horizontally along the wall). For powder inhalers, it is desirable to entrain the powder into the flow, which requires vertical motion of the particle off of the wall (since having the powder 'roll' into the mouth leads to excessive mouth–throat deposition). This should be borne in mind when reading the literature on 'removal' or 'detachment' of particles from flat walls.

Matsusaka and Masuda (1996) found that entrainment of fine powders ($d = 3.0$ μm) occurred in their experiments over several hundred seconds via the entrainment of agglomerates that initially executed a rolling movement.

In general, the nonspherical shape and poorly characterized rough surfaces of the particles and powder beds typically encountered in pharmaceutic inhalation powders limits the predictive power of typical such detachment/removal models, since adhesive and aerodynamic forces are not readily predicted on such particles. More work is needed to discern the mechanisms of detachment of pharmaceutical inhalation powders in a turbulent boundary layer.

### Example 9.10

Consider a powder entrainment region like that shown in Fig. 9.13 with a cross-sectional area of 4 mm², a development length $x = 2$ cm and inhalation flow rate of 20 l min$^{-1}$, which results in $Re_x = 1.11 \times 10^5$, which brings us just into the realm where a turbulent boundary layer may be possible if, for example, there is a high level of turbulence being convected in from upstream. However, a laminar boundary layer is also possible unless such external disturbances are very large. Assuming a lift coefficient of $C_L = 1$ for the laminar boundary layer, compare the lift force predicted for a laminar boundary assumption and a turbulent boundary layer assumption for

(a) a 3 μm particle,
(b) a 20 μm particle,
(c) a 60 μm particle.

## Solution

This problem is a matter of using Eq. (9.57) for the laminar case and either Eq. (9.63) or (9.72) for the turbulent case.

For the laminar case, the steps are the same as in the previous example, i.e. we use Eq. (9.57):

$$F_{Llam} = C_L \frac{1}{2} \rho_{fluid} U^2 \frac{\pi d^2}{4}$$

where $U$ is given by Eq. (9.50) evaluated at $y = d/2$, and we have $U_x = Q/A = 83.3 \text{ m s}^{-1}$. Performing the calculations using $v = 1.5 \times 10^{-5} \text{ m}^2 \text{ s}^{-1}$ and $\rho_{fluid} = 1.2 \text{ kg m}^{-3}$ for air, we obtain

(a) $F_{Llam} = 2.0 \times 10^{-12}$ N,
(b) $F_{Llam} = 3.9 \times 10^{-9}$ N,
(c) $F_{Llam} = 3.1 \times 10^{-7}$ N.

For the turbulent case, we must calculate $d^+$ from Eq. (9.70) and substitute this into Eq. (9.63) if $d^+ > 3$ or Eq. (9.72) if $0.3 < d^+ < 2$. Equation (9.70) gives the following values for $d^+$:

(a) $d^+ = 0.84$,
(b) $d^+ = 5.6$,
(c) $d^+ = 16.9$.

Thus, for (a) we use Eq. (9.72), while for (b) and (c) we use Eq. (9.63). Substituting in these three values of $d^+$ to the appropriate equations, we obtain

(a) $F_{Lturb} = 3.1 \times 10^{-9}$ N,
(b) $F_{Lturb} = 6.2 \times 10^{-8}$ N,
(c) $F_{Lturb} = 7.8 \times 10^{-7}$ N.

The lift force on the particle in a turbulent boundary layer is thus larger than in a laminar one by a factor of

(a) 1559 for the 3 μm particle,
(b) 16 for the 20 μm particle,
(c) 2.5 for the 60 μm particle.

As mentioned earlier, much larger instantaneous lift forces would be seen than the mean values calculated with Eq. (9.71) or (9.72) due to turbulent fluctuations.

As can be seen from the previous example, lift forces on a particle attached to a wall in a turbulent boundary layer can be considerably larger than for a laminar boundary layer, particularly for smaller particles. Indeed, the aerodynamic lift forces in the above example can readily exceed adhesion forces. However, as discussed earlier, the ergonomic need for compact devices has made it such that development lengths normally associated with values of $Re_x$ below typical turbulent transition values probably occur in many devices. However, designs that exploit the higher lift forces associated with turbulent boundary layers may be advantageous.

### 9.7.4 Entrainment by bombardment: saltation

One of the most common mechanisms for entrainment of particles from the ground into the atmosphere by wind is 'saltation' (derived from the Latin verb *saltare* which means 'to leap or dance'). This mechanism occurs because sand and other geophysical particles are often too heavy to remain entrained in the air after turbulent, wind-driven lift-off, falling back to the ground over a saltation length. When they impact the ground (at high speed since they are being swept downstream by the wind), they dislodge new particles. A large body of literature exists on saltation (recent perspectives are given by Sherman *et al.* 1998, Willetts 1998). However, saltation lengths (Maeno *et al.* 1995) are orders of magnitude larger than the lengths of powder beds encountered in inhaled pharmaceutical aerosols for the particle sizes and flow velocities expected with such aerosols. As a result, saltation like that seen in wind-driven atmospheric flows probably does not play a role in the entrainment of powder with current inhalation devices.

## 9.8 Turbulent deaggregation of agglomerates

Although shear fluidization may disperse individual carrier particles with diameters larger than a few tens of microns, such carrier particles may also carry with them smaller drug particles attached to their surface (which is indeed one purpose of the carrier particles). In addition, some inhalers direct the inhaled air directly up through the powder bed through jets (sometimes called 'gas-assist' fluidization, Dunbar *et al.* 1998), rather than over it as in shear fluidization, resulting in entrainment of large agglomerates of powder. For these reasons, many inhalers must deaggregate (i.e. breakup) the powder after entrainment, since the entrained particles are too large to escape impaction in the mouth and throat (recall we saw in Chapter 7 that particles larger than a few microns in diameter have a reasonable chance of impacting in the mouth–throat). One method for breaking up entrained powder agglomerates is through the use of turbulence, which we now examine. Note that the turbulence used for deagglomeration is typically produced by jets, grids and free shear layers, which produce turbulence more readily than wall boundary layers over the short distances typical in compact dry powder inhalers.

Before proceeding, it should be realized that exact analysis of the mechanics involved is difficult due to the complex nature of turbulence and the irregular particle shapes involved. Indeed, due to the small length scales involved (both of the particles and the turbulence) a precise analysis would require direct numerical simulation of the Navier–Stokes equations in addition to the equations governing the contact mechanics, the latter being complicated by surface roughness effects. Such an analysis is impractical at present due to its computational requirements. However, a qualitative understanding can be gained, as follows.

Consider an aggregate of particles traveling in a turbulent flow field, as shown in Fig. 9.15.

Assuming the eddies cover a range of sizes that include sizes much larger than the particle, the aggregate is then buffeted by eddies that exert aerodynamic forces on the aggregate and its individual particles. For example, as the aggregate is exposed to first one eddy, and then another, it can experience accelerations as these eddies drag the particle first one way, then another. If large enough, these accelerations may lead to

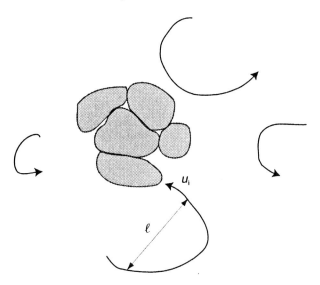

**Fig. 9.15** An aggregate of particles traveling in a turbulent flow with the large, most energetic, eddies having length $\ell$ and fluctuating velocity $u_i$.

internal forces within the aggregate that are larger than the adhesive forces, causing particles to separate off from the aggregate. Another possibility involves an eddy exerting a drag force on part of the aggregate that extends in to the eddy, causing one or more particles in the aggregate to roll, slide or lift off the aggregate. This could occur, for example, on the bottom particle of the aggregate in Fig. 9.15, which extends out into an eddy that exerts a drag and lift force on the bottom particle.

### 9.8.1 Turbulent scales

To estimate the sizes of such forces and thus estimate their potential for deaggregating agglomerates, we need to first estimate the velocity scales, length scales and time scales of the eddies comprising the turbulence. In any turbulent flow there will be a range of values to such scales. However, the large eddies contain most of the energy and the velocity, length and time scales associated with these eddies are called the integral scales (since their definition involves integrals arising from autocorrelations). We give the symbol $\ell$ to the integral length scale[3], $t_i$ to the integral time scale and the symbol $u_i$ to the integral velocity scale.

At the opposite end of the spectrum are the smallest, least energetic eddies, whose length $\eta$, time $t_K$ and velocity $v$ are referred to as the Kolmogorov scales. It should be

---

[3]Mathematically speaking, the integral length $\ell$ is defined such that the rate of turbulent energy dissipation, $\varepsilon$, is given by

$$\varepsilon \approx \frac{u_i^3}{\ell},$$

where

$$\varepsilon = -v\overline{\frac{\partial u'_m}{\partial x_n}\frac{\partial u'_n}{\partial x_m}}$$

and summation over $m$ and $n$ is implied, with the turbulent fluctuating velocity having $x$, $y$, $z$, components ($u'_1$, $u'_2$, $u'_3$) and the overbar indicates a time-averaged value (e.g. $\bar{u} = \int_0^\infty u\,dt$).

noted that for an observer traveling at the mean flow velocity, these time scales are related to length and velocity scales by

$$t_i = \ell/u_i \tag{9.73}$$

$$t_K = \eta/u_i \tag{9.74}$$

the latter resulting following because the integral scale eddies convect the Kolmogorov eddies past the particle at speed $u_i$.

If values of the integral scales are known, the Kolmogorov scales can be estimated (Tennekes and Lumley 1972) using

$$\frac{\eta}{\ell} \approx \left(\frac{u_i\ell}{\nu}\right)^{-3/4} \tag{9.75}$$

$$\frac{t_K}{t_i} \approx \left(\frac{u_i\ell}{\nu}\right)^{-3/4} \tag{9.76}$$

$$\frac{v_K}{u_i} \approx \left(\frac{u_i\ell}{\nu}\right)^{-1/4} \tag{9.77}$$

where in Eq. (9.76) the time scales are those seen by an observer moving at the mean flow velocity, and $\nu$ is the kinematic viscosity of air.

At high Reynolds numbers, a third scale which is intermediate to the integral and Kolmogorov scales, called the inertial subrange scale, is important. However, in typical dry powder inhalers, Reynolds numbers are relatively small so that the range of sizes occupied by the inertial subrange is small and we can adequately extrapolate the effect of the inertial subrange on deaggregation as being intermediate between the inertial and Kolmogorov scales.

To be precise when discussing turbulent scales, it is necessary to distinguish between the value of the turbulent time scale measured at a fixed location (the 'Eulerian' time scale) and the turbulent time scale measured by an observer that translates with the turbulence (the fluid 'Lagrangian' scale). In fact, we would like to know the values of the integral and Kolmogorov time scales as seen by a particle (which are also 'Lagrangian', but not the same as the fluid Lagrangian time scale, since the particle has its own velocity separate from the fluid). Lagrangian time scales are rarely known and difficult to obtain. Indeed, to obtain the time scale from the particle's point of view would require knowing the detailed turbulent flow field and the particle's trajectory through this field. Such data are not generally obtainable. However, for the order of magnitude type analysis that we wish to consider, we can approximate the turbulent time scales using the values obtained by an observer traveling at the mean flow velocity, which we have defined above.

In actual fact, the turbulent time scale seen by the particle depends on how the particle responds to the turbulence. If the particle's inertia is such that it rapidly acquires the velocity of each eddy that it encounters, then the time scale for the particle is approximately equal to the fluid Lagrangian time scale (since the particle essentially moves with the fluid). At the other extreme, if the particle's inertia is large, then the particle behaves like a stationary particle from the eddy's point of view, and the appropriate time scale is the Eulerian time scale. In between these two extremes, it is necessary to track the particle's motion to determine the time scale involved (Gosman and Ioannides 1983, Pozorski and Miner 1998). The key parameter distinguishing these regimes is the particle's stopping distance, $x_{\text{stop}}$ (introduced in Chapter 3). For particle stopping distances much less than the eddy size, the particle travels with the eddy so the fluid Lagrangian time scale approximates the turbulent time scale, while for stopping distances larger than the eddy size, the Eulerian time scale is more appropriate. For agglomerates seen in pharmaceutical aerosol applications, typical stopping distances are in the centimeter range, which is larger than typical eddy sizes designed to cause deaggregation in dry powder inhalers. As a result, the use of time scales obtained by an observer moving at the mean flow velocity is probably reasonable for our purposes, which allows us to sidestep the difficulty of estimating truly Lagrangian time scales.

In general, estimation of turbulent scales relies on experimental data since the equations governing turbulent flow are not readily solved analytically or computationally. Data for various geometries are given in standard reference texts, such as White (1991), as well as throughout the literature. Values of the root mean square turbulent fluctuating velocities, e.g. $\sqrt{u'^2}$, provide an estimate for the integral velocity scale (here the prime indicates a fluctuating quantity, i.e. the velocity $u$ satisfies $u = \bar{u} + u'$, where overbar indicates a time average, i.e. $\bar{u} = \int_0^\infty u\,dt$). The integral scale $\ell$ can be estimated approximately by taking $\ell$ to be equal to the largest expected dimension of the turbulent eddies (e.g. $\ell$ in a pipe is equal to the pipe diameter; $\ell$ in a boundary layer is equal to the boundary layer thickness at the current position. $\ell$ in a turbulent jet is equal to the cross-stream dimension of the jet at the current location). More accurate estimates of $\ell$ can be made from measurements of the scale $\Lambda_g = 3\ell$, where

$$\Lambda_g = \frac{\int_0^\infty \overline{u'(y)u'(y+\zeta)}\,d\zeta}{\overline{u'^2}}$$

is a two-point autocorrelation ($\zeta$ is a cross-stream distance). The reader is referred to the vast literature on turbulence for more detail on this topic (Tennekes and Lumley 1972, Hinze 1975, McComb 1990).

A round turbulent jet, shown in Fig. 9.16, produces more energetic, higher Reynolds number turbulence in comparison with other standard geometries, such as grids and

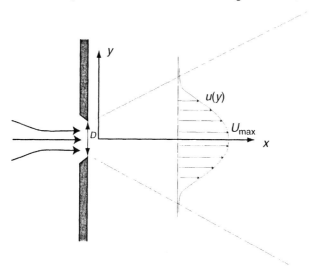

**Fig. 9.16** Schematic of round turbulent jet flow. The jet width grows linearly with distance $x$. A time-averaged velocity profile $u(y)$ is shown with maximum time-averaged velocity $U_{max}$.

mixing layers, which is why turbulent jets are one of the more commonly used sources of turbulence for deaggregation of powders. Note that at very low Reynolds numbers (of order 1) jets are laminar, but for jet velocities $> 1 \text{ m s}^{-1}$ and jet exit diameters $> 1$ mm, Reynolds numbers are large enough that turbulent, not laminar, jets are readily achieved.

Experimental data on round turbulent jets indicates that for $x/D > 20$, turbulence is present across the width of the entire jet and has length scales that can be approximated using (Tennekes and Lumley 1972, White 1991)

$$\frac{u_i}{U_{max}} \approx 0.3 \qquad (9.78)$$

$$\frac{\ell}{x} \approx 0.06 \qquad (9.79)$$

$$U_{max} \approx 6.6 U_{nozzle} \frac{D}{x} \qquad (9.80)$$

where $U_{max}$ is the average centerline velocity in the $x$-direction at the given $x$-position, $U_{nozzle}$ is the velocity of the jet at the nozzle exit ($x = 0$) and $D$ is the hole diameter. The Kolmogorov scales can be estimated by using Eqs (9.78)–(9.80) in Eqs (9.75)–(9.77). It should be noted that at distances upstream of $x/D = 20$ or so, the turbulence is confined to mixing layers at the edge of the jet and does not obey Eqs (9.78)–(9.80).

For turbulence produced by a grid (shown in Fig. 9.17), the integral scales can be estimated (Hinze 1975, Mohsen and LaRue 1990) from experimental data using

$$\frac{u_i}{U} \approx c_1 \left( \frac{x}{M} - \frac{x_0}{M} \right)^n \qquad (9.81)$$

$$\frac{\ell}{M} \approx \frac{-\sqrt{c_1}}{2n} \left( \frac{x}{M} - \frac{x_0}{M} \right)^{\frac{n+2}{2}} \qquad (9.82)$$

where $U$ is the uniform flow velocity upstream of the grid, $M$ is the width of the open space between bars of the grid, $x_0 \approx 0$ is the location of the 'virtual origin', $n \approx -1.3$ is a

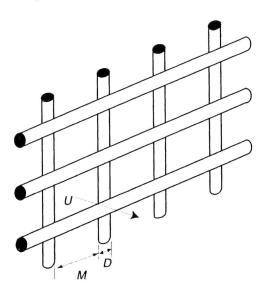

**Fig. 9.17** Fluid moving at velocity $U$ generates decaying turbulence as it passes through a grid with bars of width $D$ and spacing $D + M$.

constant, and $c_1$ is another constant that depends on the geometry of the grid (e.g. whether the bars are round, square, the geometry of the open spaces between the bars, and the ratio of $D/M$, where $D$ is the width of the bars). Typical values of $c_1$ for square-mesh grids with round bars are 0.01–0.04. Note that Eqs (9.81) and (9.82) apply only when $D/M < 0.3$ or so (otherwise jet flow instabilities occur), and are less accurate for $x/M < 40$ or so (since the turbulence is not self-similar). For estimating the integral velocity scale, the latter issue appears to be minor, since even for $x/M$ as small as 5 the value of $u_i = \sqrt{u'^2}$ differs by only a few per cent from Eq. (9.81) (Mohsen and LaRue 1990).

## Example 9.11

Estimate the integral and Kolmogorov scales 2 cm downstream from the start of a turbulent jet of exit diameter 1 mm if the flow rate is 3 l min$^{-1}$.

## Solution

The integral length scale can be determined directly from Eq. (9.79) as

$$\ell = 0.02 \times 0.06 = 0.0012 \text{ m}$$

or

$$\ell = 1.2 \text{ mm}$$

This integral scale is much larger than typical pharmaceutic inhalation powder particles, and larger even than aggregates containing hundreds of such particles.

Assuming incompressible flow, the velocity of the air at the start of the jet is obtained from the flow rate $Q = 3$ l min$^{-1}$ using

$$U_{nozzle} = Q/A_{nozzle}$$
$$= Q/(\pi D^2/4)$$
$$= (3 \text{ l min}^{-1})(10^{-3} \text{ m}^3 \text{ l}^{-1})(1 \text{ min } 60 \text{ s}^{-1})/(\pi \, 0.001^2 \text{ m}^2/4)$$
$$= 63.67 \text{ m s}^{-1}$$

This velocity is below a Mach number of 0.3, so our assumption of incompressible flow is reasonable (as discussed in Chapter 6) and we have confidence in our determined value of $U_{nozzle} = 63.67$ m s$^{-1}$. Note that the use of higher jet velocities would lead to compressible flow effects, which we have not considered here.

Substituting this value of $U_{nozzle}$ into Eq. (9.80) with $x = 0.02$ m, $D = 0.001$ m, we obtain $U_{max} = 21.0$ m s$^{-1}$ at this value of downstream location $x$. Equation (9.78) thus implies an integral velocity scale

$$u_i = 6.3 \text{ m s}^{-1}$$

The integral time scale is obtained from Eq. (9.73) as

$$t_i = \ell/u_i = 0.0012 \text{ m}/6.3 \text{ m s}^{-1} = 1.9 \times 10^{-4} \text{ s}$$

The Kolmogorov scales can now be obtained from Eqs (9.75)–(9.77), which require us to calculate the turbulent Reynolds number $Re_i = u_i\ell/\nu$. Plugging in the numbers, we have

$$Re_i = 6.3 \text{ m s}^{-1} \times 0.0012 \text{ m}/(1.5 \times 10^{-5} \text{ m}^2 \text{ s}^{-1})$$
$$= 504$$

where we have used the standard value of kinematic viscosity of air $v = 1.5 \times 10^{-5} \text{ m}^2 \text{ s}^{-1}$. Using a value of $Re_i = u_i \ell \, v = 504$ in Eqs (9.75)–(9.77) gives us the Kolmogorov scales

$$\eta = 11 \text{ μm}$$

$$t_K = 1.8 \text{ μs}$$

$$v_K = 1.3 \text{ m s}^{-1}$$

The Kolmogorov length scale here is thus only slightly larger than the desired size of a typical pharmaceutic inhalation aerosol particle.

### 9.8.2 Particle detachment from an agglomerate directly by aerodynamic forces

If values of the turbulent scales are known, it is possible to estimate whether the turbulence is able to detach particles from an aggregate by considering the simplified geometry in Fig. 9.18.

The particle shown in Fig. 9.18(b) can be removed by an eddy directly either by rolling, sliding or lifting off the agglomerate. Since the highest turbulent velocities occur with the integral scales, and the ability of an eddy to cause any of the three types of motions increases with increasing eddy velocity, the integral scales are the most likely to cause particle detachment.

We can estimate which removal mechanisms might be caused by eddies associated with the integral scales by comparing the aerodynamic forces involved with the force of adhesion. In particular, the lift and drag force are given by

$$\text{lift} = C_L \frac{1}{2} \rho_{\text{fluid}} (\pi d^2 / 4) u_i^2 \tag{9.83}$$

$$\text{drag} = C_D \frac{1}{2} \rho_{\text{fluid}} (\pi d^2 / 4) u_i^2 \tag{9.84}$$

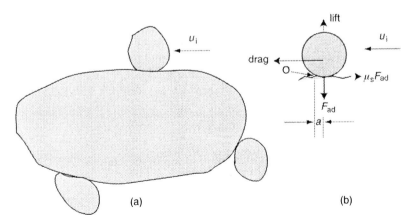

**Fig. 9.18** (a) A turbulent eddy impinging on a pharmaceutic inhalation particle that is part of an aggregate of particles. (b) Idealized version of (a) indicating the forces which can result in the particle sliding, lifting off, or rolling about point O.

where $C_L$ and $C_D$ are lift and drag coefficients for a sphere attached to a wall with flow past the wall, and depend on the particle Reynolds number $Re = u_i d/v$. Typical values of $u_i$ in dry powder inhalers are probably of the order of 1 m s$^{-1}$, while $d$ varies from a few microns up to 100 μm (for carrier particles), so that $Re$ will generally be of the order 1 or greater. For such Reynolds numbers, we saw in the last section that lift and drag coefficients for particles attached to walls in shear layers are not known, but if the velocity is assumed to vary linearly with distance away from the particle, we know from Eq. (9.47) that $C_L = 5.87$ for $Re \ll 1$, and $C_L < 1$ at higher $Re$ up to 140 in the experiments of Willetts and Naddeh (1986), while from Eq. (9.48) we know $C_D = 1.7 \times 24/Re$ for $Re \ll 1$, and we know $C_D = 24/Re$ for a particle in a uniform flow. Although we cannot give an exact value to $C_L$ or $C_D$, these data suggest $C_D \gg C_L$, and as a result the drag force is much greater than the lift force. Since the lift force must overcome the adhesion force $F_{ad}$, while the drag force must overcome only the frictional force $\mu_s F_{ad}$ (where the static coefficient of friction, $\mu_s < 1$ – see Podczeck et al. 1995), it is reasonable to suggest that sliding will occur at lower values of $u_i$ than direct lift-off. Note that a particle may slide or roll only a short distance before it detaches from the end of the agglomerate, so that sliding or rolling are feasible detachment mechanisms.

Note also that if the drag force is greater than the frictional force $\mu_s F_{ad}$, then sliding (with rolling) occurs, whereas if the drag force is less than $\mu_{roll} F_{ad}$, rolling without slip occurs (where $\mu_{roll} < \mu_s$). To examine roll without slip, it is useful to know that the torque on a sphere attached to a plane wall in a linear shear field (where the velocity is 0 at the wall and $U$ at the particle center) is given for $Re \ll 1$ by (O'Neill 1968)

$$\text{torque} = 5.93\mu U d^2 \tag{9.85}$$

Equation (9.85) is derived from the same asymptotic solution that gave Eqs (9.47) and (9.48). For roll without slip, the aerodynamic torque on the particle must exceed the moment $aF_{ad}$ where $a$ is the moment arm about point O in Fig. 9.18 (and is generally unknown, although estimates of $a$ can be made by equating $\pi a^2$ to measured contact areas given by Podczeck et al. 1996a).

## Example 9.12

A prototypical dry powder inhaler design proposes to use a turbulent jet of diameter 1 mm with flow rate 3 l min$^{-1}$ to aid deaggregation. The turbulence in the jet can be considered to combine with the powder starting at a distance 2 cm downstream from the jet exit. Estimate whether particle detachment may occur at this location directly due to aerodynamic forces of the turbulence. The aerosol consists of agglomerates containing 75 μm diameter carrier particles with attached drug particles having a diameter of 3 μm. The median adhesive force between the drug and carrier particles is approximately $10^{-11}$ N (which is typical of fine inhalation powders – see Podczeck et al. 1997), while the median adhesive force between carrier particles is approximately $3 \times 10^{-8}$ N, which is typical of lactose monohydrate carrier particles of this size (Podczeck et al. 1996a). Assume static coefficients of friction are less than 0.5.

## Solution

We calculated the integral scales for this jet in the previous example, where we found the integral velocity scale at this location was $u_i = 6.3$ m s$^{-1}$ and the integral length scale was $\ell = 1.2$ mm. The integral length scale is large enough compared to the particle sizes that treating the flow as in Fig. 9.18(b) is not unreasonable for rough estimation purposes (if the integral length scale was much smaller than the particles involved, such an idealization becomes less reasonable). Calculating the particle Reynolds number, we have $Re = u_i d/v = 1.26$ for a 3 micron particle, and $Re = 31.5$ for a 75 $\mu$m particle. These are large enough that Eqs (9.47), (9.48) and (9.84) are of questionable accuracy. Using instead a lift coefficient of $C_L = 1$ (by roughly interpolating between Eq. (9.47) and the data of Willetts and Naddeh (1986)), we obtain a lift force from Eq. (9.83) of

$$\text{lift} = C_L \frac{1}{2}\rho_{\text{air}}(\pi d^2/4)u_i^2$$
$$= 0.5(1.2 \text{ kg m}^{-3})(\pi(3.0 \times 10^{-6} \text{ m})^2/4)(6.3 \text{ m s}^{-1})^2$$
$$= 1.6 \times 10^{-10} \text{ N for a 3 } \mu\text{m particle}$$

Similarly, we obtain

$$\text{lift} = 10^{-7} \text{ N for a 75 } \mu\text{m particle}$$

To estimate the drag force and torque, we must resort to extrapolation since we have no data on the drag on a particle attached to a wall at the given Reynolds numbers. A rough estimate of the drag force can be had by using Eq. (9.48) in Eq. (9.85), from which we obtain

$$\text{drag} = 1.7(24/1.26)0.5(1.2 \text{ kg m}^{-3})(\pi(3.0 \times 10^{-6} \text{ m})^2/4)(6.3 \text{ m s}^{-1})^2$$
$$\text{drag} = 5.2 \times 10^{-9} \text{ N for a 3 } \mu\text{m particle}$$

Similarly we obtain

$$\text{drag} = 10^{-7} \text{ N for a 75 } \mu\text{m particle}$$

For spheres in uniform flow we know that extrapolating $C_D = 24/Re$ to Reynolds numbers above 1 underestimates the drag coefficient (by a factor of approximately $1 + 0.15 Re^{0.687}$ as we saw in Chapter 3, which is a factor of 3 for $Re = 40$); if similar underestimation occurs by extrapolating Eq. (9.48) the above estimates for drag may be underestimated, but we cannot know this without data in the present parameter range.

Comparing to the median adhesion forces, we see that the lift force is larger than the adhesion force of a 3 $\mu$m particle to a carrier particle ($1.6 \times 10^{-10}$ N vs. $10^{-11}$ N) or between carrier particles ($10^{-7}$ N vs. $3 \times 10^{-8}$ N). Thus, some detachment of 3 $\mu$m particles from the carrier particles, as well as separation of carrier particles from each other, may occur by direct lift-off of the particles by integral scale eddies.

If we examine the drag force due to an integral scale eddy ($5 \times 10^{-9}$ for 3 $\mu$m particle, $10^{-7}$ for 75 $\mu$m particle), we see that it is greater than the frictional force (which is $< 0.5 F_{ad}$ assuming a coefficient of friction less than 0.5 as stated) for both size particles, so that we expect that particles which are not lifted off will slide (and roll, since the frictional and drag forces cause a couple), i.e. such particles will execute roll with slip.

In summary, we expect the integral scales of this turbulent jet to cause particle detachment from agglomerates by direct lift-off, as well as by rolling with slip.

## 9.8.3 Particle detachment from an agglomerate induced by turbulent transient accelerations

In addition to removing particles directly by having eddies acting on individual particles, it is also possible for particles to be detached by eddies that act on the entire agglomerate. Indeed, in typical dry powder inhalers the integral length scales may be larger than 1 mm, so that such eddies may completely engulf even large agglomerates. As an agglomerate is buffeted by first one such integral scale eddy and then another, the agglomerate will experience accelerations. For particles on the edges of the agglomerate, sudden accelerations of the agglomerate can result in these edge particles sliding or rolling off the agglomerate. To examine whether such behavior is possible, consider the following order of magnitude analysis.

From Chapter 3, we know that a particle of diameter $d$ that is suddenly exposed to a flow with velocity $u_t$ will have a velocity given by[4]

$$|\mathbf{v}| = u_t\left(1 - e^{-\frac{t}{\tau}}\right) \tag{9.86}$$

in which the particle relaxation time is

$$\tau = \rho_{particle}d^2 C_c / 18\mu \tag{9.87}$$

($d$ is particle diameter, $\mu$ is dynamic viscosity of the fluid and $C_c$ is the Cunningham slip factor). Treating the velocity $u_t$ as a turbulent velocity scale, the particle will be exposed to this velocity only over the turbulent time scale $t_t$, so the particle will achieve a velocity

$$|\mathbf{v}| = u_t\left(1 - e^{-\frac{t_t}{\tau}}\right) \tag{9.88}$$

in the time $t_t$. The longest time scale in a turbulent flow is the integral scale, defined for our purposes by Eq. (9.73). Integral length scales in typical dry powder inhalers that rely on turbulence for deaggregation are on the order of 1 mm, and integral velocities are on the order of a few meters per second, so that turbulent time scales on the order of a few tenths of a millisecond can be expected. However, relaxation times $\tau$ for agglomerates (which may have diameters up to hundreds of microns) are several milliseconds. As a result, we can approximate the exponential in Eq. (9.88) by assuming $t_t \ll \tau$ to yield

$$|\mathbf{v}| \approx u_t \frac{t_t}{\tau} \tag{9.89}$$

The acceleration, $a$, experienced by the particle in this time can be approximated as

$$a \approx \frac{|\mathbf{v}|}{t_t} \tag{9.90}$$

Combining Eqs (9.89) and (9.90), we obtain an estimate for the acceleration experienced by the particle:

$$a \approx \frac{u_t}{\tau} \tag{9.91}$$

---

[4]Actually, Eq. (9.86) is valid only for particle Reynolds numbers $\ll 1$ (since Stokes drag, with drag coefficient $C_D = 24/Re$ was used in deriving this equation). However, we wish only to perform an order of magnitude estimation, for which the correction to this equation that accounts for Reynolds numbers up to perhaps 100 that might be seen by large agglomerates is relatively minor from such a point of view.

For a given agglomerate with relaxation time $\tau$, the largest accelerations will occur with the largest turbulent velocity scale $u_l$, which means the integral scales are most likely to result in particle detachment due to the accelerations they induce in agglomerates.

To estimate whether these transient accelerations induced by integral scale eddies are large enough to result in particle detachment, we can compare them to the adhesive force $F_{ad}$ which may keep the particle from detaching directly (in a direction normal to the contact surface), or the friction force $\mu_s F_{ad}$ which may keep the particle from starting to slide off along the contact surface (we would also need to consider rolling without slip if the accelerations are not large enough to overcome the frictional force). If the acceleration is directly opposite to the adhesive force, detachment can occur if

$$ma > F_{ad} \tag{9.92}$$

while if the acceleration is directly opposite the friction force, the particle will start to slide (and possibly detach) if

$$ma > \mu_s F_{ad} \tag{9.93}$$

We can estimate a typical acceleration from Eq. (9.91) using the integral velocity scale $u_i$ for the turbulent velocity scale $u_t$ in this equation. Substituting this into Eqs (9.92) and (9.93), using Eq. (9.87) for particle relaxation time (where we use the diameter $d_{agg}$ and density $\rho_{agg}$ of the agglomerate in Eq. (9.87)), assuming a Cunningham slip factor $C_c = 1$, and writing the particle mass as $m = \pi \rho_{particle} d^3/6$, we obtain

$$\frac{3\mu\pi\rho_{particle}d^3u_i}{\rho_{agg}d_{agg}^2} > F \tag{9.94}$$

where $F$ is either $F_{ad}$ or $\mu_s F_{ad}$.

Equation (9.94) provides a crude estimate of the condition that should be satisfied if accelerations induced by turbulence are to detach particles from agglomerates. The left-hand side of Eq. (9.94) can be thought of as a 'force' associated with the transient motion induced by the turbulence, and when this force exceeds the adhesive/frictional force, particle detachment occurs. As seen in the following example, this 'force' is typically smaller than the aerodynamic (drag or lift) detachment forces that we considered in the previous section. As a result, transient accelerations due to agglomerates being buffeted about by turbulent eddies are probably not the dominant turbulence-induced mode of deaggregation in dry powder inhalers.

## Example 9.13

For powder deaggregation by jet turbulence considered in the previous two examples, estimate whether transient accelerations induced by turbulence might detach

(a) a drug particle from a single carrier particle;
(b) a carrier particle from an agglomerate of carrier particles with diameter 300 μm.

Recall that drug particles had a diameter of 3 μm, and carrier particles had a diameter of 75 μm.

## Solution

We can use Eq. (9.94) for an approximate condition for transient accelerations due to turbulence to cause particle detachment. This equation is very approximate, so we can also approximate the particle and aggregate densities as being the same. The integral velocity scale for the turbulent jet under consideration was already calculated in a previous example as $u_i = 6.3 \text{ m s}^{-1}$. With this value, and using a dynamic viscosity $\mu = 1.8 \times 10^{-5} \text{ kg m}^{-1} \text{s}^{-1}$ for air, Eq. (9.94) becomes

$$\frac{1.07 \times 10^{-3} d^3}{d_{agg}^2} > F \qquad (9.95)$$

where $d$ has units of meters. For part (a), we substitute in $d = 3 \times 10^{-6}$ m (i.e. 3 μm, the stated size of the drug particles in the previous example) for the particle size and $d_{agg} = 75 \times 10^{-6}$ as the aggregate size (which is the stated size of a single carrier particle in the previous example and is the approximate aggregate size if we are interested in the detachment of a single drug particle from an aggregate consisting of a single carrier particle with attached drug particle). Equation (9.95) then evaluates to

$$F < 5 \times 10^{-12} \text{ N}$$

According to the data given in the previous example (where the adhesive force between a drug particle and a carrier particle $F_{ad}$ was $10^{-11}$ N), both the adhesive and frictional forces $F$ are of the same order as the turbulent transient acceleration 'force' given by $m|a| = 5 \times 10^{-12}$ N (which is the left-hand side of Eq. (9.94) or (9.95)). Note also that the drag and lift forces of the turbulence on the drug particle calculated in the previous example ($5 \times 10^{-9}$ N and $1.6 \times 10^{-10}$ N, respectively) are much larger than the turbulent transient acceleration 'force' of $5 \times 10^{-12}$ N. Thus, although some drug particle detachment due to transient accelerations of the agglomerate (induced by the turbulence) may occur, it is probable that aerodynamic detachment by turbulence dominates in the present case.

For part (b), we substitute $d = 75 \times 10^{-6}$ μm and $d_{agg} = 300 \times 10^{-6}$ μm to obtain the criteria for detachment as

$$F < 5 \times 10^{-9} \text{ N}$$

The median adhesive force between carrier particles was given as $3 \times 10^{-8}$ N, so that we do not expect significant deaggregation of the given agglomerate size due to turbulent transient accelerations. As just discussed with drug particle detachment, the 'force' associated with transient accelerations induced by the turbulence is much smaller than the drag or lift forces on a single carrier particle, so that we also expect the dominant mode of detachment of carrier particles from agglomerates to be caused by aerodynamic forces induced by eddies impinging on the agglomerate.

## 9.9 Particle detachment by mechanical acceleration: impaction and vibration

An alternative approach to creating rapid accelerations that lead to agglomerate breakup is to induce these accelerations mechanically, usually by having the particles impact at relatively high speed on a solid surface (resulting in rapid deceleration during the impaction), or by vibrating the particles during entrainment from the dosing container. For such mechanical accelerations to detach particles from agglomerates, we can proceed as we did when examining turbulence-induced accelerations, so that Eqs (9.92) or (9.93) give estimates of the needed mechanical accelerations. For vibration-induced accelerations, let us assume the powder bed is vibrated at a frequency $f$ and the displacement of the bed is $\Delta x$, so that for harmonic vibration the position of the bed is given by

$$x = \Delta x \sin(2\pi f t) \tag{9.96}$$

The acceleration $a = d^2x/dt^2$ is then given by

$$a = -4\pi^2 f^2 \Delta x \sin(2\pi f t) \tag{9.97}$$

An estimate for the magnitude of mechanical acceleration experienced by agglomerates is thus

$$a \approx 4\pi^2 f^2 \Delta x \tag{9.98}$$

For this acceleration to result in particle detachment, $ma$ should be greater than the adhesive/frictional force $F$, i.e.

$$ma > F \tag{9.99}$$

Combining Eqs (9.98) and (9.99), an approximate condition for detachment of a particle of mass $m$ attached to an agglomerate with force $F$ ($F = F_{ad}$ for direct detachment, $F = \mu_s F_{ad}$ for detachment by sliding) is

$$m4\pi^2 f^2 \Delta x > F \tag{9.100}$$

If instead the particle's acceleration is caused by the agglomerate impacting a solid surface (e.g. the bars of a grid or a wall) at speed $v_0$, the acceleration $a$ can be estimated as

$$a \approx \frac{v_0}{\Delta t} \tag{9.101}$$

where $\Delta t$ is the collision time. Values of $\Delta t$ are difficult to estimate, particularly for agglomerates consisting of several particles of the same size in contact with each other, since sliding of these particles occurs during collision (Boerefijn et al. 1998) resulting in highly inelastic behavior, so that simple estimates based on elastic rebound of an elastic sphere are inappropriate. Values of $\Delta t$ on the order of a few microseconds are seen with large agglomerates (300 μm diameter, 2000 particles) of 9–11 μm lactose monohydrate particles in the study by Ning et al. (1997) where impact velocities were a few meters per second.

For a particle of mass $m$ on the outside of an agglomerate, a criterion for detachment by impaction can be written based on Eqs (9.99) and (9.101) as

$$\frac{mv_0}{\Delta t} > F \tag{9.102}$$

## Example 9.14

A design being considered for a dry powder inhaler uses a grid to cause breakup of powder after entrainment. The grid consists of a square-mesh with round bars of diameter $D = 0.15$ mm having spacing 0.9 mm. The grid fills a round tube of diameter

1 cm and the design flow rate through the inhaler is $100 \, l \, min^{-1}$. For the powder considered in the previous two examples, suggest whether the dominant mode of breakup is likely to be:

(a) turbulence-induced aerodynamic forces,
(b) turbulence-induced transient acceleration,
(c) impaction on the grid surfaces.

## Solution

We need to first calculate the integral length and velocity scales for this grid, which can be done using Eqs (9.81) and (9.82):

$$\frac{u_i}{U} \approx c_1 \left(\frac{x}{M} - \frac{x_0}{M}\right)^n \tag{9.81}$$

$$\frac{\ell}{M} \approx \frac{-\sqrt{c_1}}{2n} \left(\frac{x}{M} - \frac{x_0}{M}\right)^{\frac{n+2}{2}} \tag{9.82}$$

with $c_1 \approx 0.25$ and $n \approx -1.3$. For the present grid, $D + M = 0.9$ mm and $D = 0.15$ mm, so $M = 0.75$ mm (implying $D/M = 0.2$ which is in the range of validity of $D/M < 0.3$ for Eqs (9.81) and (9.82)). To use Eq. (9.81), we need the flow velocity $U$, which is simply the flow rate of $100 \, l \, min^{-1}$ divided by the cross-sectional area of the 1 cm tube containing the grid, which gives $U = 21.2 \, m \, s^{-1}$. Although Eqs (9.81) and (9.82) show that the integral scales vary with distance downstream of the grid, the largest values of $u_i$ (and therefore the highest deaggregation forces due to the turbulence, since both turbulent aerodynamic and acceleration forces increase with $u_i$) occur closest to the grid. Realizing that for distances closer that about 5 $M$ Eq. (9.81) is invalid, we find that typical values of $u_i$ for distances $x = 4$–$40$ mm are near $u_i \approx 0.3 \, m \, s^{-1}$, while Eq. (9.82) suggests the integral length scale has values near 0.3 mm.

To calculate the aerodynamic forces associated with detachment by an integral eddy (which, having a length scale of 0.3 mm is probably larger than many agglomerates) as shown in Fig. 9.18, we use Eqs (9.83) and (9.84)

$$\text{lift} = C_L \frac{1}{2} \rho_{\text{fluid}} (\pi d^2/4) u_i^2 \tag{9.83}$$

$$\text{drag} = C_D \frac{1}{2} \rho_{\text{fluid}} (\pi d^2/4) u_i^2 \tag{9.84}$$

The particle Reynolds number associated with a velocity $u_i$ is given by $Re = u_i d/\nu$, which gives $Re = 0.1$ for $d = 3 \, \mu m$. This is small enough that Eqs (9.47) and (9.48) are reasonable (although we do not know if the velocity profile is linear as assumed in Eqs (9.47) and (9.48), so these equations must be considered as approximations only). From Eqs (9.47) and (9.48), we have $C_L \approx 5.87$ and $C_D \approx 1.7 \times 24/Re$, from which Eqs (9.83) and (9.84) give us

$$\text{lift} = (5.87)(1/2)(1.2 \, kg \, m^{-3})[\pi(3 \times 10^{-6} \, m)^2/4](0.3 \, m \, s^{-1})^2$$
$$= 2.2 \times 10^{-12} \, N$$

$$\text{drag} = (1.7 \times 24/0.2)(1/2)(1.2 \, kg \, m^{-3})[\pi(3 \times 10^{-6} \, m)^2/4](0.3 \, m \, s^{-1})^2$$
$$= 1.5 \times 10^{-10} \, N$$

Recall that we were told the median adhesion force $F_{ad} = 10^{-11}$ N and the coefficient of friction was less than 0.5. The lift force is much less than $F_{ad}$, so direct lift-off of drug particles from carrier particles is not expected. The drag force however is considerably larger than the friction force $\mu_s F_{ad}$ so that drug particles may slide/roll off carrier particles due to aerodynamic drag of impinging turbulent eddies.

If we recalculate the lift and drag using a 75 μm particle, we find both the lift and drag are less than adhesion force of $3 \times 10^{-8}$ N given in the previous example, suggesting that turbulent aerodynamic forces probably will not directly cause many carrier particles to separate from each other.

To consider mechanism (b), i.e. detachment by turbulence-induced accelerations, we use Eq. (9.94). Using a particle size $d = 3 \times 10^{-6}$ m, $u_i = 0.3$ m s$^{-1}$, $\mu = 1.8 \times 10^{-5}$ kg m$^{-1}$ s$^{-1}$, assuming the aggregate size $d_{agg}$ is approximately equal to the stated 75 μm carrier particle size (i.e. an aggregate consists of one carrier particle with small drug particles attached), and assuming the drug particle and aggregate particle densities are approximately equal, Eq. (9.94) becomes

$$F < 2 \times 10^{-13} \text{ N}$$

Here $F$ represents either the given adhesion ($F_{ad} = 10^{-11}$ N) or frictional forces ($< F_{ad}$), both of which are much larger than $2 \times 10^{-13}$ N, so that we do not expect turbulent transient accelerations to detach many drug particles from the carrier particles. Note also that if we recalculate Eq. (9.94) using a carrier particle size $d = 75$ μm, and use an agglomerate size of 300 μm, we obtain $F < 3 \times 10^{-10}$ N for carrier particle detachment from a 300 μm agglomerate. We were given $F_{ad} = 3 \times 10^{-8}$ N as the adhesive force between carrier particles, so that it is unlikely that turbulent transient accelerations can deagglomerate aggregates of carrier particles.

To examine whether impaction on the bars of the grid may result in particle detachment, we should first examine whether we think the particles will even hit the bars of the grid. For this purpose, we need to estimate the fraction of particles that will hit the bars of the grid. For this purpose, we can calculate the area $\phi_b$ the bars block off from the flow, being given (refer to Fig. 9.17) by

$$\phi_b = 1 - \frac{M^2}{(M+D)^2} \tag{9.103}$$

Using the given values of $M = 0.75$ mm, $D = 0.15$ mm, we have $\phi_b = 31\%$. The collection efficiency of a cylinder in a flow is known to depend on the Stokes number (Marple et al. 1993). The Stokes number in our case is given (see Chapter 3) as

$$Stk = U\rho_{particle}d^2 C_c/18\mu D$$

Using the flow velocity of $U = 21$ m s$^{-1}$ (which we obtained from the flow rate and tube cross-section), $D = 0.15$ mm, $d = 75 \times 10^{-6}$ m for a carrier particle, we obtain a Stokes number $> 1000$ (depending on the density of the particles, which we haven't been given, but which is typically of order 1000 kg m$^{-3}$ or greater). At Stokes numbers above 10 or so, collection efficiencies on cylinders in a uniform flow are 100% (Marple et al. 1993). Thus, all of the particles in the area $\phi_b$ will impact, so we expect 31% of the particles to undergo impaction on the grid bars. We should now calculate whether we think impaction is likely to lead to particle detachments from agglomerates. To do this we use Eq. (9.102) with $v_0 = 21$ m s$^{-1}$ being the flow velocity approaching the grid. We do not know the value of the collision time $\Delta t$ in Eq. (9.102), but it is likely $\Delta t < 100 \ \mu s$ (Ning *et al.* 1997), so that the left-hand side of Eq. (9.102) $> 10^{-9}$ N for a 3 μm particle and greater than $10^{-6}$ N for particles larger than 75 μm in diameter, both of which are many times greater than the given adhesion or friction forces. Thus, we expect particles impacting on the grid to result in drug and carrier particles detaching from agglomerates.

In summary, our calculations suggest that of the three mechanisms, (a) turbulence-induced aerodynamic forces, (b) turbulence-induced transient acceleration and (c) impaction on the grid surfaces may occur, the forces in (a) and (b) are considerably weaker than (c). As a result, we can speculate that (c) may dominate, although only 31% of the particles are expected to be exposed to (c) (some of which may adhere to the bars of the grid, which is an additional design consideration we have not examined). We thus speculate that (c) is probably a dominant mechanism of deaggregation in this geometry, however, we must bear in mind that this analysis is very approximate due to our limited understanding of the details of the mechanics involved.

## 9.10  Concluding remarks

The mechanics of dry powder inhalers are complex and not very well understood, largely because of the complexity of the fluid mechanics and adhesion mechanics for irregularly shaped, rough particles exposed to a fluid flow that may be turbulent. In addition, the short length and time scales involved in the dynamics of powder uptake and deaggregation complicate measurements of the transient mechanics involved and largely limit measurements to the initial and final states of the powder. However, we can develop a basic understanding of some of the factors involved by considering simplified geometries, as we have seen in this chapter. From this understanding, we see that powder entrainment occurs by aerodynamic forces on the particles in the powder bed, and that turbulence plays a role in deaggregating entrained powder agglomerates, with the dominant action of turbulence probably occurring through turbulent aerodynamic forces on agglomerates. Impaction of particles on solid surfaces is also a likely deaggregation mechanism.

Many different dry powder inhaler designs are available, and to cover the mechanics of all of these in appropriate detail is beyond the scope of this text, although the mechanics involved in many of these cases can be estimated by modifying concepts given in this chapter. There remains much room for future research though, particularly aimed at developing predictive mechanistic methods for the powder behavior. Because of the transient nature of dry powder inhaler mechanics (involving the rapid entrainment and deaggregation of a short burst of powder), the little we do know about powders in standard steady geometries, such as fluidized beds, does not contribute greatly to aiding our predictive understanding. Clearly, much challenging research lies ahead.

## References

Anderson, J. H. (1996) The effect of additives on the tribocharging of electrophoretic toners, *J. Electrostatics* 37:197–209.

Bailey, A. G. (1993) Charging of solids and powders, *J. Electrostatics* **30**:167–180.

Batchelor, G. K. (1967) *An Introduction to Fluid Dynamics*, Cambridge University Press, Cambridge.

Bennett, F. S., Carter, P. A., Rowley, G. and Dandiker, Y. (1999) Modification of electrostatic charge on inhaled carrier lactose particles by addition of fine particles, *Drug Dev. Ind. Pharm.* **25**:99-103.

Boerefijn, R., Ning, Z. and Ghadiri, M. (1998) Disintegration of weak lactose agglomerates for inhalation applications, *Int. J. Pharm.* **172**:199-209.

Bowling, R. A. (1988) A theoretical review of particle adhesion, in *Particles on Surfaces 1: Detection, Adhesion and Removal*, ed. K. L. Mittal, Plenum Press, New York.

Burnham, N. A., Colton, R. J. and Pollock, H. M. (1992) Work-function anisotropies as an origin of long-range surface forces, *Phys. Rev. Lett.* **69**:144-147.

Byron, P. R., Peart, J. and Staniforth, J. N. (1997) Aerosol electrostatics I: Properties of fine powders before and after aerosolization by dry powder inhalers, *Pharm. Res.* **14**:698-705.

Carter, P. A., Cassidy, O. E., Rowley, G. and Merrifield, D. R. (1998) Triboelectrification of fractionated crystalline and spray-dried lactose, *Pharm. Pharmacol. Commun.* **4**:111-115.

Cichocki, B. and Jones, R. B. (1998) Image representation of a spherical particle near a hard wall, *Physica A* **258**:273-302.

Crowe, C., Sommerfeld, M. and Tsuji, Y. (1998) *Multiphase Flows with Droplets and Particles*, CRC Press, Boca Raton, Florida.

Derjaguin, B. V., Muller, V. M. and Toporov, Y. P. (1975) Effect of contact deformations on the adhesion of particles, *J. Colloid Interface Sci.* **53**:314-326.

Desai, T. R., Li, D., Finlay, W. H. and Wong, J. P. (2001) Determination of surface free energy of interactive dry powder liposome formulations using capillary penetration technique, colloids and surfaces B, in press.

Dunbar, C. A., Hickey, A. J. and Holzner, P. (1998) Dispersion and characterization of pharmaceutical dry powder aerosols, *KONA, Powder and Particle* **16**:7-45.

Goldman, A. J., Cox, R. G. and Brenner, H. (1967) Slow viscous motion of a sphere parallel to a plane wall – II. Couette flow, *Chem. Eng. Sci.* **22**:653-660.

Gosman, A. D. and Ioaniddes, E. (1983) Aspects of computer simulation of liquid-fueled combusters, *J. Energy* **7**:482-490.

Grundke, K., Bogumil, T., Gietzelt, T., Jacobasch, H.-J., Kwok, D. Y. and Neumann, A. W. (1996) Wetting measurements on smooth, rough and porous solid surfaces, *Progr. Colloid Polym. Sci.* **101**:58-68.

Hall, D. (1988) Measurements of the mean force on a particle near a boundary in turbulent flow, *J. Fluid Mech.* **187**:451-466.

Hamaker, H. C. (1937) The London–Van der Waals attraction between spherical particles, *Physica* **4**:1059-1072.

Hinze, J. O. (1975) *Turbulence*, McGraw-Hill, New York.

Israelachvili, J. (1992) *Intermolecular and Surface Forces*, Academic Press, London.

Johnson, K. L., Kendall, K. and Roberts, A. D. (1971) Surface energy and the contact of elastic solids, *Proc. Roy. Soc. London A* **324**:301-313.

King, M. R. and Leighton, D. T. (1997) Measurement of the inertial lift on a moving sphere in contact with a plane wall in a shear flow, *Phys. Fluids* **9**:1248-1255.

Krishnan, G. P. and Leighton, D. T. (1995) Inertial lift on a moving sphere in contact with a plane wall in a shear flow, *Phys. Fluids* **7**:2538-2545.

Krupp, H. (1967) Particle adhesion theory and experiment, *Adv. Colloid Interface Sci.* **1**:111-239.

Larhrib, H., Zeng, X. M., Martin, G. P., Marriott, C. and Pritchard, J. (1999) The use of different grades of lactose as a carrier for aerosolised salbutamol sulphate, *Int. J. Pharm.* **191**:1-14.

Lazaridis, M., Drossinos, Y. and Georgopoulus, P. G. (1998) Turbulent resuspension of small nondeformable particles, *J. Colloid Interface Sci.* **204**:24-32.

Leighton, D. and Acrivos, A. (1985) The lift on a small sphere touching a plane in the presence of a simple shear flow, *ZAMP* **36**:174-178.

Liftshitz, E. M. (1956) *Soviet Phys. JETP* (Engl. Transl.) **2**:73-83.

Maeno, N., Nishimura, K. and Sugiura, K. (1995) Grain size dependence of eolian saltation lengths during snow drifting, *Geophys. Res. Lett.* **22**:2009-2012.

Marple, V. A., Rubow, K. L. and Olson, B. A. (1993) Inertial, gravitational, centrifugal, and thermal collection efficiencies, in *Aerosol Measurement*, ed. K. Willeke and P. A. Baron, Van Nostrand Reinhold, New York.

Matsusaka, S. and Masuda, H. (1996) Particle reentrainment from a fine powder layer in a turbulent air flow, *Aerosol Sci. Technol.* **24**:69–84.

McComb, W. D. (1990) *The Physics of Fluid Turbulence*, Oxford University Press, Oxford.

Mei, R. (1992) An approximate expression for the shear lift on a spherical particle at finite Reynolds number, *Int. J. Multiphase Flow* **18**:145–147.

Mohsen, S. M. and LaRue, J. C. (1990) The decay power law in grid-generated turbulence, *J. Fluid Mech.* **249**:195–214.

Mollinger, A. M. and Nieuwstadt, F. T. M. (1996) Measurement of the lift force on a particle fixed to the wall in the viscous sublayer of a fully developed turbulent boundary layer, *J. Fluid Mech.* **316**:285–306.

Muller, V. M., Yushchenki, S. and Derjaguin, B. V. (1983) General theoretical consideration of the influence of surface forces on contact deformations and the reciprocal adhesion of elastic spherical particles, *J. Colloid Interface Sci.* **92**:92–101.

Ning, Z., Boerefijn, R., Ghadir, M. and Thornton, C. (1997) Distinct element simulation of impact breakage of lactose agglomerates, *Adv. Powder Technol.* **8**:15–37.

O'Neill, M. E. (1968) A sphere in contact with a plane wall in a slow linear shear flow, *Chem. Eng. Sci.* **23**:1293–1298.

Orr, F. M., Scriven, L. E. and Rivas, A. P. (1975) Pendular rings between solids: meniscus properties and capillary force, *J. Fluid Mech.* **67**:723–742.

Padmadisastra, Y., Kennedy, R. A. and Stewart, P. J. (1994) Solid bridge formation in sulphonamide–Emdex interactive systems, *Int. J. Pharm.* **112**:55–63.

Panton, R. L. (1996) *Incompressible Flow*, Wiley, New York.

Podczeck, F. (1997) The relationship between particulate properties of carrier materials and the adhesion force of drug particles in interactive powder mixtures, *J. Adhesion Sci. Technol.* **11**:1089–1104.

Podczeck, F. (1999) Investigations into the reduction of powder adhesion to stainless steel surfaces by surface modification to aid capsule filling, *Int. J. Pharm.* **178**:93–100.

Podczeck, F., Newton, J. M. and James, M. B. (1995) The assessment of particle friction of a drug substance and a drug carrier substance, *J. Material Sci.* **30**:6083–6089.

Podczeck, F., Newton, J. M. and James, M. B. (1996a) The estimation of the true area of contact between microscopic particles and a flat surface in adhesion contact, *J. Appl. Phys.* **79**:1458–1463.

Podczeck, P., Newton, J. M. and James, M. B. (1996b) The influence of constant and changing relative humidity of the air on the autoadhesion force between pharmaceutical powder particles, *Int. J. Pharm.* **145**:221–229.

Podczeck, F., Newton, J. M. and James, M. B. (1997) Influence of relative humidity of storage air on the adhesion and autoadhesion of micronized particles to particulate and compacted powder surfaces, *J. Coll. Interface Sci.* **187**:484–491.

Pollock, H. M., Burnham, N. A. and Colton, R. J. (1995) Attractive forces between micron-sized particles: a patch charge model, *J. Adhesion* **51**:71–86.

Pozorski, J. and Minier, J.-P. (1998) On the Lagrangian turbulent dispersion models based on the Langein equation, *Int. J. Multiphase Flow* **24**:913–945.

Reeks, M. W., Reed, J. and Hall, D. (1988) On the resuspension of small particles by a turbulent flow, *J. Phys. D: Appl. Phys.* **21**:574–589.

Rietema, K. (1991) *The Dynamics of Fine Powders*, Elsevier, London.

Rumpf, H. (1977) Particle adhesion, in *Agglomeration 77*, ed. K. V. S. Sastry, Am. Inst. Min. Met. Pet. Eng. Inc., New York.

Schlichting, H. (1979) *Boundary-layer Theory*, McGraw-Hill, New York.

Sherman, D. J., Jackson, D. W. T., Namikas, S. L. and Wang, J. (1998) Wind-blown sand on beaches: an evaluation of models, *Geomorphology* **22**:113–133.

Soltani, M. and Ahmadi, G. (1994) On particle adhesion and removal mechanisms in turbulent flows, *J. Adhesion Sci. Technol.* **8**:763–785.

Soltani, M. and Ahmadi, G. (1995) Particle detachment from rough surfaces in turbulent flows, *J. Adhesion* **51**:105–123.

Tennekes, H. and Lumley, J. L. (1972) *A First Course in Turbulence*, MIT Press, Cambridge, MA.

Visser, J. (1972) On Hamaker constants: a comparison between Hamaker constants and Lifshitz–van der Waals constant, *Adv. Colloid Interface Sci.* **3**:331–363.

Wang, H.-C. (1990) Effects of inceptive motion on particle detachment from surfaces, *Aerosol Sci. Technol.* **13**:386–393.

Wang, Q., Squires, K. D., Chen, M. and McLaughlin, J. B. (1997) On the role of the lift force in turbulence simulations of particle deposition, *Int. J. Multiphase Flow* **23**:749–763.

White, F. M. (1991) *Viscous Fluid Flow*, McGraw-Hill, New York.

White, F. M. (1999) *Fluid Mechanics*, 4th edition, McGraw-Hill, Boston.

Willetts, B. (1998) Aeolian and fluvial grain transport, *Phil. Trans. R. Soc. Lond. A* **356**:2497–2513.

Willetts, B. B. and Naddeh, K. F. (1986) Measurement of lift on spheres fixed in low Reynolds number flows, *J. Hydraulic Res.* **24**:425–435.

Xie, H.-Y. (1997) The role of interparticle forces in the fluidization of fine particles, *Powder Technol.* **94**:99–108.

Zimon, A. D. (1982) *Adhesion of Dust and Powder*, Consultants Bureau (Plenum), New York.

Ziskind, G., Fichman, M. and Gutfinger, C. (1995) Resuspension of particulates from surfaces to turbulent flows – review and analysis, *J. Aerosol Sci.* **26**:613–644.

Ziskind, G., Fichman, M. and Gutfinger, C. (1997) Adhesion moment model for estimating particle detachment from a surface, *J. Aerosol Sci.* **28**:623–634.

# 10
# Metered Dose Propellant Inhalers

The most commonly used delivery device for inhaled pharmaceutical aerosols worldwide is currently the propellant (pressurized) metered dose inhaler (usually abbreviated pMDI or simply MDI). Devices of this type have existed since the mid-1950s (Thiel 1996) and have changed surprisingly little in their conceptual design since that time. Since previous authors have addressed their basic use (Morén *et al.* 1993, Hickey 1996), and our focus here is on the fundamental mechanics, only a brief overview of their basic design will be given.

The basic design of pMDIs relies on aerosol propellant technology, in which a high vapor-pressure substance (the propellant) contained under pressure in a canister is released, sending liquid propellant at relatively high speed from the canister. Drug is either suspended (colloidally) or dissolved (in solution) homogeneously in the propellant in the canister (with the drug usually occupying less than a few tenths of 1% by weight). By metering the release of a given volume of propellant (typically on the order of 50–100 µl), a metered dose of drug (typically less than a few hundred micrograms) is delivered in the propellant volume. Breakup of the propellant into liquid droplets that vaporize rapidly leaves the residual nonvolatile drug in the form of aerosol powder particles that are of suitable size for inhalation into the lung.

Figure 10.1 shows the basic geometry of a typical pressurized metered dose inhaler.

During inhalation, the outlet valve is open, but the metering chamber valve is closed, resulting in the release of the propellant/drug mixture that was contained in the metering chamber. At the end of delivery (usually a few tenths of a second), the outlet valve closes and the metering chamber opens to allow the next dose to enter the metering chamber. Valve operation occurs automatically by the passive opening and closing of passages that occurs as the patient presses the pMDI container down (e.g. so that an opening into the expansion chamber slides from outside the container into the metering chamber, 'opening' the outlet valve). Coordination of valve operation with inhalation is important in order to achieve good delivery to the lung, and is a significant issue in the design and use of metered dose inhalers.

In common with nebulizers and dry powder inhalers, the detailed mechanics of the aerosol formation process in propellant driven metered dose inhalers remains relatively poorly understood. Although successful design is possible using empirical data (much of which is not in the public domain due to its proprietary nature), quantitative prediction from a fundamental viewpoint remains elusive due to the complex, unsteady, multiphase fluid dynamics that occurs in these devices. However, certain aspects of aerosol formation in metered dose inhalers can be understood in an introductory manner, which is the topic of this chapter.

It should be noted that although the aerosol formation process is important to pMDIs (since it affects the size of the residual drug particles for inhalation), this process cannot

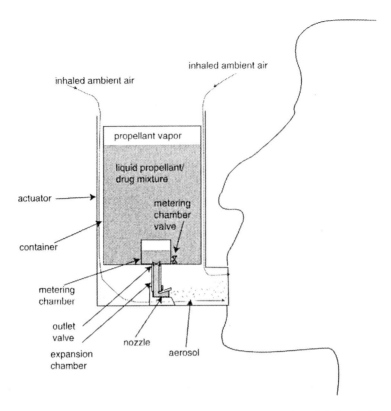

inhaled ambient air

inhaled ambient air

propellant vapor

liquid propellant/
drug mixture

actuator

metering
chamber
valve

container

metering
chamber

outlet
valve

nozzle

expansion
chamber

aerosol

**Fig. 10.1** Simplified schematic of a typical pressurized metered dose inhaler design.

occur without a drug-propellant system that remains stable in the canister over time (e.g. settling or flocculation of suspended drug particles can negatively impact drug dosing). This often requires consideration of colloidal stability in a nonaqueous medium (if the drug is suspended in the propellant), with the use of surfactants to control interfacial interactions, or cosolvents to aid in the dissolution of drug in propellant (for solution pMDIs). Such considerations are beyond the scope of this text. The interested reader is referred to Johnson (1996) for an introduction to some of the considerations involved.

## 10.1 Propellant cavitation

The flow of liquid propellant out of the metering chamber valve, through the expansion chamber and out the actuator nozzle constriction can be viewed in classical engineering terms as the transient release of a fluid from a pressurized vessel. However, this flow is complicated by the possibility that the propellant may partly vaporize as it travels through the expansion chamber and nozzle due to the high vapor pressure of the propellant. This vaporization is usually refered to as 'cavitation' or 'flashing', which is a subject all in itself (Brennen 1995), so that our discussion here is necessarily brief. However, to understand the dynamics of aerosol formation in pMDIs, it is useful to examine cavitation.

Cavitation arises in essence because the liquid propellant is exposed to such a large pressure gradient force as it flows through the expansion chamber and nozzle that the intermolecular attractive forces between fluid molecules are overcome (in typical pMDIs, pressure gradients approaching GPa m$^{-1}$ can occur). The fluid is essentially 'torn apart' by this pressure gradient, leaving cavities that are filled with propellant vapor. The process is somewhat analogous to a solid that undergoes tensions above its tensile strength, where the solid fails at its weakest locations. For liquids exposed to large tensions at temperatures below approximately $0.9T_c$, where $T_c$ is the critical point temperature ($T_c = 101.03°C$ for HFA 134a, $T_c = 101.77°C$ for HFA 227), flashing occurs due to heterogeneous cavitation, meaning that cavities develop only at so-called 'nucleation sites' that are present as naturally occurring small vapor pockets or 'cavities' (and are analogous to the weakest locations where a solid fails when in tension). Such nucleation sites are nearly always present, even at pressures well above that where cavitation occurs (Brennen 1995). In metered dose inhalers, nucleation sites are probably present at the irregular surfaces of suspended drug particles, and at the surface of the walls of the expansion chamber and actuator passage. The vapor pockets at these sites are present while the propellant is in its saturated liquid form inside the metering chamber and canister. Nucleation sites are ubiquitous in most liquid flows unless special care is taken to remove them.

To examine how nucleation sites can result in tensile failure of the liquid, it is useful to examine the fate of a small vapor cavity (at a nucleation site), as shown in Fig. 10.2.

When the drug particle in Fig. 10.2 resides in the metering chamber prior to its release, the pressure, $p_b$, in the bubble cavity will balance the force of the pressure $p$ of the liquid on the bubble's interface and the surface tension force $\gamma_{LV}$, as shown in Fig. 10.3.

From Fig. 10.3, in order for the bubble to be in static equilibrium we must have the forces balanced, i.e.

$$p\pi R^2 = p_b\pi R^2 - 2\pi R\gamma_{LV} \sin(\pi - \theta) \qquad (10.1)$$

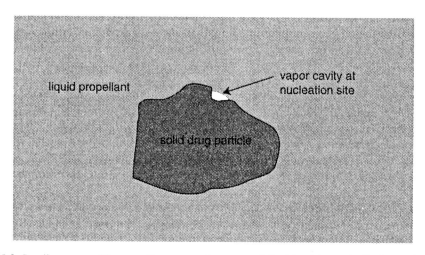

**Fig. 10.2** Small vapor cavities are often present in saturated liquids, either as individual submicron diameter bubbles at walls or at the surfaces of solid particulates (as shown here occurring on a drug particle in a suspension pMDI), or as individual free bubbles (not shown).

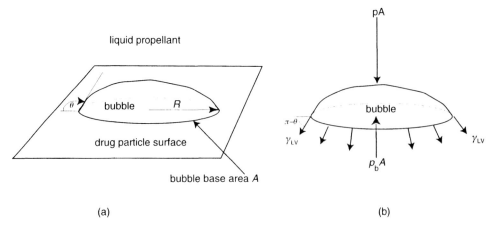

**Fig. 10.3** (a) Enlarged schematic of the vapor cavity/bubble in Fig. 10.2 (b) Free body diagram of the bubble showing the forces acting on the bubble, where $\gamma_{LV}$ is the surface tension of the propellant liquid/vapor interface and $\theta$ is the contact angle, defined in Chapter 9.

where we have assumed for simplicity that the area $A$ in Fig. 10.3 is a circle of radius $R$. Equation (10.1) can be simplified to read

$$p = p_b - \frac{2\gamma_{LV}\sin\theta}{R} \tag{10.2}$$

When the propellant is at rest in the metering chamber, the pressure $p_b$ in the bubble will be the saturated vapor pressure $p_s$ that we saw in Chapter 4. Thus, we can write Eq. (10.2) as

$$p = p_s - \frac{2\gamma_{LV}\sin\theta}{R} \tag{10.3}$$

When the pressure in the liquid exterior to the bubble falls below the pressure $p$ given by Eq. (10.3) (due to the propellant flowing through the expansion chamber and nozzle), the bubble will no longer be in equilibrium and its surface will expand, resulting in growth of the bubble. In other words, the condition for bubble growth (i.e. cavitation) is

$$p_s - p > \frac{2\gamma_{LV}\sin\theta}{R} \tag{10.4}$$

For very small bubble cavities (i.e. very small $R$), the right-hand side of Eq. (10.4) becomes very large and for a given vapor pressure $p_s$, Eq. (10.4) cannot be satisfied. Thus, there is a minimum bubble size, called the critical bubble size $R_c$, that must be present in order for heterogeneous cavitation to be possible. This bubble size can be calculated using Eq. (10.3) or (10.4) if the liquid pressure $p$ and the vapor pressure $p_s$ are known.

Notice that the critical radius is dependent on the surface tension of the vapor/liquid interface and the contact angle. Thus, the presence of surfactants in metered dose inhalers, which can alter both surface tension and contact angle (as well as altering the evaporation rate at liquid/vapor surfaces, as discussed in Chapter 4), can be expected to alter the behavior of metered dose inhaler formulations, as is indeed observed (Clark 1996).

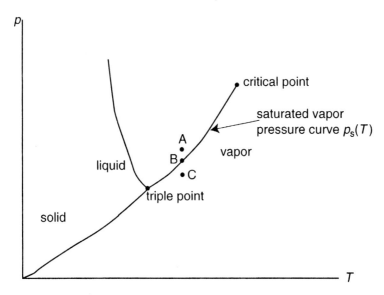

**Fig. 10.4** Pressure–temperature phase diagram. Liquid at state A is referred to as a subcooled or compressed liquid, B is a saturated liquid/vapor, and a liquid at C is unstable and referred to as a superheated liquid.

From the point of view of thermodynamic state, cavitation occurs when the pressure in a liquid is reduced suddenly so that the state of the liquid is suddenly below the saturated liquid/vapor surface in a $p$–$T$ phase diagram, as shown in Fig. 10.4.

Prior to opening the outlet valve, the propellant in a metered dose inhaler is in a saturated liquid/vapor state (e.g. state B in Fig. 10.4), with the pressure in the canister being given by the saturated vapor pressure $p_s(T)$. When the outlet valve is opened, the propellant is exposed to a sudden decrease in pressure as it travels out of the expansion chamber, and is no longer in equilibrium on the saturated liquid/vapor line, but instead drops to point C in Fig. 10.4. However, the propellant is still a liquid at this point, not having had time to reach its equilibrium vapor state, and is of course unstable. Because state C can also be reached theoretically by increasing the temperature, a liquid at state C is called superheated (although a better name in our case where C is reached by reducing the pressure might be 'subpressurized'). Cavitation results as the liquid vaporizes at the nucleation sites.

## Example 10.1

Calculate the critical bubble radius $R_c$ in order for heterogeneous cavitation to be possible for propellant HFA 134a in a pMDI as it flows through the expansion chamber and actuator nozzle immediately after the opening of the metering chamber valve. Assume the propellant's saturated vapor pressure obeys the following empirical relation:

$$p_s(T) = 6.021795 \times 10^6 \exp[-2714.4749/T] \text{ N m}^{-2} \tag{10.5}$$

while the surface tension $\gamma_{LV}$ obeys the following linear fit to experimental data:

$$\gamma_{LV}(T) = 0.05295764 - 0.0001502222T \tag{10.6}$$

Assume the liquid propellant pressure is $p = 101\,320$ Pa (i.e. atmospheric pressure), and assume a constant contact angle of $\theta = 135°$ for the propellant vapor/liquid/solid interface at a nucleation site. Plot the critical radius vs. temperature.

### Solution

From Eq. (10.4) we obtain the critical radius as

$$R_c = \frac{2\gamma_{LV}(T)\sin\theta}{p_s(T) - p} \tag{10.7}$$

We are given the functions $\gamma_{LV}(T)$, $p_s(T)$ and are told $p = 101\,320$ Pa, while $\theta = 135°$. Substituting these into Eq. (10.7), we obtain $R_c(T)$, a plot of which is shown in Fig. 10.5. At 20°C, we find $R_c = 27$ nm, so that as long as there are vapor pockets of this dimension or larger in the propellant in the expansion chamber and actuator nozzle (either on the irregular surfaces of suspended drug particles or at walls, or floating free as bubbles), cavitation is possible. It seems entirely possible that vapor cavities of at least this size are present in HFA 134a metered dose inhalers, so that heterogeneous cavitation through explosive growth of these pockets likely induces cavitation in propellant as it exits such inhalers. However, data on nucleation sites is difficult to obtain and the author is unaware of published data on this issue specific to meter dose inhaler formulations.

Notice in Fig. 10.5 that the critical size of the bubbles needed for heterogeneous cavitation increases rapidly with decreasing temperature, due to the rapid decrease in vapor pressure with temperature. Indeed, below 251 K ($-22°C$) in Fig. 10.5, nucleation bubbles larger than 1 μm in radius are needed for cavitation to occur, which is probably an unlikely situation (especially if the majority of nucleation sites are supplied by suspended particles in suspension pMDIs, since the suspended particles themselves are not much larger than this). Thus, the dynamics of the spray formation process are probably very different if the inhaler has been cooled prior to use (e.g. by having been stored at cold, outside temperatures), which partly explains the usual recommendation that these inhalers be kept at room temperature.

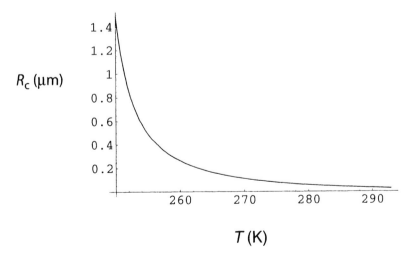

$R_c$ (μm)

$T$ (K)

**Fig. 10.5** Critical radius for cavitation bubble growth as a function of temperature for HFA 134a.

## 10.2 Fluid dynamics in the expansion chamber and nozzle

The above example suggests that cavitation occurs in the flow of propellant through the expansion chamber and nozzle during normal operation of typical pMDIs. Equation (10.4) can be used as a criterion to estimate whether such cavitation may occur (and suggests it may not occur at low temperatures, depending on the size of the naturally occurring bubbles at nucleation sites). However, once cavities appear in the propellant, they can grow rapidly and coalesce, leading to 'flash evaporation'.

The early part of this process (called incipient cavitation) in which the fluid consists of liquid propellant with dispersed, rapidly growing bubbles (occupying only a low vapor volume fraction) can be examined using the equations of bubble dynamics (Brennen 1995, Elias and Chambré 2000) derived from consideration of the Navier–Stokes equations. However, solution of the equations for bubble growth requires specification of the pressure in the liquid phase, which itself is a forbidding task in our case for several reasons. First, typical Reynolds numbers ($Re = \rho v D / \mu$) are several hundred thousand (velocities ($v$) in the expansion chambers of MDIs are greater than 100 m s$^{-1}$ (Dunbar et al. 1997a), while a typical internal dimension $D$ is 1 mm; liquid propellant densities $\rho$ are near $10^3$ kg m$^{-3}$, while dynamic viscosities $\mu$ are several hundred μPa s, giving high $Re$). At these Reynolds numbers turbulent flow can be expected in the pipe-like geometry of the expansion chamber.

Turbulence always complicates the fluid dynamics, but in our case its effect is dauntingly complex because of the pressure fluctuations associated with turbulent eddies. These localized pressure fluctuations can dramatically affect the behavior of cavitating bubbles (O'Hern 1990, Rood 1991), since bubble sizes during incipient cavitation can easily be submicron in diameter, and so they are readily contained within eddies. Bubbles in different eddies will behave differently, implying that we must resolve the turbulence itself, a task that is difficult at the high Reynolds numbers expected. Progress in modeling the effect of turbulence on cavitation may allow capturing of the first stages of cavitation dynamics in metered dose inhalers.

However, a second complicating factor is our lack of knowledge of the number and size of the nucleation sites in the liquid propellant (which is difficult information to obtain experimentally due to the small sizes of nucleation sites). Nucleation site information supplies the initial condition for the resulting cavitation, and without this information the details of the cavitation process are not readily obtained.

Even if we could capture the first stages of the cavitation process in the metering chamber and upper parts of the expansion chamber, our main interest is in the final aerosol produced. However, prediction of the behavior of the propellant at the stages between incipient cavitation and production of the propellant aerosol at the nozzle exit is difficult because of the highly convoluted, three-dimensional, transient liquid/vapor interface occurring in this process.

Experimental measurements (Domnick and Durst 1995) of incipient cavitation with the propellant CFC 12 flowing through a 2:1 constriction (5 mm constricted height) in a rectangular channel (15:1 aspect ratio) at Reynolds numbers of 50 000–130 000, show that bubble growth occurs in recirculation regions inside the constriction, as depicted in Fig. 10.6.

The lowest pressure in this flow occurs in the cross-section where the recirculation regions and vena contracta are present. Nucleation bubbles that enter the recirculation region are exposed to this reduced pressure for a longer time than bubbles that stream

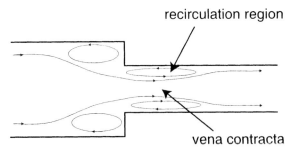

**Fig. 10.6** Mean streamlines of the flow in a constriction examined by Domnick and Durst (1995).

through and which are not captured by the recirculation region. The increased residence time of bubbles in the recirculation region allows such bubbles to undergo large growth, causing the volume of the recirculation region to grow until it obstructs enough of the channel that the incoming liquid flushes a large portion of the recirculation region downstream. Bubbles in the recirculation region, which is reduced in size after this flushing, then build up the size of the recirculation region, and the cycle occurs again. The result is a rapidly oscillating bubbly flow downstream (at a frequency of 500–800 Hz).

The geometry of the flow in the expansion chamber and nozzle of a typical pMDI is somewhat different than in Fig. 10.6, as is seen in Fig. 10.7. However, the presence of recirculation regions in the flow, both in the sump and the nozzle, would allow

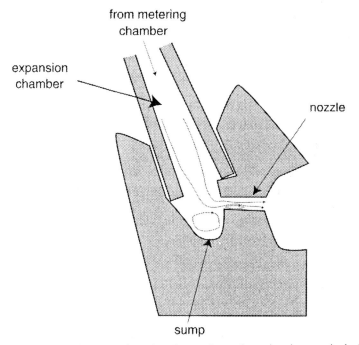

**Fig. 10.7** Enlarged view of the expansion chamber and nozzle region in a typical pMDI, after Dunbar *et al.* (1997b).

preferential bubble growth in these regions. Indeed, Dunbar (1997), and Dunbar *et al.* (1997a,b) suggest that a vena contracta present in the nozzle produces similar multiphase flow behavior to that observed by Domnick and Durst (1995) and is responsible for the oscillation at approximately 700 Hz of the aerosol cloud issuing from an HFA 134a pMDI (Dunbar *et al.* 1997b).

Although the above discussion helps us in our general understanding, the flow in the expansion chamber and nozzle is a high Reynolds number, turbulent, cavitating, multiphase flow, the detailed dynamics of which are beyond the abilities of current theoretical or numerical models of such flows. Dunbar *et al.* (1997a) summarize applicable simplified approaches to analyzing this flow, but find that only the empirical correlation of Clark (1991) allows reasonable estimation of the droplet sizes exiting the nozzle from the pMDI they considered. This correlation gives the mass median diameter (*MMD*) of the droplets at the nozzle exit as

$$MMD = \frac{8.02}{Y_{vap}^{0.56}\left[\dfrac{p_{ec} - p_{amb}}{p_{amb}}\right]} \tag{10.8}$$

where $Y_{vap}$ is the vapor mass fraction in the expansion chamber, $p_{ec}$ is the pressure in the expansion chamber and $p_{amb}$ is ambient atmospheric pressure. Estimates of $Y_{vap}$ and $p_{ec}$ can be made using adiabatic, isentropic one-dimensional analysis of the flow (Solomon *et al.* 1985, Clark 1991, Dunbar 1996, Dunbar *et al.* 1997a, Schmidt and Corradini 1997) if the discharge coefficients of the nozzle and the outlet valve are known. The transient, multiphase, compressible nature of the flow complicates matters, with the flow reaching sonic velocities for part of the discharge period (although the speed of sound in a mixed vapor–liquid flow can be dramatically lower than in the pure vapor, and depends strongly on the vapor volume fraction – see Wallis 1969, Clark 1991, Brennen 1995).

It should be noted that a simple estimate of the vapor mass fraction exiting the nozzle can be obtained (Dunbar *et al.* 1997a) from a control volume analysis of the energy in the expansion chamber and actuator nozzle, using the energy equation given in Chapter 8 (Eq. 8.36):

$$\int_S q\,dS - \int_S \rho h\mathbf{v}\cdot\hat{n}\,dS = \frac{d}{dt}\int_V \rho\hat{u}\,dV \tag{10.9}$$

where differences in kinetic and gravitational energy between the inlets and outlets of the expansion chamber and nozzle, as well as friction, have been neglected. Here $V$ is the control volume surrounding the propellant in the expansion chamber and actuator nozzle, $S$ is the surface of this volume, $\rho$ is the fluid density, $h$ is the specific enthalpy and $\hat{u}$ is the internal energy. If we assume the first and third terms are negligible compared to the middle term (which implies an adiabatic, quasi-steady process), this equation can be approximated as

$$(\rho_{vap}h_g + \rho_l h_f)|_2 = \rho_l h_l|_1 \tag{10.9}$$

where subscript 2 indicates the entrance to the expansion chamber and subscript 1 indicates the exit of the actuator nozzle, where $\rho_{vap}$ and $\rho_l$ at station 2 are the mass concentrations of vapor and liquid in the fluid exiting the nozzle, while $\rho_l$ at station 1 is the density of the liquid propellant as it enters the expansion chamber. In Eq. (10.9), $h_f$ and $h_g$ are the specific enthalpies of pure saturated liquid and vapor propellant, respectively.

Assuming a volume fraction occupied by the liquid droplets at the nozzle $\ll 1$, we can approximate $\rho_{vap}$ as $\rho_g$, the density of pure propellant vapor and $\rho_{l_2}$ as $\rho_f$, the density of pure liquid propellant, both at the temperature at the nozzle exit. Equation (10.9) can thus be written

$$\rho_g h_g(T) + \rho_f h_f(T) = \rho_f h_f(T_{mc}) \tag{10.10}$$

where $T$ is the temperature of the propellant at the nozzle exit, and $T_{mc}$ is the temperature of the liquid propellant as it exits the metering chamber into the expansion chamber. Neglecting the temperature variation of $\rho_f$ with temperature, and assuming $\rho_g \ll \rho_f$, Eq. (10.10) can be approximated as

$$Y_{vap} = \frac{h_f(T_{mc}) - h_f(T)}{L} \tag{10.11}$$

Here, $Y_{vap}$ is the vapor mass fraction at the nozzle exit (i.e. $Y_{vap} = \rho_{vap}/\rho$) and $L$ is the latent heat of vaporization $L = h_g - h_f$, evaluated at the temperature $T$ of the propellant exiting the nozzle.

If the droplets exiting the nozzle are assumed to have reached their quasi-steady evaporation temperature, i.e. $T = T_s$ (where $T_s$ is their 'wet bulb' temperature and can be calculated using the equations of droplet evaporation from Chapter 4), then Eq. (10.11) provides an estimate for the vapor mass fraction in the aerosol exiting the nozzle.

## Example 10.2

Calculate the fraction of HFA 134a propellant that exits the nozzle of a pMDI as droplets and as vapor. Approximate the droplet 'wet bulb' temperature by assuming the droplets are entrained in 20°C room air (i.e. zero propellant mass fraction and room temperature and pressure). Assume the temperature of the liquid HFA 134a in the metering chamber is 20°C. Assume linear variation of transport properties with temperature as given in Example 4.10 at the end of Chapter 4, and for pure liquid HFA 134a the specific enthalpy can be approximated as (ASHRAE 1997)

$$h_l(T) = 1000(149.2 + 1.2847T) \tag{10.12}$$

## Solution

The most difficult part of this calculation is obtaining the droplet temperature $T$, which we calculated in Example 4.10 at the end of Chapter 4. This involved using the droplet evaporation equations developed in Chapter 4, where we needed to include the effect of Stefan flow. The result found for the droplet wet bulb temperature was $T = 211$ K. From Eq. (10.12), at this temperature the enthalpy of liquid HFA 134a is $h_f(T) = 1.23 \times 10^5$ J kg$^{-1}$, while at the given metering chamber temperature of 20°C $h_f(T) = 2.27 \times 10^5$ J kg$^{-1}$. From Example 4.10 in Chapter 4, the latent heat of vaporization was approximated as

$$L = 1000 \times (388.3988 - 0.7025714T) \text{ J kg}^{-1}$$

so that $L(211 \text{ K}) = 2.40 \times 10^5$ J kg$^{-1}$.

Substituting these values into Eq. (10.11) we obtain the vapor mass fraction at the nozzle exit as

$$Y_{vap} = 0.44$$

The volume fraction can be shown to be related to the mass fraction by

$$\alpha_{vap} = \frac{Y_{vap}}{\frac{\rho_g}{\rho_f}(1 - Y_{vap}) + Y_{vap}} \tag{10.13}$$

which gives the volume fraction of the vapor as 99.92% (where we have used vapor and liquid propellant densities of $\rho_g = 0.885$ kg m$^{-3}$ and $\rho_f = 1471$ kg m$^{-3}$ at 211 K, see ASHRAE (1997)).

We thus see that the aerosol exiting the pMDI has only a small fraction of its volume occupied by droplets, as is seen experimentally (Dunbar et al. 1997b).

It should be noted that the temperature in the metering chamber decreases as it empties, since the vapor in the chamber expands to lower densities, causing it to cool (Clark 1991, Dunbar et al. 1997a). In addition, liquid propellant in the metering chamber may evaporate (with resultant cooling) as the propellant in the metering chamber maintains its saturated vapor/liquid state while the pressure in the canister drops (due to release of liquid propellant). The temperature of the metering chamber as it empties can be estimated using a control volume energy analysis, similar to that used in Chapter 8 to estimate the cooling of nebulizers (although inclusion of kinetic energy losses associated with the high speed propellant exiting the outlet valve is necessary). However, thermal resistances associated with heat transfer from the bulk canister into the metering chamber are not known. This difficulty is removed if an adiabatic assumption is made, and the temperature of the metering chamber temperature can then be estimated (Dunbar et al. 1997a). Note that if the propellant in the metering chamber is assumed to remain in a saturated liquid/vapor state, the temperature in the metering chamber will not drop below the saturated vapor pressure at ambient pressure (since the pressure in the canister will not drop below ambient pressure), which is $-26°C$ at 101.3 kPa. Thus, the metering chamber temperature is bracketed by room temperature and $-26°C$.

Cooling of the metering chamber will result in cooler propellant exiting the metering chamber, giving less vaporization of propellant in the expansion chamber and nozzle. For example, redoing the previous example with a metering chamber temperature of $-20°C$ (253.15 K) instead of $+20°C$, reduces the vapor mass fraction to 26% from its value of 44% (the wet bulb temperature drops to 203 K), although the volume fraction is still very high at 99.8% so that the propellant is still very much an aerosol (and not a bubbly liquid jet). However, a major effect of reduced vapor mass fraction as the metering chamber cools is increased droplet size (Dunbar et al. 1997a), as suggested by Eq. (10.8) where changing the vapor mass fraction from 44% to 26% results in the droplet size going from 6 µm to 29 µm. This suggests that propellant droplet sizes exiting the nozzle of a pMDI should increase from the start to finish of the actuation of a pMDI (as the metering chamber cools), as is indeed seen experimentally (Dunbar et al. 1997a).

## 10.3 Post-nozzle droplet breakup due to gradual aerodynamic loading

At the exit of the nozzle, the propellant droplets are expelled at high speed into a coflowing ambient air stream being drawn in through the mouthpiece by the patient's inhalation effort. Droplets with high relative velocity can undergo breakup into smaller secondary droplets due to the aerodynamic forces acting on the droplet, as discussed in Chapter 8. In the parlance of Chapter 8, the droplets from a pMDI undergo gradual aerodynamic loading and their possible breakup into secondary droplets can be analyzed using the correlations of Shraiber *et al.* (1996) given in Chapter 8.

### Example 10.3

Redo the analysis of Example 8.2 where a nebulized water droplet was assumed to undergo aerodynamic loading, instead performing the calculations for HFA 134a droplets as follows:

(a) 10 μm droplet diameter, 30 m s$^{-1}$ initial relative velocity (which is the approximate exit velocity measured by various authors – see Hickey and Evans (1996));
(b) 10 μm droplet diameter, 200 m s$^{-1}$ initial relative velocity (which is the approximate predicted initial velocity of HFA 134a droplets in the HFA 134a pMDI examined by Dunbar *et al.* (1997a));
(c) 30 μm droplet diameter, 30 m s$^{-1}$ initial relative velocity;
(d) 30 μm droplet diameter, 200 m s$^{-1}$ initial relative velocity.

For simplicity, assume the density and viscosity of the continuous phase surrounding the droplet is that of pure air at the wet bulb temperature of 211 K (associated with 20°C surrounding ambient air, as calculated in Example 4.10), i.e. $\rho_{air} = 1.67$ kg m$^{-3}$, $\mu_{air} = 1.4 \times 10^{-5}$ kg (m$^{-1}$ s$^{-1}$). Assume the droplet surface tension is $\sigma = 0.021$ N m$^{-1}$, the density of liquid HFA 134a at 211 K is 1471 kg m$^{-3}$ and its viscosity is $7.15 \times 10^{-4}$ kg m$^{-1}$ s$^{-1}$ (ASHRAE 1997). To keep the analysis manageable, neglect droplet size changes associated with evaporation of the droplet prior to its breakup time.

### Solution

This is a matter of redoing Example 8.2 but now using the given fluid properties associated with HFA 134a and air at a temperature of 211 K. Following that example, the Rayleigh–Taylor time scale, given by

$$\tau_0 \approx d(\rho_l/\rho_g)^{1/2}/U_0 \tag{10.14}$$

has the following values for each of the four cases:

(a) $\tau_0 = 10 \times 10^{-6}$ m$(1418/1.67)^{1/2}/30$ m s$^{-1} = 9.9 \times 10^{-6}$ s;
(b) $\tau_0 = 10 \times 10^{-6}$ m$(1418/1.67)^{1/2}/200$ m s$^{-1} = 1.5 \times 10^{-6}$ s;
(c) $\tau_0 = 30 \times 10^{-6}$ m$(1418/1.67)^{1/2}/30$ m s$^{-1} = 3.0 \times 10^{-5}$ s;
(d) $\tau_0 = 30 \times 10^{-6}$ m$(1418/1.67)^{1/2}/200$ m s$^{-1} = 4.5 \times 10^{-6}$ s.

Following Example 8.2, we substitute these values into Newton's second law ($m\,dU/dt$ = drag force) with an assumed drag coefficient of $C_d = 5$, and integrate to obtain droplet relative velocities vs. time as

(a) $U(t) = 30/(1 + 12\,735.5t)$ m s$^{-1}$                  (10.15)
(b) $U(t) = 200/(1 + 84\,904t)$ m s$^{-1}$                  (10.16)
(c) $U(t) = 30/(1 + 4245.2t)$ m s$^{-1}$                  (10.17)
(d) $U(t) = 200/(1 + 28\,301t)$ m s$^{-1}$                  (10.18)

Using the empirical correlation for natural period of droplet oscillation given in Chapter 8

$$\tau_n = 0.83\rho_l d^2 Oh/\mu_l \qquad (10.19)$$

where the Ohnesorge number $Oh = \mu_l/[\rho_l d\sigma]^{1/2}$ gives

(a), (b) $Oh = 0.041$, $\tau_n = 6.9 \times 10^{-6}$ s                  (10.20)
(c), (d) $Oh = 0.023$, $\tau_n = 3.6 \times 10^{-5}$ s                  (10.21)

The parameter $H$ defined by Shraiber *et al.* (1996) is

$$H = \frac{1}{\tau_n} = \int_0^{t_c} \frac{\rho_g U^2 d}{\sigma}\,dt \qquad (10.22)$$

where $t_c$ is the time at which breakup occurs. Substituting Eqs (10.15)–(10.21) into Eq. (10.22), we obtain

(a) $H \approx 2.3 \times 10^{12}t_c/(2.24 \times 10^7 + 2.85 \times 10^{11}t_c)$           (10.23)
(b) $H \approx 1.02 \times 10^{14}t_c/(2.24 \times 10^7 + 1.90 \times 10^{12}t_c)$           (10.24)
(c) $H \approx 1.32 \times 10^{12}t_c/(2.24 \times 10^7 + 9.50 \times 10^{10}t_c)$           (10.25)
(d) $H \approx 5.90 \times 10^{13}t_c/(2.24 \times 10^7 + 6.33 \times 10^{11}t_c)$           (10.26)

We must now substitute each of these into the Shraiber *et al.* (1996) correlation

$$\rho_g dU^2(t_c)/\sigma = 4 + (12 + \ln Oh)\exp[-(0.03 - 0.024\ln Oh)H(t_c)] \qquad (10.27)$$

and solve for $t_c$. Doing so, we find

(a) Eq. (10.27) has no real-valued solution, indicating droplet breakup does not occur
(b) $t_c = 2.0 \times 10^{-5}$ s                  (10.28)
(c) Eq. (10.27) again has no solution, indicating droplet breakup does not occur
(d) $t_c = 1.37 \times 10^{-4}$ s                  (10.29)

Thus, we find that when the propellant droplets have a relative velocity of 30 m s$^{-1}$, droplet breakup due to aerodynamic loading is not expected. However, at 200 m s$^{-1}$ (which is the predicted velocity of HFA 134a droplets in the HFA 134a pMDI examined by Dunbar *et al.* (1997a)), both 10 and 30 µm droplets are predicted to undergo breakup into secondary droplets by aerodynamic forces.

It is worth examining the distance, $s$, the droplets are expected to travel before they break up in the time $t_c$, since if this distance is too large the droplets can be expected to deposit in the oropharynx before they undergo breakup. We can calculate the distance $s$ by integrating $s = dU/dt$ where $U(t)$ is given by Eqs (10.15)–(10.18). For the cases where we expect droplet breakup (i.e. cases (b) and (d)) the inhalation velocity is much less than the droplet relative velocity of 200 m s$^{-1}$, so the relative velocity $U$ can be approximated as the absolute velocity, from which we obtain

(b) $s = 0.0024 \ln(2.24 \times 10^7 + 1.90 \times 10^{12}t)$          (10.30)

(d) $s = 0.0071 \ln(2.24 \times 10^7 + 6.33 \times 10^{11}t)$          (10.31)

Substituting $t_c$ from Eq. (10.28) into (10.29) and $t_c$ from Eq. (10.29) into (10.31), we obtain

(b) $s = 4.2$ cm

(d) $s = 13$ cm

When placed directly in the mouth, the distance from the nozzle of a pMDI to the back of the throat is considerably less than 13 cm, so that we would expect the 30 μm droplet to impact at the back of the mouth–throat before it undergoes breakup. If a spacer or add-on device is used (as discussed in the next section), the droplet may have 13 cm to travel and so may undergo breakup in this case.

For case (b), the breakup distance for the 10 μm droplet is predicted to be only 4.2 cm, which is less than the distance to the back of the mouth–throat, so this droplet may undergo secondary breakup.

Our analysis thus suggests that only the 10 μm droplet has a chance of undergoing breakup before impaction if the pMDI is placed directly in the mouth.

It is interesting to note that for the 10 μm droplet, recalculating the numbers above indicates that droplet breakup does not occur if the initial relative velocity of the droplet is less than 150 m s$^{-1}$. For the 30 μm droplet, droplet breakup does not occur below a relative velocity of 71 m s$^{-1}$. Thus, very high droplet exit velocities are needed if aerodynamic breakup of droplets is to be expected.

The above example is approximate, since it neglects droplet evaporation prior to breakup and the effect of surfactants on surface properties, as well as extrapolating the correlation of Shraiber et al. (1996) to values of $H \geq 12$. However, it does suggest that droplet breakup may occur with pMDI sprays at the highest exit velocities (upwards of 100 m s$^{-1}$) suggested in the literature, but not at the lower velocities (below 50 m s$^{-1}$) observed by other researchers.

Because of the short duration, high velocity and optically dense nature of pMDI sprays near the nozzle, the extent to which drop breakup into secondary droplets by aerodynamic forces occurs in current pMDI sprays has not been studied experimentally and remains a topic for future research.

## 10.4 Post-nozzle droplet evaporation

The evaporation of droplets after their exit from the nozzle is governed by the equations of Chapter 4. We saw there that the inclusion of Stefan flow in these equations is necessary to produce reasonable estimates of evaporation rates for typical pMDI propellants. Indeed, if Stefan flow is included, and the droplet sizes after any aerodynamic breakup are known, estimates can be made of the subsequent droplet sizes by solving the equations of Chapter 4 (with Stefan flow included), and these estimates are in reasonable agreement with experimental measurements (Dunbar et al. 1997a). A complicating factor in such an analysis is the need to solve for the motion of the continuous phase turbulent gas jet (propellant + air) downstream of the nozzle. Dunbar *et al.* (1997) find that solving the Reynolds-averaged Navier–Stokes equations (with the standard $k$–$\varepsilon$ turbulence model) and the usual Gosman and Ionnadies (1983) approach to capture the effect of turbulence on transport of the droplets, provides reasonable agreement with experimental data on droplet sizes.

It should be noted that the exposure of pMDI droplets to high water vapor mass fractions (i.e. high humidity) as they are evaporating can reduce their evaporation rate, possibly by interacting with surfactants in the formulation to alter the surface physics (Lange and Finlay 2000). This effect explains the increased impaction of droplets in add-on devices used in ventilated settings where the pMDI is actuated into warm, saturated air supplied by a ventilator (Lange and Finlay 2000). Mechanistic models of this effect for inclusion in the equations of droplet evaporation have not been presented, to the author's knowledge.

The droplet evaporation in the aerosol plume downstream of the nozzle is two-way coupled (i.e. the droplet evaporation alters the surrounding gas, and vice versa – see Chapter 4), which complicates the analysis somewhat (see Chapter 4). However, this coupling is largely due to the sensitivity of the droplet evaporation process to the continuous phase temperature, since droplet evaporation is relatively insensitive to the mass fraction of propellant in the air. In other words, it is cooling of the air around the droplets (caused by the droplets themselves), not the presence of evaporated propellant (coming off the droplets) that causes the conditions in the air around the droplets to primarily affect their evaporation rate. Thus, if droplet evaporation rates are to be increased downstream of the nozzle (in order to reduce droplet sizes to avoid impaction in the mouth–throat), ways of transporting heat more effectively to the droplets should be more effective than methods focusing on reducing propellant concentrations in the surrounding air.

## 10.5 Add-on devices

Droplet evaporation can be enhanced by allowing the droplets more time to evaporate before being inhaled, yielding smaller inhaled particle sizes and less mouth–throat deposition. In addition, if the distance from the pMDI nozzle to the back of the mouth–throat (the oropharynx) is increased, the velocity of the aerosol is reduced due to jet entrainment, again reducing impaction in the mouth–throat by lowering the Stokes number. Both of these factors are invoked by the many add-on or accessory devices that are sometimes used with pMDIs (such devices having different names, including 'spacers' for those devices which merely add distance between the pMDI nozzle and the mouth, while the term 'holding-chamber' is normally reserved for chambers into which the aerosol is fired and then allowed to slowly settle until the patient inhales through a valve). Such devices also aid in coordinating patient inhalation with firing of the pMDI.

Aerosol mechanics in add-on devices are governed by the equations of Chapter 3 on particle motion, with the equations of Chapter 4 on droplet evaporation and the Navier–Stokes equations governing the fluid motion of a turbulent gas (propellant + air) jet exiting into these devices. However, electrostatic charge on pMDI droplets (Peart *et al.* 1998) complicates the behavior of the droplets, particularly since the electrostatic charge of the surface of the add-on device affects the electric field in the device (O'Callaghan *et al.* 1994, Piérart *et al.* 1999, or van der Veen and van der Zee 1999). The mechanism of this charge production has not been well characterized – charge on the pMDI aerosol particles may arise from triboelectric effects as the liquid propellant flows across the plastic walls of the expansion chamber, or possibly for suspension pMDIs from contact charging of the suspended particulates when they impact walls. Our present poor understanding of the electrostatic charge generation process with pMDI droplets, combined with the relatively complicated fluid dynamics of a transient jet in a confined chamber and the complicating effect of excipients (such as surfactants or cosolvents) on droplet evaporation have all hindered the modeling of the dynamics of add-on devices, although empirical models have been presented (Zak *et al*. 1999). Most of the work on add-on devices has been experimental, aimed at measuring the fraction and particle size of aerosol inhaled from various add-on device designs compared to pMDI alone. A large body of literature exists on this topic (Amirav and Newhouse 1997, Bisgaard 1997, Finlay and Zuberbuhler 1999, Lange and Finlay 2000).

## 10.6 Concluding remarks

The mechanics of propellant metered dose inhalers is complex, involving a transient, cavitating turbulent fluid (with solid particulates in the case of a suspension pMDI) that flashes into rapidly evaporating droplets. Although certain aspects of the mechanics are understood, particularly droplet evaporation after exiting the pMDI nozzle (using the Stefan flow equations of Chapter 4), an understanding of the detailed mechanics of actual pMDI formulations remains elusive due to the difficulty of performing experiments at the small length scales and short time scales involved. In addition, mechanistic modeling is hampered by the complex phenomena involved that remain poorly understood, including electrostatic charge generation, the effect of surfactants and water vapor on evaporation rates, and the effect of turbulence on cavitation. Clearly, much challenging research lies ahead.

## References

Amirav, I. and Newhouse, M. T. (1997) Metered-dose inhaler accessory devices in acute asthma. Efficacy and comparison with nebulizers: a literature review, *Arch. Pedatr. Adolesc. Med.* **151**:876–882.

ASHRAE (1997) *1997 American Society of Heating, Refrigerating and Air-Conditioning Engineers Handbook: Fundamentals*, ASHRAE, Atlanta, Georgia.

Bisgaard, H. (1997) Delivery of inhaled medication to children, *J. Asthma* **34**:443–467.

Brennen, C. E. (1995) *Cavitation and Bubble Dynamics*, Oxford University Press, New York.

Clark, A. R. (1991) Metered atomization for respiratory drug delivery, PhD thesis, Loughborough, University of Technology.

Clark, A. R. (1996) MDIs: physics of aerosol formation, *J. Aerosol Med.* **9S**:19–26.

Domnick, J. and Durst, F. (1995) Measurement of bubble size, velocity and concentration in flashing flow behind a sudden constriction, *Int. J. Multiphase Flow* **21**:1047–1062.

Dunbar, C. A. (1996) An experimental and theoretical investigation of the spray issued from a pressurized metered-dose inhaler, PhD thesis, Manchester University.

Dunbar, C. A. (1997) Atomization mechanisms of the pMDI, *Particulate Sci. Technol.* **15**:253–271.

Dunbar, C. A., Watkins, A. P. and Miller, J. F. (1997a) Theoretical investigation of the spray from a pressurized metered-dose inhaler, *Atomization and Sprays* **7**:417–436.

Dunbar, C. A., Watkins, A. P. and Miller, J. F. (1997b) An experimental investigation of the spray issued from a pMDI using laser diagnostic techniques. *J. Aerosol Med.* **10**:351–368.

Elias, E. and Chambré, P. L. (2000) Bubble transport in flashing flow. *Int. J. Multiphase Flow* **26**:191–206.

Finlay, W. H. and Zuberbuhler, P. (1999) In vitro comparison of salbutamol hydrofluoroalkane (Airomir) metered dose inhaler aerosols inhaled during pediatric tidal breathing from five valved holding chambers. *J. Aerosol Med.* **12**:285–291.

Gosman, A. D. and Ioannides, E. (1983) Aspects of computer simulation of liquid-fueled combustors. *J. Energy* **7**:482–490.

Hickey, A. J. (1996) *Inhalation Aerosols: Physical and Biological Basis for Therapy*, Marcel Dekker, New York.

Hickey, A. J. and Evans, R. M. (1996) Aerosol generation from propellant-driven metered dose inhalers, in *Inhalation Aerosols: Physical and Biological Basis for Therapy*, ed. A. J. Hickey, Marcel Dekker, New York, pp. 417–439.

Johnson, K. A. (1996) Interfacial phenomena and phase behavior in metered dose inhaler formulations, in *Inhalation Aerosols: Physical and Biological Basis for Therapy*, ed. A. J. Hickey, Marcel Dekker, New York.

Lange, C. F. and Finlay, W. H. (2000) Overcoming the adverse effect of humidity in aerosol delivery via pMDIs during mechanical ventilation. *Am. J. Crit. Resp. Care Med.* **161**:1614–1618.

Morén, F., Dolovich, M. B., Newhouse, M. T. and Newman, S. P. (1993) *Aerosols in Medicine. Principles, Diagnosis and Therapy.* 2nd edition. Elsevier, Amsterdam.

O'Callaghan, C., Lynch, J., Cant, M. and Robertson, C. (1994) Improvement in sodium cromoglycate delivery from a spacer device by use of an antistatic lining, immediate inhalation and avoiding multiple actuations of drug. *Thorax* **48**:603–606.

O'Hern, T. J. (1990) An experimental investigation of turbulent shear flow cavitation. *J. Fluid Mech.* **215**:365–391.

Peart, J., Magyar, C. and Byron, P. R. (1998) Aerosol electrostatics – metered dose inhalers (MDIs): reformulation and device design issues, in *Respiratory Drug Delivery VI*, ed. R. N. Dalby, P. R. Byron and S. J. Farr, Interpharm, Buffalo Grove, IL, pp. 227–233.

Piérart, F., Wildhaber, J. H., Vrancken, I., Devadason, S. G. and Le Souëf, P. N. (1999) Washing plastic spacers in household detergent reduces electrostatic charge and greatly improves delivery. *Eur. Respir. J.* **13**:673–678.

Rood, E. P. (1991) Review – mechanisms of cavitation inception. *J. Fluids Eng.* **113**:163–175.

Schmidt, D. P. and Corradini, M. L. (1997) Analytical prediction of the exit flow of cavitating orifices. *Atom. and Sprays* **7**:603–616.

Shraiber, A. A., Podvysotsky, A. M. and Dubrovsky, V. V. (1996) Deformation and breakup of drops by aerodynamic forces. *Atomization and Sprays* **6**:667–692.

Solomon, A. S. P., Rupprecht, S. D., Chen, L. D. and Faeth, G. M. (1985) Flow and atomization in flashing injectors. *Atomisation Spray Technol.* **1**:53–76.

Thiel, C. G. (1996) From Susie's question to CFC-free: an inventor's perspective on forty years of MDI development and regulation, in *Respiratory Drug Delivery V*, ed. R. N. Dalby, P. R. Byron and S. J. Farr, Interpharm Press, Buffalo Grove, IL, pp. 115–123.

van der Veen, M. J. and van der Zee, J. S. (1999) Aerosol recovery from large-volume reservoir delivery systems is highly dependent on the static properties of the reservoir. *Eur. Respir. J.* **13**:668–672.

Wallis, G. B. (1969) *One-dimensional Two-phase Flow*, McGraw-Hill, New York.

White, F. M. (1999) *Fluid Mechanics*, 4th edition, McGraw-Hill, Boston.

Zak, M., Madsen, J., Berg, E., Bülow, J. and Bisgaard, H. (1999) A mathematical model of aerosol holding chambers. *J. Aerosol Med.* **12**:187–196.

# Index

Note: Figures and Tables are indicated (in this index) by *italic page numbers*, footnotes by suffix 'n'. Bold page numbers refer to main discussion.

Printed in the United Kingdom
by Lightning Source UK Ltd.
117954UK00001B/66